NMR IMAGING IN BIOMEDICINE

Advances in
MAGNETIC RESONANCE

EDITED BY

JOHN S. WAUGH

DEPARTMENT OF CHEMISTRY
MASSACHUSETTS INSTITUTE OF TECHNOLOGY
CAMBRIDGE, MASSACHUSETTS

NMR IMAGING IN BIOMEDICINE

SUPPLEMENT 2
Advances in Magnetic Resonance

P. MANSFIELD

DEPARTMENT OF PHYSICS
UNIVERSITY OF NOTTINGHAM
NOTTINGHAM, ENGLAND

P. G. MORRIS

THE NMR CENTRE
NATIONAL INSTITUTE FOR MEDICAL RESEARCH
LONDON, ENGLAND

1982

ACADEMIC PRESS

A Subsidiary of Harcourt Brace Jovanovich, Publishers

New York   London
Paris   San Diego   San Francisco   São Paulo   Sydney   Tokyo   Toronto

ACADEMIC PRESS, INC.
111 Fifth Avenue, New York, New York 10003

United Kingdom Edition published by
ACADEMIC PRESS, INC. (LONDON) LTD.
24/28 Oval Road, London NW1 7DX

LIBRARY OF CONGRESS CATALOG CARD NUMBER: 65–26774

ISBN 0–12–025562–6

PRINTED IN THE UNITED STATES OF AMERICA

83 84 85   9 8 7 6 5 4 3 2

Contents

1. General Introduction

2. Water in Biological Systems

3. Basic Imaging Principles

4. Classification and Description of NMR Imaging Methods

5. Comparison of Imaging Methods

6. Imaging Regimes

7. Potential Use in Medicine

8. Some Hardware Considerations

9. Biomagnetic Effects

10. Conclusion

Acknowledgments

We wish to thank all those authors of articles on imaging and related topics who have kindly given their permission to reproduce figures from their cited works. In the case of photographic reproductions, we are especially indebted to those friends and colleagues who have readily supplied original photographs. We also thank the various publishers for their permission to use or reproduce figures from their journals.

We are indebted to the following members of our NMR imaging group for reading through and commenting on the manuscript: V. Bangert, R. J. Ordidge, I. L. Pykett, and R. R. Rzedzian. We also thank all members of the group for many clarifying discussions on numerous aspects of the book.

We are extremely grateful to a number of other friends and colleagues for critically reading selected sections of the book and suggesting improvements to the text. In particular we wish to thank the following people: Professor R. E. Coupland and Professor P. H. Fentem of the Queen's Medical Centre, University of Nottingham, who made many valuable suggestions, Dr. P. S. Allen and Dr. W. Derbyshire of the Department of Physics, University of Nottingham, Dr. R. D. Saunders of NRPB, Dr. A. T. Barker and Mr. J. R. Polson of the Department of Medical Physics, Royal Hallamshire Hospital, University of Sheffield, and Dr. M. Burle and Dr. I. L. Young of Thorn-EMI, whose comments were most helpful. We are also grateful to Professor D. R. Wilkie, University of London, for his kind remarks.

We also wish to express our sincere appreciation of the unflagging efforts of Mrs. M. Newsum-Smith in the exacting task of typing the many drafts of the manuscript and Mr. D. R. S. Cameron for the drawing of original diagrams and extensive photographic work.

Preface

When this work began, it was intended to be simply a chapter on the state of the emerging subject of NMR imaging. Since then, some of the most dramatic changes in the subject with respect to innovative proposals and experimental accomplishment have occurred. A detailed explanation of all the imaging methods soon turned the chapter into a book. As such, we have tried to cover all material, making the work as comprehensive and up to date as possible. Inevitably, some of the more recent publications could not be mentioned in the text. However, where relevant, we have included some additional discussion or comment in numbered Notes which, for convenience, are collected in a single section entitled Notes Added in Proof at the end of the book.

Although originally intended as a review, much of the material presented in this book is original work of the authors. Helpful suggestions have been received from many colleagues and friends (see Acknowledgments). While the book has undoubtedly benefited from their comments, errors in the text must remain the responsibility of the authors. We would be grateful, therefore, for readers' comments and corrections.

Many of us who pioneered this new subject and others who have contributed over the years have firmly believed all along that NMR imaging would eventually emerge as a useful clinical diagnostic aid. The general improvements in imaging quality have surpassed our most optimistic expectations, allowing full clinical trials to be conducted in several hospital centers in Britain and the United States, using prototype commercial machines. It is still too soon to say with authority where NMR imaging will fit in. But examples demonstrating its superiority to computerized tomographic X-ray imaging now exist, thus casting the die for its use in hitherto difficult diagnostic areas. Many more uses, already on the horizon, will emerge, and they could eventually lead to a reappraisal of NMR imaging, transforming it from what we currently believe it to be, namely, a supplementary or secondary investigative technique, into a primary imaging modality. Only time and effort will tell, but the prognosis seems excellent.

P. MANSFIELD
P. G. MORRIS

1. General Introduction

1.1. Historical Background of NMR

The subject of nuclear magnetic resonance (NMR) has grown rapidly from its inception in 1946 into a sophisticated technique with very wide applications in both physical and organic chemistry, a whole range of analytical techniques in a broad variety of disciplines including, of course, solid-state physics and, more recently, biophysics and biochemistry. Many of these applications were really concerned with investigations of microscopic interactions in materials at the molecular and atomic level. There has also been considerable interest in microscopic molecular motions in liquids and gases, a subject which has been widely studied by various NMR techniques. An important and growing application of standard NMR techniques is the study of water content in a variety of materials, including foodstuffs and fuels. Several commercial instruments are currently available which are specifically designed to measure water content and also what one might refer to as mobile proton content in, for example, saturated and unsaturated margarine and related foodstuffs. These latter applications of NMR are really macroscopic applications rather than the microscopic studies that have been the basis of NMR for so many years.

Even in what we might regard as conventional NMR techniques, for example, the study of chemical shifts and their anisotropies in solids and liquid crystals, we have seen a growing sophistication of NMR apparatus designed to produce complicated pulse sequences and to automatically record the NMR signals and process them on-line. There is little doubt that this type of experiment has been made possible through the tremendous advances made in microelectronics, in particular, the advent of the minicomputer and, more recently, the microprocessor (although, because of the structure of the microprocessor, the operating speed is usually some-what slower than a minicomputer and it therefore has a limited application). The enormous flexibility implied by computer control experiments, from the point of view of spectrometer control, data acquisition, and display, etc., has enabled hitherto unthinkable experiments, in terms of complexity, to be successfully attempted. Indeed, new applications of computer-controlled spectrometers continue to be devised. The ever-increasing speed of data acquisition and data processing which can be achieved currently with array processors, for example, suggests that we can expect still more innovation in NMR experiments.

Of course, computers, both minicomputers and microprocessors, have an enormous influence on other disciplines. We have in mind, in particular, the outstanding advances made in medical imaging using computerized tomographic X-ray techniques, phased-array ultrasonic techniques, and, more recently, positron emission techniques. It is doubtful whether any of these highly sophisticated imaging modalities would have evolved without the advent of the laboratory minicomputer.

1.2. NMR Imaging Background

Although reported only at an anecdotal level, it is known that some of the early pioneers of NMR had used themselves as specimens in some early biological experiments. For example, as long ago as thirty years, Edward Purcell is reputed to have placed his head in a suitable NMR coil and magnet in an effort to try to observe differences in NMR signal shape or line shape caused by his thinking intensively or concentrating hard on a specific task as opposed to, as far as possible, making the mind blank. Erwin Hahn also tried experiments of a similar nature on his own head. In all cases there appeared to be no difference between the observed signals when the brain was idling as opposed to when the brain was concentrating. No effects, adverse or otherwise, were reported. Although these experiments could in no way be interpreted as imaging experiments, or indeed, forerunners of NMR imaging experiments, it is clear that the early practitioners of NMR had strongly in their minds the possibility of applying NMR to the study of biological systems. Indeed, with the passage of time this has happened, although the interest and emphasis in biological NMR has been until recently almost exclusively concerned with molecular interactions. The first published accounts of the new techniques of NMR imaging were in 1973 and since then development has been very rapid in particular techniques as well as in the number and range of techniques up until the beginning of clinical trials in 1980. It is a remarkable fact that all the development and, indeed, the initial clinical trials have been achieved in universities, and it is to their lasting credit, as well as, of course, the funding agencies, that this development was allowed to occur in this manner. As originators of some of the ideas in NMR imaging, it is most satisfying for us to see ideas born and developed to the stage of being practically useful, in a university environment.

1.3. Commercial Exploitation

Naturally, the activities in various universities have attracted a great deal of interest from potential commercial instrument manufacturers. Although it would indeed be wrong and misleading to say that industry as a whole has played no role in the development of NMR imaging, this role has so far been minimal. Most manufacturers of X-ray CT scanning equipment have shown a great deal of interest in the development of NMR imaging, and we feel sure that in the very near future commercial machines capable of producing medical images for hospital use and clinical evaluation will be available. In addition to the commercial instrument makers, there has been over the last few years a growing interest shown by potential users of NMR imaging machines in hospital centers and medical institutions. Several institutions in the United States, for instance, are known to be seeking funds to build their own imaging machines, and this can only help accelerate the approach to perfection necessary in a working clinical environment.

1.4. Range of Applications

It is clear from the preliminary work done in evaluating NMR imaging that there will undoubtedly be a wide range of applications for the technique. Foremost among these will be, of course, the application to medicine. Indeed, much of the work at universities has been aimed at emphasizing the imaging or pictorial representation of data. However, there is a growing feeling that NMR imaging will find wider application in areas other than the study of morphology of the body.

Since protons are the most abundant and most sensitive nuclei in biological systems, there has naturally been a preoccupation with the study of mobile protons in the distributed water, fat, and oil in living tissue. However, there is evidence already that the resonances from other nuclei (for example, phosphorus and, possibly, sodium) contain important information on the metabolic processes or chemistry which is taking place in living tissues. There are also new imaging developments on the horizon. For example, it now seems likely that fluid velocity can be measured by modifications of the imaging process, and with the developments of high-speed imaging techniques, there is now a strong possibility that in the near future we may see moving pictures. With these new and exciting developments it would seem that NMR imaging will find its true role not in the study of anatomy, but rather in the study of physiological functions of the body. Since there is no known hazard associated with NMR imaging, this makes the technique attractive from the point of view of prolonged time course studies, which would be difficult, impossible, or hazardous by X-ray techniques, but which would fit in well with the emerging physiological role of NMR.

Medical imaging systems for human beings involve large magnets and costly installation. However, other useful applications of NMR imaging lie in the study of small biological systems, for example, small animals and excised and perfused organs. These applications are, of course, at the macroscopic level, but it is envisaged that NMR machines capable of examining material at the microscopic level will become available in the not too distant future. An NMR microscope will be capable of examining tissue at the cellular level and could be of value both in evaluating tissue in biopsy studies and as a research tool for studying purely biological problems. Cell dynamics, diffusion processes, drug incorporation, and cell nutrition are all possible candidates for study at the cellular level using an NMR microscope. Of course, these studies are not limited exclusively to animal tissue, and we emphasize the importance at both the macroscopic and microscopic level of NMR as a new means of examining a whole range of botanical problems.

A related topic of application for NMR imaging is food science and technology. Interesting problems which could be tackled include the measurement and control of water content, although this can be achieved without NMR imaging *per se*, process control, and nondestructive examination of the contents of, for example, cooked pies containing meat and other types of filling.

Another potentially important application of NMR imaging is the area of building science and technology. Water plays an important, positive role in the building industry from the basic materials point of view, for example: cements, concretes, clays, and various types of plaster. It also has a negative aspect in terms of seepage into building materials and the problem of rising dampness. Nondestructive testing of such building materials when subjected to various conditions of dampness could well lead to new methods of dampness proofing.

Other applications of NMR, NMR imaging, or variants of the techniques, could find application in the study of fuels, water content in fuels, and the prospecting and discovery of oil. Yet other important applications, particularly with respect to water-content monitoring could arise in the paper industry, and even in libraries and museums for the estimation of water content in old manuscripts and books.

An important biological application of NMR, which has already been exploited, is the development of oil-bearing seeds by the examination of the oil content in certain seeds such as corn or maize. This type of nondestructive selection has led to new and high-oil-yielding seed varieties in the United States and elsewhere.

Most of the discussion of NMR imaging and, indeed, its application to date has been in the study of mobile protons in biological tissue. However, by the combination of certain imaging techniques, together with multiple-pulse techniques designed to remove the strong dipolar interaction in solids, it seems feasible, at least in principle, that the imaging of solid materials or solid regions within materials is possible. This automatically leads to new possibilities for these combined imaging experiments in solids. For example, materials testing on a range of polymers would be feasible. Ultrasonic techniques which have proved their worth for the detection of flaws and cracks in metals turn out not to be so valuable in the search for flaws and faults in materials made from plastics. For example, modern helicopter blades are fabricated from laminated plastic material bonded with polymer resins. Pockets of nonhomogenized plasticizer cause localized weaknesses which can lead to the complete destruction of the blades.

Yet, other applications of NMR imaging, or at least techniques related to NMR imaging, are what one might call warp field detectors. The disturbance of a spatially uniform magnetic field by the presence of a ferromagnetic material or even weakly paramagnetic materials has long been known. Indeed, the idea has already been exploited in the development and use of NMR sensors for the detection of buried objects. The apparatus is very sensitive to inhomogeneous field distributions and has been used primarily as a method of locating archeological remains buried up to a few feet below the surface of the earth. It is not hard to see how the idea may be simply extended to two- or three-dimensional imaging by having a set of detectors placed strategically on a grid or lattice.

1.5. The Origins of NMR Imaging

NMR imaging has caused a certain degree of contention on several fronts, and perhaps most important is the question of who invented the subject. From the point of view of the learned literature there is no doubt that the subject was first discussed openly in 1973 by Lauterbur and, independently, by Mansfield and Grannell. Both groups had taken the important conceptual step of realizing that it was necessary to use linear magnetic field gradients in order to get spatially differentiated discrimination of the nuclear magnetic response of a distributed system. However, a patent filed in 1972 and published in 1974 revealed that Damadian had previously considered the possibility of making T_1 measurements *in vivo* in the human body. The precise way in which this would be achieved was never really defined in a quantitative sense. In subsequent publications of Damadian, a method of making both spin density and relaxation time measurements evolved, based on the reception of a nuclear signal from a localized region of the specimen defined by degrading the static magnetic field everywhere except at the localized point of interest. The idea of using uniform magnetic field gradients did not figure in his imaging scheme and all the advantages of input data multiplexing are thrown away. Nevertheless, one can obtain images by this point scan method, although the scan time is inevitably long.

The second point of contention relates to the data-gathering efficiency of the various systems. In other words, which is the fastest imaging system and which is the slowest? As we shall see in later chapters, those methods which rely on linear magnetic field gradients are, generally speaking, faster than the point imaging methods. We shall also see later that there is a growing variety of NMR imaging techniques which rely on the use of linear magnetic field gradients of one sort or another, and these have a range of data-gathering efficiencies which depend on the precise details of the technique itself. We shall also see later that the most efficient and fastest imaging technique is, generally speaking, the most difficult to achieve in practice. Although these practicalities may ultimately be overcome, the cost of overcoming them may be prohibitive in a commercially viable machine. The converse situation is also true, namely, that the slower imaging techniques are generally relatively easy to build from the technological point of view and are correspondingly cheaper. However, in medicine one cannot afford to be slow. As previously discussed, the emphasis is moving from anatomical or morphological information to physiological information. In this case, imaging speed is of the essence. In reality, the machines which are finally built and used clinically are most likely to be a compromise between speed on the one hand and cost on the other.

Although true NMR imaging is able to achieve all that single-point methods can do in terms of chemical analysis, imaging, etc., the reverse is not the case in general. That is to say, the simultaneous collection of data from all parts of the specimen, whether it be a plane or set of planes, is impossible, in general, by point-imaging methods.

A third area of contention with a slightly humorous aspect is the range of names which have been used to describe the various imaging techniques.

The simplest and most descriptive name which we favor, and which is perhaps the most widely used at the moment is NMR imaging. Equally descriptive are the titles of spin imaging and spin mapping, although we prefer to reserve the term mapping for the graphical representation of other NMR parameters (for example, the spatial variations of spin relaxation times, diffusion constants, and velocities). A more esoteric name for imaging is zeugmatography, based on the Greek word *zeugma*, meaning yoke. Because of the growing use of the word tomography in the medical X-ray world, it has been suggested that a suitable description of NMR imaging could simply be NMR tomography. For the acronymologist we have the subtle alternating acronym formed from the words field-focused nuclear magnetic resonance to give the word FONAR. A recent adaptation of the FONAR technique to the study of high-resolution phosphorus signals from a localized region within a biological specimen has been named topical NMR from the Greek *topos*, meaning a point or region.

1.6. Scope and Content

This book is intended to be useful to students of NMR imaging and the NMR purist, as well as to medical and biological scientists. One of the attractions of this new subject is its interdisciplinary character. The range of activities involved comprises physics, chemistry, biology, biochemistry, biophysics, medical physics, anatomy, physiology, digital electronics, rf electronics, computing, computer control, data handling, data acquisition, and display. Few new subjects can boast so wide a background. It would seem that there is something for everybody and of course this is the attraction and strength of this new field.

The work presented here started out as a simple review but has now grown into a book in which large parts represent original work of the authors. Substantial parts of the book are written from a fundamental point of view and assume very little knowledge of either the theory or practice of NMR on the part of the reader. For example, Chapter 2 assumes very little background knowledge of NMR and could be read by most newcomers to the field with very little difficulty. Chapter 3 attempts to introduce the basic theoretical ideas of NMR and their application to NMR imaging. This chapter, though of a fundamental character and containing some new material, could well be skipped over on a first reading. However, it is necessary to look at this chapter in some detail if a full understanding of the principles which underlie the techniques described in subsequent chapters is to be gained.

In Chapter 4 we attempt to describe in some detail the range of imaging techniques currently available. Full mathematical analysis of the various techniques is given and in some cases, original material presented. For the non-NMR specialist, much of the analysis may be skipped over on first reading, effort being concentrated purely on the descriptive aspects of the various techniques. Chapter 5 is concerned entirely with the mathematical comparison of the various techniques described in Chapter 4, and is largely review material. The comparison is made both in terms of imaging speed and data-acquisition rate. On this basis the optimum technique is that one which has both the fastest imaging time and the highest data-acquisition rate. Other parameters not taken into account in this chapter are cost and technical feasibility. On this basis, as we have mentioned previously, the most efficient imaging system is not usually the cheapest imaging system. Indeed, it is most likely that the converse is true. In Chapter 6 we attempt to define a number of practical ranges or imaging regimes in terms of sensitivity, sample size, and operating frequency. Much of this chapter is original work.

Chapter 7 discusses potential applications of NMR imaging in medicine. This chapter is entirely nonmathematical, and after a brief perusal of Chapter 4 which concerns the various imaging techniques, could be read without difficulty by the nonexpert contemplating entry into the new field. After a suitably brief look at Chapters 2 and 4, Chapter 7 might well be a good starting place for radiologists, medical physicists, neuroradiologists, anatomists, and physiologists interested in the new imaging modality. Chapter 8 is designed to give an introductory overview of the apparatus requirements for those interested in the instrumentation of NMR imaging machines. Much of the contents will be familiar to the NMR expert, although there is some original work on coil design and optimization of the signal-to-noise performance of imaging systems. The majority of this chapter, however, will undoubtedly appeal to the postgraduate student of NMR imaging, the medical physicist, those engaged in the design and construction of NMR imaging machines, and last, those perhaps who, though not directly concerned with the design and construction of machines, wish to know more about the building blocks and the basic working principles of such machines. Chapter 9 is concerned entirely with biomagnetic effects. Biomagnetic effects are categorized into three main groups; namely, the effects of static magnetic fields, the effects of relatively slowly varying time-dependent fields, both uniform and nonuniform, and, finally, rf magnetic fields. Discussion of the effects of static magnetic fields is largely in the nature of a review, although there is an element of original work here. A discussion of rf biomagnetic effects is centered around rf heating and is therefore largely a review. However, discussion of the measurement of power deposition in living subjects is new work. The discussion of induced currents in a living subject caused by switched magnetic field gradients is entirely original work of the authors. Also included in this chapter is a review of the current recommendations of a number of institutions and bodies concerning the safe levels of exposure of personnel to both static and time-dependent magnetic fields. The advent of NMR imaging has led government committees in a number of countries to look closely at the possible hazards of NMR imaging to patients exposed to this new imaging modality. The recommendations contained in the *Notes of Guidance* for NMR clinical imaging issued by the National Radiological Protection Board in Britain are also included.

2. Water in Biological Systems

2.1. Introduction

Water plays an essential role in all biological systems, although its presence is often taken for granted and therefore ignored. This is true, for example, in medical diagnosis, except in the more obvious conditions like edema or swelling around an infected region. The reason for this is twofold. First, water is the medium in which most of the body chemistry or metabolism takes place. As the vehicle by which the essential ingredients for reaction within the cell are brought together, water is often considered to be of only secondary importance. The other reason why water has so far played a minor role in medical diagnosis is simply the fact that until the advent of NMR imaging, there really was no way of studying it *in vivo*.

Now that there are ways of studying the characteristics of water and its distribution, this situation is likely to change radically. For as we now know, water in biological tissue carries with it information about its cellular and extracellular environment. Biochemical analytical techniques have been developed and refined over the years to test for the presence of very small quantities of organic or inorganic material in the body fluids and are the established initial diagnostic methods based on direct chemical assay. NMR allows small quantities of material to be studied either directly by high-resolution techniques or indirectly by looking at the effect of these dissolved constituents on the properties of water.

Of course, we should make it quite clear that we do not envisage all chemical assays being replaced by NMR tests. Indeed, that would be most unlikely, simply because we know that many soluble organic compounds do not significantly change the NMR characteristics of water. What we do suspect is that with the continued study of water, of its distribution and state within the body as a whole and in specific organs, a new diagnostic dimension will emerge. Even if it does not become a general indicator of the overall state

10

of health in a patient, it could well emerge as a specific pointer to the disease states of organs and tissues within the body.

We have referred to water distribution and the state of water. The distribution of water among the various tissues and organs has been studied widely in the past on post mortem and biopsy specimens and we shall discuss some of these results in this chapter. By the state of water we mean several things; for example, the number of phases of water, *vide infra*, in tissue and their effect on the spin–lattice relaxation times in the laboratory frame T_1, the rotating frame $T_{1\rho}$, the spin–spin relaxation time T_2, and the self-diffusion coefficient D. We also have in mind the effects of dissolved nutrients and salts on these parameters.

Most of the work to date on water in biological systems has, perforce, been performed on material *in vitro* by a variety of techniques, including NMR, and is extensively discussed and reviewed elsewhere.[1-8] However, many of the effects that we now hint at can only be performed *in vivo*: for example, the localized effect on T_1, T_2, and D of oxygenation of blood and the diffusion of O_2 and nutrients through the capillaries and lymph to the cells.

The distribution of water within the body is discussed in Section 2.2. The relaxation times T_1 and T_2 and their relationship to water content are treated in Section 2.3 and we conclude with a discussion of diffusion in Section 2.4.

2.2. Water Content

2.2.1. GENERAL

The human body as a whole contains on average about 55% by weight of water, the remainder comprising protein, fat, and inorganic material.[9-11]

[1] F. Franks (Ed.), "Water: A Comprehensive Treatise," Vols. 1–4. Plenum, New York, 1972.

[2] R. Cooke and R. Wien, *Biophys. J.* **11**, 1002 (1971).

[3] R. Cooke and I. D. Kuntz, *Annu. Rev. Biophys. Eng.* **3**, 95 (1974).

[4] H. A. Resing, A. N. Garroway, and K. R. Foster, ACS Symposium No. 34, "Magnetic Resonance in Colloid and Interface Science," p. 516 (1976).

[5] K. R. Foster, H. A. Resing, and A. N. Garroway, *Science* **194**, 324 (1976).

[6] K. J. Packer, *Phil. Trans. R. Soc. London Ser. B* **278**, 59 (1977).

[7] F. Franks, *Phil. Trans. R. Soc. London Ser. B* **278**, 33 (1977).

[8] S. B. Ahmad, K. J. Packer, and J. M. Ramsden, *Mol. Phys.* **33**, 857 (1977).

[9] P. L. Altman and D. S. Dittmer (Eds.), "Biology Data Book," Vol. 3, p. 1986. Fed. of Amer. Soc. for Exp. Biol., Bethesda, Maryland, 1973.

[10] K. Diem and C. Leutner (Eds.), "Documenta Geigy Scientific Tables," p. 517. Gregory, Basel, 1970.

[11] P. L. Altman and D. S. Dittmer (Eds.), "Biology Data Book 2," p. 1206. Fed. of Amer. Soc. for Exp. Biol., Bethesda, Maryland, 1973.

The water content is ordinarily measured by a drying and weighing operation. The drying is done over a period of a day or so at between 80–100°C so as not to drive off other volatile material like fat or oil. The water content determined in this manner is close to the total water content of the tissue. However, strongly bound surface-water layers within cells attached to protein or membrane structures may still be retained in the tissues at these temperatures. There is thus an intrinsic uncertainty in the absolute determination of water in biological materials. Similarly, the determination of water content by NMR imaging is really only sensitive to the mobile proton centers. Proteins and tightly bound water will give solid-like signals which would go undetected in NMR images. Nevertheless, it would be surprising to obtain an exact correlation between water content measured by weighing and by NMR. With this in mind, therefore, we should regard the tissue water contents discussed in this section as simply a guide to that which one might expect to see in actual NMR images. We also stress that NMR images will reveal other sources of mobile protons, particularly those contained in adipose or fatty tissue. Fortunately, the spin–lattice relaxation times of water and fat are generally quite different, so that areas containing these constituents may be differentiated on grounds other than signal intensity.

2.2.2. Tissue Dependence

Table 2.1 shows the water content of various human tissues determined by the drying and weighing method. The corresponding water contents of other mammalian vertebrates, e.g., cattle, dogs, rabbits, and rats, are broadly similar.

We have used some of the water-content values from Table 2.1 together with morphological detail taken from a delta scan CT cross-sectional image to produce the water-content image[12] shown in Fig. 2.1a. This hand-produced image comprises a 64^2 array and indicates the differentiation and resolution that one might expect from pure water-content images. The liver, kidney, vertebral column, aorta, ribs, spleen, muscle, and skin are clearly differentiated in this figure. A similar hand-produced cross-sectional image of the head is shown in Fig. 2.1b. This clearly reveals the cerebrospinal fluid in the lateral ventricles and differentiation of white and gray cerebral tissue as well as bone and epidermis. For actual NMR images, see Chapter 7.

2.2.3. Age Dependence

The average values of water content cannot be regarded as absolute measures, since it is found that additional factors affecting water distribution

---

[12] P. Mansfield and I. L. Pykett, *J. Magn. Reson.* **29,** 355 (1978).

TABLE 2.1

NORMAL TISSUE WATER CONTENT IN ADULT HUMANS EXPRESSED AS A
PERCENTAGE PER KILO OF FAT-FREE TISSUE

Tissue	Water content	Ref.	Tissue	Water content	Ref.
Skeletal muscle	79.2	10	Teeth:		
Heart:			Enamel	3.0	10
Left ventricle	78.2	10	Dentine	10.0	10
Right ventricle	80.2	10	Femur cortex	12.2	10
Auricles	81.2	10	Placenta,		
Septum	79.2	10	20–40 weeks	86.6	10
Liver	71.1	10	Fluids:		
Kidney	81.0	10	Blood plasma	93	11
Spleen	79.0	10	Pleural fluid	98	11
Brain:			Peritoneal fluid	95–99	11
White matter	84.3	10	Cerebrospinal		
	84	11	fluid	96–98.8	11
Gray matter	70.6	10	Saliva	99.4	11
	70–74	11	Gastric juice	99.4–99.5	11
Spinal chord	64.4	10	Pancreatic juice	98.7	11
	82	11	Urine	98.7	11
Peripheral nerve	55.7	10	Sweat	99–99.5	11
	56	11	Semen	91.5	11
Myelin	40	11			
Epidermis	64.5	10			

and concentration are the state of health, nutrition, drugs, climate, and age.[13] The values of water content quoted are for normal adults in the age range 20–40 yr. With age, there is a progressive dehydration: the body finds it more difficult to retain fluid. As a result, statistically speaking, the total water content drops from 55% body weight at 40 yr at about 2.5% per decade to around a steady 47% at the age of 70. Whether this overall percentage drop in water content reflects equally among the various tissues and organs as described in Table 2.1 remains to be seen. Our feeling is that the whole subject of water content and water balance needs careful study and appraisal in normal healthy subjects by NMR, bearing in mind the factors of climate, drug incorporation, as well as diet and age. This is really a prerequisite to establishing the norm before any meaningful diagnosis of ill health can be attempted.

[13] See Altman and Dittmer,[9] p. 204; also p. 2041.

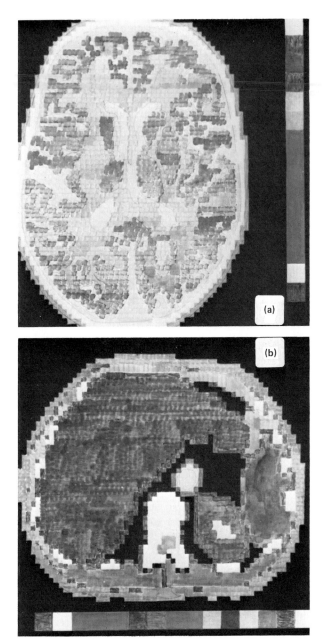

FIG. 2.1.   Hand-produced pictures showing the distribution of water throughout human anatomical sections. Each picture comprises a matrix of 64^2 points: (a) cross section through the head; (b) cross section through the upper abdomen at about L1. [See text for further details; from P. Mansfield and I. L. Pykett, *J. Magn Reson* **29**, 355 (1978).]

2.3. Relaxation Times

2.3.1. INTRODUCTION

In this section we limit ourselves to discussion of the proton relaxation rates in biological tissues. This is an extremely important subject in its own right and we seek here only to give a feeling for the orders of magnitude of the relaxation rates, their frequency dependence, and some of the proposed relaxation mechanisms. We refer the interested reader to the series of books edited by Franks[1] and a recent review article on water dynamics in heterogeneous systems by Packer.[6] In addition, the literature is reviewed biannually by Derbyshire[14,15] who includes NMR measurements in foodstuffs, plants, seeds, and other heterogeneous systems. These may well be of interest to future NMR imagers, but are given only brief attention here.

The conventional view of the role of water in heterogeneous systems has recently been challenged[16] and is currently the subject of some controversy.[17] The "structured water" proponents hold that interaction of macromolecules and ions with the cellular water imposes on the latter a kind of semicrystallinity. This accounts for the observed NMR relaxation parameters T_1, $T_{1\rho}$, and T_2, and the diffusion coefficient D, which have values intermediate between those expected for liquids and solids.

On the other hand, the traditional approach which we follow in the discussion below, views the cytoplasm as a dilute solution of ions and macromolecules in water. The magnitude of the diffusion coefficient [about one-half and equal to the free water value for translational (see Table 2.4) and rotational motion, respectively] can largely be explained in terms of a barrier effect of the macromolecules. However, it is necessary to introduce a minimum of two water phases, undergoing exchange, to explain the relaxation results. The theoretical analysis is based on the work of Zimmerman and Brittin[18] and we summarize their main results below.

2.3.2. THE EXCHANGE MODEL

In the general case of n exchanging phases the relaxation times have only been evaluated in the fast and slow exchange limits. If P_i denotes the fraction of water present in the ith phase, T_i denotes a general relaxation time (T_1, T_2, or $T_{1\rho}$), and $M(t)$ denotes a normalized magnetization, then

[14] W. Derbyshire, *Chem. Soc. Spec. Period. Rep.*, Nuclear Magnetic Resonance, **5**, 264 (1976).

[15] W. Derbyshire, *Chem. Soc. Spec. Period. Rep.*, Nuclear Magnetic Resonance, **7**, 193 (1978).

[16] G. N. Ling, "A Physical Theory of the Living State. The Association-Induction Hypothesis." Blaisdell, New York, 1962.

[17] G. B. Kolata, *Science* **192**, 1220 (1976).

[18] J. R. Zimmerman and W. E. Brittin, *J. Phys. Chem.* **61**, 1328 (1957).

for the slow exchange case ($T_i \ll$ lifetime in the ith phase),

$$M(t) = \sum_{i=1}^{n} P_i e^{-t/T_i}. \qquad (2.1)$$

Each phase thus relaxes independently with time constant T_i. Such a situation would pertain, for instance, in the presence of compartmentalization, where the water is physically barred from diffusing between the various regions.

For the case of rapid exchange ($T_i \gg$ lifetime in the ith phase),

$$M(t) = \exp\left[-t\bigg/\left(\sum_{i=1}^{n} P_i/T_i\right)\right]. \qquad (2.2a)$$

In this regime the whole system relaxes with a single characteristic time T given by

$$1/T = \sum_{i=1}^{n} P_i/T_i. \qquad (2.2b)$$

In the special case that $n = 2$, an exact expression can be derived [see Zimmerman and Brittin,[18] Eq. (40)]. This reduces to the above results in the limiting cases, but also allows the region of intermediate exchange to be studied.

Most analyses of NMR relaxation data from tissue water make the assumption of two phases known as "bound" and "free." The majority of the water is considered to be in the free state and has NMR properties similar to those of normal water (or, more correctly, a solution of the appropriate cellular salts). The remaining fraction ($\leq 10\%$) is considered to be "bound" to the macromolecules. The connotation of the word "bound" varies with the particular model being considered; it may imply that the water is rigidly or irrotationally bound to the macromolecules, or that the water molecules undergo restricted, possibly anisotropic, motion.

The main evidence for the presence of two water phases lies in the observation of a nonfreezing water component in a great variety of biological systems.[19,20] Many researchers have identified this component directly with the bound phase,[21,22] although this assignment is not universal.[5]

The magnetic interaction leading to relaxation is generally considered to be dipolar in origin, although a number of others have been suggested, for instance, paramagnetic centers bound to the macromolecules,[23] chemical

[19] I. D. Kuntz, T. S. Brassfield, G. D. Law, and G. V. Purcell, *Science* **163**, 1329 (1969).
[20] I. D. Kuntz, *J. Am. Chem. Soc.* **93**, 514 (1971).
[21] P. S. Belton, K. J. Packer, and T. C. Sellwood, *Biochim. Biophys. Acta* **304**, 56 (1973).
[22] D. R. Woodhouse, W. Derbyshire, and P. Lillford, *J. Magn. Reson.* **19**, 267 (1975).
[23] D. P. Hollis, J. S. Economou, L. C. Parks, J. C. Eggleston, L. A. Saryan, and J. L. Czeisler, *Cancer Res.* **33**, 2156 (1973).

exchange, and anisotropy of local magnetic fields.[2] The correlation times, which have been proposed to explain the relaxation data, fall into three categories: first, a free-phase correlation time τ_f in the region of 10^{-12} sec; second, a bound-phase correlation time τ_b, in the region of 10^{-9}–10^{-8} sec, and last, a characteristic time for exchange τ_e lying in the range 10^{-6}–10^{-5} sec.

2.3.3. SPIN–LATTICE RELAXATION

The observation that spin–lattice recovery is at least approximately exponential in most biological samples leads one to believe that this process is in the fast-exchange regime ($T_1 \gg \tau_e$) and is governed by Eq. (2.2b). For a two-phase system this can be rewritten as

$$1/T_1 = (1/T_1)_b P_b + (1/T_1)_f (1 - P_b), \tag{2.3a}$$

where $(1/T_1)_b$, $(1/T_1)_f$ denote the spin–lattice relaxation rates in bound and free phases, respectively, and P_b denotes the fraction of bound water.

For typical NMR frequencies (in the range 1–100 MHz, say), the dominant relaxation process is due to the molecules in the bound phase $[(1/T_1)_b \gg (1/T_1)_f]$ and the important correlation time is τ_b. This may be characteristic of the Brownian motion of the macromolecules to which the water is rigidly bound,[24] as it is in some protein solutions,[25] or else it may reflect the restricted motions of the water molecules themselves. Equation (2.3a) may be rearranged as

$$1/T_1 = (1/T_1)_f + P_b\{(1/T_1)_b - (1/T_1)_f\}. \tag{2.3b}$$

From this it is apparent that provided $(1/T_1)_b$ is not a function of water concentration, the observed relaxation rate is proportional to the fraction of bound water P_b, or alternatively, to the concentration of macromolecules. Such behavior is well known in protein systems[26] and has also been observed in some biological tissues.[2] Over a sufficiently narrow composition range, T_1 should be approximately proportional to water content and this dependence has been reported in animal tissues by a number of authors.[27-31]

[24] R. R. Knispel, R. T. Thompson, and M. M. Pintar, *J. Magn. Reson.* **14,** 44 (1974).

[25] S. H. Koenig and W. E. Schillinger, *J. Biol. Chem.* **244,** 3283 (1969).

[26] O. K. Daszkiewicz, J. W. Hennel, B. Lubas, and T. W. Szczepkowski, *Nature (London)* **200,** 1006 (1963).

[27] I. C. Kiricuta, Jr., D. Demco, and V. Simplaceanu, *Arch. Geschwülstforsch.* **42,** 226 (1973).

[28] I. C. Kiricuta, Jr. and V. Simplaceanu, *Cancer Res.* **35,** 1164 (1975).

[29] W. R. Inch, J. A. McCredie, R. R. Knispel, R. T. Thompson, and M. M. Pintar, *J. Natl. Cancer Inst.* **52,** 353 (1974).

[30] L. A. Saryan, D. P. Hollis, J. S. Economou, and J. C. Eggleston, *J. Natl. Cancer Inst.* **52,** 599 (1974).

[31] W. Bovee, P. Huisman, and J. Schmidt, *J. Natl. Cancer Inst.* **52,** 595 (1974).

A qualitative indication of this correlation can be seen by comparing the rat-tissue data of Fig. 2.2 (at, say, 100 MHz) with the tissue-water contents of Table 2.1. For instance, liver has a low water content and a short T_1 (high relaxation rate) due to the relatively large fraction of water which is bound, brain and muscle have high water contents, and T_1's and kidney and spleen are intermediate. A more quantitative example is given in Fig. 2.3 which shows the data of Hollis et al.[32] both for water content (Fig. 2.3a) and T_1 (Fig. 2.3b), in various normal and malignant rat tissues and is valuable as a reference table of T_1 and water content. Although not strictly applicable to human imaging, the values are close enough to serve as a useful guide. Similar work on rat tissue has been reported by Kiricuta and

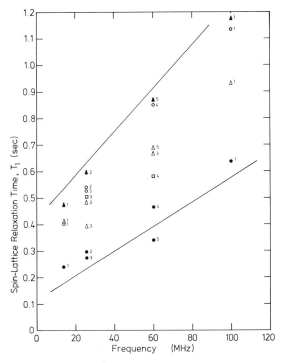

FIG. 2.2.   The frequency dependence of spin–lattice relaxation times for various rat tissues: ▲, brain; △, kidney; ●, liver; ○, muscle; □, spleen. Superscripts denote the origin of results: [1] R. E. Block and G. P. Maxwell, *J. Magn. Reson.* **14**, 329 (1974); [2] R. Damadian, *Science* **171**, 1151 (1971); [3] D. P. Hollis, L. A. Saryan, J. C. Eggleston, and H. P. Morris, *J. Natl. Cancer Inst.* **54**, 1469 (1975); [4] W. Bovee, P. Huisman, and J. Smidt, *J. Natl. Cancer Inst.* **52**, 595 (1974); [5] I.-C. Kiricuta, Jr. and V. Simplaceanu, *Cancer Res.* **35**, 1164 (1975).

[32] D. P. Hollis, L. A. Saryan, J. C. Eggleston, and H. P. Morris, *J. Natl. Cancer Inst.* **54**, 1469 (1975).

Simplaceanu.[28] Water contents were measured by weighing and drying the tissues at 100°C for 24 hr. Their results for both T_1 and T_2 are shown in Fig. 2.4a and b and show explicitly the observed relationship between T_1, T_2, and water content for a number of rat tissues. A particularly clear illustration of the linear dependence of T_1 on water content is given in a study by

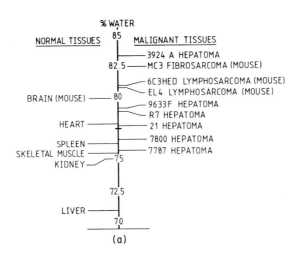

(a)

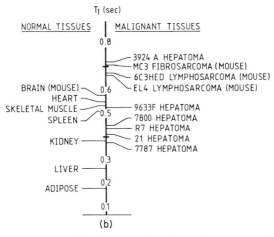

(b)

FIG. 2.3. (a) Water content of various normal rat tissues (except mouse brain), and malignant tumors from rats (Morris hepatomas) and mice. (b) Spin–lattice relaxation times T_1 at 24 MHz for various normal rat tissues (except mouse brain) and malignant tumors from rats (Morris hepatomas) and mice. [Both figures taken from D. P. Hollis, L. A. Saryan, and J. C. Eggleston. *J. Natl. Cancer Inst.* **54**, 1469 (1975).]

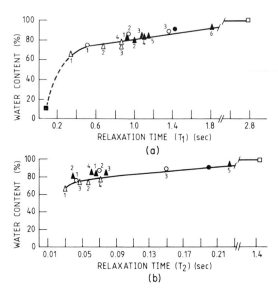

FIG. 2.4.   (a) Percent water content versus T_1 for various normal and malignant tissues. (b) Percent water content versus T_2 for various normal and malignant tissues. In both graphs, each point represents the average of values for each tissue type: △, normal tissue [(1) liver, (2) kidney, (3) heart, (4) brain]; ▲, malignant tissue [(1) lymph node metastasis of Walker 256 carcinoma, (2) Ehrlich solid tumor, (3) Walker 256 carcinoma, (4) Ehrlich ascites (cells), (5) Guérin T_5, (6) Ehrlich ascites (cells and liquid)]; ○, immature tissue [(1) liver, (2) heart, (3) brain]; ●, embryo; □, distilled water; ■, dried spleen. [From I. O. Kiricuta and V. Simplaceanu, *Cancer Res.* **35**, 1164 (1975).]

Lauterbur et al.[33] of edematous dog lung. (This work was undertaken in order to evaluate the possible use of NMR imaging methods to detect pulmonary extravascular water resulting from systemic or direct lung trauma). These interesting results show quite strikingly what one might describe as a relaxation time amplification effect embodied in our empirical relationship

$$\rho = 0.65 + 0.126T_1 \tag{2.4}$$

obtained from Fig. 2.4a, where T_1 is measured in seconds and ρ is the fractional water content. In its differential form we obtain

$$\Delta T_1 = 7.94\,\Delta\rho \tag{2.5}$$

showing that due to the magnitude of $(1/T_1)_b$, small changes in water content, typically of a few percent, can give rise to large variations in T_1. This

[33] P. C. Lauterbur, J. A. Frank, and M. J. Jacobson, *Dig. Int. Conf. Med. Phys., 4th,* Physics in Canada, **32**, Abstract 33.9 (1976).

is the basis behind the desire to form NMR images which reflect a T_1 distribution rather than simply spin density. It may be possible to enhance such tissue discrimination by the use of paramagnetic contrast agents such as Mn^{2+} [34] (see Section 2.3.4), allowing one to distinguish between normal, infarcted, and ischemic regions of the heart, for example.

Whereas the water content is generally a good guide to tissue relaxation rates, their strict linear dependence should not be taken too seriously. One might expect that this would hold in a single organ, within narrow physiological limits, but it seems hardly reasonable to expect the relaxation processes to be identical in brain and muscle fiber, for instance! We would not, therefore, dismiss water-content images out of hand. Indeed, our view is that all the NMR parameters, even though they are to a varying extent interrelated, may have some eventual role to play in tissue typing and diagnosis of disease states.

Figure 2.2, which contains data from a number of different sources in the literature, is intended to give the reader a guide to the range and frequency dependence of T_1 values to be expected in healthy animal tissue. The variation in results at 24.3 MHz and at 60 MHz reflects both normal sample variability and the difference in preparative techniques employed by various workers. A more extensive T_1 study of rabbit tissue at 24 MHz has recently been published by Mallard et al.[35]

It has been shown that T_1 values for a particular organ generally do not show a great variation with the species of animal.[36] Thus the data of Fig. 2.2 should give a good indication of the values to be expected in humans. That this is approximately correct can be seen by comparing the rat tissue data at 100 MHz with Table 2.2, which lists T_1 values at 100 MHz for human samples (taken from the results of Damadian et al.[37]).

With regard to the frequency dependence, in the simplest case of isotropic motion characterized by a single correlation time τ_b, the relaxation rate obeys the well-known modified Bloembergen, Purcell, and Pound (BPP)[38,39] expression

$$\left(\frac{1}{T_1}\right)_b \simeq B \left\{ \frac{\tau_b}{1 + \omega_0{}^2\tau_b{}^2} + \frac{4\tau_b}{1 + 4\omega_0{}^2\tau_b{}^2} \right\}, \tag{2.6}$$

[34] P. C. Lauterbur, M. Helena Mendonca Dias, and A. M. Rudin, Private Communication (1978).

[35] J. Mallard, J. M. S. Hutchison, W. A. Edelstein, C. R. Ling, M. A. Foster, and G. Johnson, Phil. Trans. R. Soc. London Ser. B 289.

[36] G. L. Cottam, A. Vasek, and D. Lusted, Res. Commun. Chem. Pathol. Pharmacol. 4, 495 (1972).

[37] R. Damadian, K. Zaner, D. Hor, and R. DiMaio, Proc. Natl. Acad. Sci. U.S.A. 71, 1471 (1974).

[38] N. Bloembergen, E. M. Purcell, and R. V. Pound, Phys. Rev. 73, 679 (1948).

[39] R. Kubo and K. Tomita, J. Phys. Soc. Jpn. 9, 888 (1954).

TABLE 2.2

T_1 Relaxation Times at 100 MHz in Normal and Malignant Human Tissues[a]

Tissue	$T_{1\,tumor}$	$T_{1\,normal}$	Probability that difference in means are not significant
Breast	1.080 ± 0.08 (13)	0.367 ± 0.079 (5)	0.52 × 10⁻⁴
Skin	1.047 ± 0.108 (4)	0.616 ± 0.019 (9)	0.55 × 10⁻⁴
Muscle:			
Malignant	1.413 ± 0.082 (7)	1.023 ± 0.029 (17)	0.50 × 10⁻⁵
Benign	1.307 ± 0.1535 (2)		
Esophagus	1.04 (1)	0.804 ± 0.108 (5)	
Stomach	1.238 ± 0.109 (3)	0.765 ± 0.075 (8)	0.40 × 10⁻²
Intestinal tract	1.122 ± 0.04 (15)	0.641 ± 0.080 (8)[b]	0.27 × 10⁻⁵
		0.641 ± 0.043 (12)[c]	
Liver	0.832 ± 0.012 (2)	0.570 ± 0.029 (14)	
Spleen	1.113 ± 0.006 (2)	0.701 ± 0.045 (17)	
Lung	1.110 ± 0.057 (12)	0.788 ± 0.063 (5)	0.25 × 10⁻²
Lymphatic	1.004 ± 0.056 (14)	0.720 ± 0.076 (6)	0.52 × 10⁻²
Bone	1.027 ± 0.152 (6)	0.554 ± 0.027 (10)	0.74 × 10⁻²
Bladder	1.241 ± 0.165 (3)	0.891 ± 0.061 (4)	0.36 × 10⁻¹
Thyroid	1.072 (1)	0.882 ± 0.045 (7)	
Nerve	1.204 (1)	0.557 ± 0.158 (2)	
Adipose	2.047 (1)	0.279 ± 0.008 (5)	
Ovary	1.282 ± 0.118 (2)	0.989 ± 0.047 (5)	
Uterus:			
Malignant	1.393 ± 0.176 (2)	0.924 ± 0.038 (4)	
Benign	0.973 (1)		
Cervix	1.101 (1)	0.827 ± 0.026 (4)	
Testes	1.223 (1)	1.200 ± 0.048 (4)	
Prostate	1.110 (1)	0.803 ± 0.014 (2)	
Adrenal	0.683 (1)	0.608 ± 0.020 (5)	
Peritoneum	1.529 (1)	0.476 (1)	
Malignant			
melanomas	0.724 ± 0.147 (6)		
Tongue	1.288 (1)		
Pericardial layer			
(mesothelioma)	0.758 (1)		
Kidney		0.862 ± 0.033 (13)	
Brain		0.998 ± 0.016 (8)	
Pancreas		0.605 ± 0.036 (10)	
Heart		0.906 ± 0.046 (9)	

[a] Probability values are reported for series with sample size ≥ 3. Errors reported are standard error of the mean (SEM). Number of cases analyzed are indicated in parenthesis. [From R. Damadian et al., Proc. Natl. Acad. Sci. U.S.A. **71**, 1471 (1974).]

[b] Small bowel.

[c] Colon.

where B is a measure of the (usually dipolar) interaction strength and ω_0 is the Larmor frequency. For $\omega_0\tau_b \gtrsim 1$, $(1/T_1)_b$ should vary roughly as ω_0^{-2} and Knispel et al.[24] observe this sort of dependence in mouse tissues over the 17–45 MHz region. Note that on the basis of this model, the divergence of T_1 values in Fig. 2.2 at high frequency would seem to indicate a variation in τ_b between the different organs, lending some weight to the view that the tissue T_1 variation is not simply a matter of water content. This divergence also calls into question the often-made assumption that tissue T_1 discrimination improves at lower Larmor frequencies, being optimal at about 2 MHz.[40]

In order to fit adequately the T_1 temperature and frequency variations many workers have found it necessary to introduce a distribution of corre-

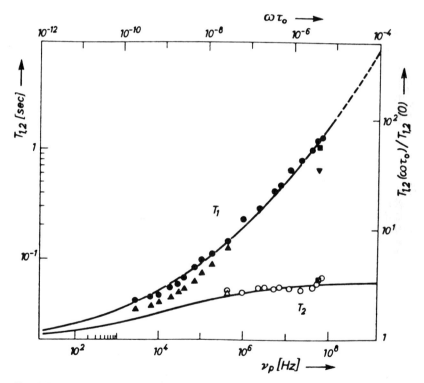

FIG. 2.5. Proton spin–lattice (T_1) and spin–spin (T_2) relaxation times in frog muscle as a function of frequency. (a) *Rana esculenta:* ●, T_1 at 25°C; ▲, T_1 at 0°C; ○, T_2 at 25°C; △, T_2 at 0°C. (b) *Rana pipiens:* ■, T_1 at 25°C; ▼, T_1 at 0°C; □, T_2 at 0 and 25°C. (c) Curves are theoretical fits to Eqs. (3a)–(3c) of Reference 41 with $\tau_0 = 1.1 \times 10^{-14}$ sec, $\alpha = 0.15$, and $T_1(0) = T_2(0) = 1.8 \times 10^{-2}$ sec. [From G. Held, F. Noack, V. Pollak, and B. Melton, Z. *Naturforsch.* **28c**, 59 (1973).]

[40] J. G. Diegel and M. M. Pintar, *J. Natl. Cancer Inst.* **55**, 725 (1975).

lation times.[21,41-44] Figure 2.5, taken from the work of Held *et al.*[41] on frog muscle, is a good example of what can be achieved. Fast exchange between the various sites ensures, of course, that the recovery remains exponential. The logarithmic distributions, which are generally used, unfortunately afford little physical insight into the system and are often incapable of simultaneously fitting the T_2 dispersion data.[44] In addition, Woessner and Snowden[45] have demonstrated the presence of anisotropic motion in heterogeneous systems.

Finally, we remark that some tissues, for instance, breast,[46-48] have markedly nonexponential T_1 recoveries. In the case of human breast tissue the recovery can be resolved into two exponential components with T_1's of about 200 and 900 msec at 60 MHz[47] which are ascribed to fat and glandular tissue, respectively. This situation clearly corresponds to the slow exchange regime of Eq. (2.1). Nonexponentiality, particularly in the early part of the recovery, can also arise from cross-relaxation effects as has been recently demonstrated by Edzes and Samulski,[49,50] (The bound-water protons are believed to be the relaxation sinks for the macromolecular protons.)

2.3.4. PARAMAGNETIC IMPURITY EFFECT ON T_1

Much of the early work on T_1 investigations in various pure liquids was vitiated by the effect of minute traces of paramagnetic material picked up either in the manufacturing process or inadvertently introduced in sample preparation. When it was finally realized, liquid samples were commonly triply distilled and all oxygen removed by the so-called "freeze, pump, thaw" method. The T_1 of tap water, for example, may vary from 0.5 to 1 sec, but by careful distillation, etc., the true T_1 is found to be 2.72 sec.

The theory of impurity effects on T_1 was first formulated by Bloembergen[51] for paramagnetic centers in crystalline solids. In biological tissue we may ignore the rather small amount of bound water, which is thought to behave

[41] G. Held, F. Noack, V. Pollak, and B. Melton, *Z. Naturforsch.* **28C**, 59 (1973).

[42] R. K. Outhred and E. P. George, *Biophys. J.* **13**, 83 (1973).

[43] R. K. Outhred and E. P. George, *Biophys. J.* **13**, 97 (1973).

[44] I. D. Duff and W. Derbyshire, *J. Magn. Reson.* **15**, 310 (1974).

[45] D. E. Woessner and B. S. Snowden, Jr., *J. Chem. Phys.* **50**, 1516 (1969).

[46] D. Medina, C. F. Hazlewood, G. G. Cleveland, D. C. Chang, H. J. Spjut, and R. Moyers, *J. Natl. Cancer Inst.* **54**, 813 (1975).

[47] W. M. M. J. Bovee, K. W. Getreuer, J. Smidt, and J. Lindeman, *J. Natl. Cancer Inst.* **61**, 53 (1978).

[48] J. A. Koutcher, M. Goldsmith, and R. Damadian, *Cancer* **41**, 174 (1978).

[49] H. T. Edzes and E. T. Samulski, *Nature (London)* **265**, 521 (1977).

[50] H. T. Edzes and E. T. Samulski, *J. Magn. Reson.* **31**, 207 (1978).

[51] N. Bloembergen, *Physica* **15**, 386 (1949).

more like an ice structure. In this case, paramagnetic impurities may be considered to interact essentially with the mobile protons of the "normal" cytoplasmic and extracellular water.

The effect of paramagnetic impurities on proton relaxation times has been studied both experimentally and theoretically.[52,53] The simplified model, valid at room temperature, in which the protons in water and the electronic spins of the impurity ion are taken to interact by translational modulation of the dipole–dipole interaction, give for the impurity-dominated spin–lattice relaxation rate

$$1/T_{1i} = (16/15)\pi^2\gamma^2\langle\mu_i^2\rangle N_i(D_i/6\pi a_i), \qquad (2.7)$$

where N_i is the number of ionic relaxation centers per cubic centimeter, $\langle\mu_i^2\rangle$ is the average ionic magnetic moment, D_i the ionic self-diffusion coefficient, a_i the ionic radius assumed to be spherical, and γ the magnetogyric ratio. In this formulation, a simplifying assumption is made that D_i and a_i are the same as for a water molecule; account can be taken when this is not the case. We assume that the classical Stokes relationship for diffusing spheres

$$D = kT/(6\pi a\eta) \qquad (2.8)$$

pertains where a is the spherical radius, η the viscosity, T the absolute temperature, and k Boltzmann's constant.

Equation (2.7) holds well for ferric Fe^{3+} ions and cupric Cu^{2+} ions, but not so well for manganese Mn^{2+} ions, and this is thought to arise from a strong proton–electron exchange interaction.

From the imaging standpoint, especially with possible contrast agents in mind, the most important aspect of Eq. (2.7) is the reciprocal relationship between T_{1i} and N_i. For ferric ions, for example, the product $T_{1i}N_i = 10^{17}$ where T_{1i} is measured in seconds. A similar result obtains for Cu^{2+} ions in glycerol.[54] An important conclusion can be drawn from this result. Namely, that to significantly reduce biological T_1 values below about 0.5 sec, doping concentrations of the order of ten parts per million are required.

2.3.5. SPIN–SPIN RELAXATION

It has been known for many years that the spin–spin relaxation times T_2 in biological systems are about fifty times less than in pure water.[55] In con-

[52] A. Abragam, "The Principles of Nuclear Magnetism." Oxford Univ. Press (Clarendon) London and New York, 1961.

[53] I. Solomon, *Phys. Rev.* **99**, 559 (1955).

[54] P. Mansfield and J. G. Powles, *Proc. Col. Ampere, 11th, Eindhoven, Holland* p. 194. North-Holland, Amsterdam, 1963.

[55] E. Odeblad, B. Bhar, and G. Lindstrom, *Arch. Biochem. Biophys.* **63**, 221 (1956).

trast with spin–lattice relaxation, the spin–spin process is often nonexponential. Normally, this is attributed to compartmentalization effects and the free induction decay (FID) is resolved into a series of exponentials.[5,56,57] Typical results, taken from the work of Foster et al.[5] on the giant barnacle muscle, are shown in Table 2.3.

The microsecond fraction is attributed to rigid macromolecular protons, whereas the millisecond fraction is attributed to lipid and mobile macromolecular protons. This latter assignment is based on the observation that

TABLE 2.3

SUMMARY OF TRANSVERSE RELAXATION DATA FOR SINGLE BARNACLE MUSCLE CELLS
AT ROOM TEMPERATURE[a]

Proton fraction designation	T_2 (msec)	Relative amplitude	Tentative proton source
Protein	~ 0.02	~ 0.2	Tissue protein
Millisecond	0.75 ± 0.25	0.033 ± 0.006^b	Relatively mobile protons
	0.3–4	0.056 ± 0.014^c	in tissue protein and lipid
		0.062 ± 0.003^d	
Major	35	0.91–0.94	Intracellular water
	45^e		
	55^f		
	20^g		
Extracellular	400	0.03	Extracellular water

[a] The relative amplitudes are normalized so that they sum to unity for the three fractions with slowest relaxation. [From K. R. Foster et al., Science 194, 324 (1976).]

[b] The relaxation data from nondeuterated tissue between 100 μsec and 50 msec could be adequately fitted by a sum of two exponential decay functions, corresponding to the millisecond and major proton fractions. The values given here are the averaged results from several fibers, assuming a two-exponential decay function.

[c] The relaxation data from nondeuterated fibers between 100 μsec and 50 msec were also fitted to a sum of three decay functions, where the two shortest decay times (corresponding to the millisecond proton fraction) were arbitrarily set equal to those calculated for the heavily deuterated tissue. The averaged results are given here.

[d] Extrapolated from the amplitude of the millisecond fraction in deuterated tissue samples, assuming that none of the millisecond protons exchange.

[e] Rotating frame relaxation time $T_{1\rho}$ in an effective field $H_1 = 2.4$ G, measured using the Carr–Purcell/Meiboom–Gill pulse sequence with pulse spacing $2\tau = 50$ μsec.

[f] Same as footnote e but at 5°C.

[g] Deuteron T_2 at 25°C; the barnacle was deuterated in vivo. All other data are from nondeuterated tissues.

[56] P. S. Belton, R. R. Jackson, and K. J. Packer, Biochim. Biophys. Acta 286, 16 (1972).

[57] R. T. Pearson, I. D. Duff, W. Derbyshire, and J. M. V. Blanshard, Biochim. Biophys. Acta 362, 188 (1974).

no exchange with D_2O occurs regardless of treatment. It is therefore not feasible to associate this millisecond fraction with a bound-water phase. The bulk of the tissue water (91–94%) has a T_2 of about 35 msec and a smaller fraction ($\sim 3\%$) a T_2 of 400 msec. These fractions are attributed to intracellular and extracellular water, respectively. The percentage of extracellular water is, in this case, extremely small due to the use of a single cell. However, the relaxation times are typical.

The process leading to relaxation is believed to be dipolar coupling modulated by exchange and the important correlation time is therefore τ_e. Knispel et al.[24] originally suggested that this referred to the time for proton exchange but later preferred to associate it with the lifetime of a molecule in the bound phase.[58] The predominant dipolar interaction is the intramolecular one,[5] but interactions with the macromolecular protons may also be important.[50,59]

Since typically $\omega_0\tau_e \gg 1$, the main relaxation term will be the zero-frequency component which is proportional to τ_e. The T_2 frequency dependence is therefore fairly flat in the region $\omega_0 > 1$ MHz and this is illustrated in Fig. 2.5 for the case of frog gastrocnemius muscle.

Whereas the overall nonexponentiality of the FID is generally resolved into a series of exponential components corresponding to slowly or nonexchanging phases, some of this behavior may arise from a distribution of exchange times[58] or from dipolar interactions which remain unaveraged due to the anisotropic nature of the motion.[45]

The range of T_1 values (~ 20–100 msec for the major water component at standard NMR frequencies in biological tissue) is important from the imaging standpoint as it sets a lower bound on the field gradient strengths and an upper bound on the resolution of some methods, notably echo-planar imaging. This may well be an important restriction if one envisages small-scale imaging of seeds, say, where the T_2's are rather shorter. The ratio $(T_2/T_1)^{1/2}$ is also of interest where steady-state free precession methods are used (for example, the sensitive[60] or multisensitive[61] point method), since it determines the maximum observable transverse magnetization (see Section 6.3.7). (In the case of Fig. 2.5 this ratio varies between 0.35 at 1 MHz and 0.14 at 100 MHz). These and other points are taken up in greater detail in Chapter 6 where the application of imaging techniques to a variety of biological systems is discussed. Finally, also of interest from the imaging viewpoint, $T_{1\rho}$ shows a strong dispersion[24,58] in the frequency (ω_1) region where $\omega_1\tau_e \sim 1$.

[58] J. G. Diegel and M. M. Pintar, Biophys. J. 15, 855 (1975).
[59] D. C. Chang and D. E. Woessner, Science 198, 1180 (1977).
[60] W. S. Hinshaw, J. Appl. Phys. 47, 3709 (1976).
[61] E. R. Andrew, P. A. Bottomley, W. S. Hinshaw, G. N. Holland, W. S. Moore, and C. Simaroj, Phys. Med. Biol. 22, 971 (1977).

2.3.6. Tumor Discrimination

The observation by Hazlewood and Nichols[62] that the proton linewidths in immature skeletal muscle were less than those in mature muscle led Damadian[63] to the conclusion that relaxation times should be greater in dedifferentiated or neoplastic tissue. He demonstrated this for the case of two malignant rat tumors, Walker sarcoma and Novikoff hepatoma, which had T_1 relaxation times at 24 MHz of 736 and 826 msec, respectively, compared with a range in normal tissues extending from 257 to 595 msec (see Fig. 2.2 which includes the latter results). T_2 showed a similar discrimination.

Confirmation of these results followed rapidly both in laboratory animals[23,31,36,64-68] and in humans.[23,29,36,46,69-72] For example, Hazlewood et al.[64] distinguished tumor, preneoplastic, and normal states in murine mammary glands on the basis of T_1, T_2, and the diffusion constant D. Iijima and Fujii[68] distinguished solid murine tumors, and Hollis et al.[23] were able to distinguish Morris hepatomas (slow growing and therefore considered to be minimally deviated) and a number of other experimental and human malignant tumors.

In 1973, Damadian and co-workers[37,69,70] undertook the most extensive examination (106 samples) of normal and malignant human tissues. These results demonstrated that T_1 values are elevated relative to the corresponding normal tissue except in the isolated case of melanoma. (See Table 2.2 which is taken from their work[37].) The discrimination is, however, not always as clear as it is with the large, fast-growing experimental tumors which are often used in animal studies. In these cases T_1 values are generally elevated with respect to *all* normal tissues. For instance, Saryan et al.[30] found for mice at 24.3 MHz that T_1 values for healthy tissues were between 186 and 526 msec, whereas tumor T_1's lay in the 593–847 msec range. Thus, in the human case, diagnostic problems could arise in the case of metastases.[32]

[62] C. F. Hazlewood and B. L. Nichols, *Physiologist* **12**, 251 (1969).

[63] R. Damadian, *Science* **171**, 1151 (1971).

[64] C. F. Hazlewood, D. C. Chang, D. Medina, G. Cleveland, and B. L. Nichols, *Proc. Natl. Acad. Sci. U.S.A.* **69**, 1478 (1972).

[65] H. E. Frey, R. R. Knispel, J. Kruuv, A. R. Sharp, R. T. Thompson, and M. M. Pintar, *J. Natl. Cancer Inst.* **49**, 903 (1972).

[66] I. D. Weisman, L. H. Bennett, L. R. Maxwell, M. W. Woods, and D. Burk, *Science* **178**, 1288 (1972).

[67] D. P. Hollis, L. A. Saryan, and H. P. Morris, *Johns Hopkins Med. J.* **131**, 441 (1972).

[68] N. Iijima and N. Fujii, *Jeol. News Ser. A* **9**(4), 5 (1972).

[69] R. Damadian, K. Zaner, D. Hor, T. DiMaio, L. Minkoff, and M. Goldsmith, *Ann. N.Y., Acad. Sci.* **222**, 1048 (1973).

[70] R. Damadian, K. Zaner, D. Hor, and T. DiMaio, *Physiol. Chem. Phys.* **5**, 381 (1973).

[71] M. Schara, M. Sēntjurc, M. Auersperg, and R. Golouh, *Br. J. Cancer* **29**, 483 (1974).

[72] J. C. Eggleston, L. A. Saryan, and D. P. Hollis, *Cancer Res.* **35**, 1326 (1975).

Frey et al.[65] showed that T_1's are elevated in the nontumorous organs of tumorous mice thereby demonstrating the presence of a tumor systemic effect. This result means that apparently unaffected organs from tumorous hosts cannot be taken as reliable control tissues. Nevertheless, recent careful studies, using graphical rather than null T_1 determinations and employing large numbers of samples, have demonstrated that many human tumors can be reliably distinguished, for instance, breast,[47,48] gastrointestinal,[73] and lung.[74]

Koutcher et al.[48] introduced the concept of a malignancy index in the course of these investigations. This combines two or more NMR parameters to yield a figure which gives improved cancer discrimination. It may well be that a similar index will be of value in a medical imaging system, when the results of a number of scans could be combined to yield a map of malignancy index rather than simply of spin density or relaxation time.

There has been much speculation concerning the origin of the T_1 elevation in cancerous tissue. Many authors believe it to be simply a matter of water content,[27-32] whereas others consider that it reflects changes in the macromolecular structure.[75-79] In this regard, it has been shown[79] that T_1 varies as a function of cell cycle in synchronized populations of HELA cells. It is clear, however, that water concentration must be a major factor.

It has been suggested[80] that T_1 elevation is related to growth rate, and higher water contents and correspondingly longer T_1's have been observed in fetal tissue, regenerating liver,[29] and in growing tumors.[32] Elevated T_1's are also seen in a variety of human diseases and are certainly not cancer specific.[72] There is therefore likely to be a wide range of currently prevalent illnesses which are amenable to NMR imaging investigations.

We have described here only proton NMR experiments. Other nuclei, such as phosphorus, may also be of diagnostic value.[81,82]

As a final cautionary note, we add that, with the exception of the experiment of Weisman et al.,[66] all the above work was performed in vitro. The

[73] M. Goldsmith, J. Koutcher, and R. Damadian, Cancer 41, 183 (1978).

[74] M. Goldsmith, J. A. Koutcher, and R. Damadian, Br. J. Cancer 36, 235 (1977).

[75] S. S. Ranade, R. S. Chaughule, S. R. Kasturi, J. S. Nadkarni, G. V. Talwalkar, U. V. Wagh, K. S. Korgaonkar, and R. Vijayaraghavan, Indian J. Biochem. Biophys. 12, 229 (1975).

[76] S. S. Ranade, S. S. Shah, K. S. Korgaonadr, S. R. Kasturi, R. S. Chaughule, and R. Vijayaraghavan, Physiol. Chem. Phys. 8, 131 (1976).

[77] S. R. Kasturi, S. S. Ranade, and S. S. Shah, Proc. Indian Acad. Sci. 84B, 60 (1976).

[78] C. F. Hazlewood, G. Cleveland, and D. Medina, J. Natl. Cancer Inst. 52, 1849 (1974).

[79] P. T. Beall and C. F. Hazlewood, Science 192, 904 (1976).

[80] P. Carver, Biophys. Soc. 13, 331a (1973).

[81] K. S. Zaner and R. Damadian, Science 189, 729 (1975).

[82] D. I. Hoult, S. J. W. Busby, D. G. Gadian, G. K. Radda, R. E. Richards, and P. J. Seeley, Nature (London) 252, 285 (1974).

effects of the rigor process and the state of the animal prior to death can give rise to pronounced changes in the relaxation time.[57] The normal assumption is that changes are only of the order of 10%. However, we cannot be certain of this until parameters measured by NMR imaging become available.

2.4. Diffusion

The use of pulsed NMR spin–echo techniques has long been established as a valuable method of measuring the translational self-diffusion coefficient D of mobile spins in a static magnetic field gradient.[83,84] Variants of the spin–echo technique have also been developed, using pulsed field gradients.[85,86] It would thus appear feasible, in principle, to measure the spatially dependent self-diffusion coefficient $D(x, y, z)$ by combining diffusion techniques with imaging experiments. As with other NMR parameters, it would seem reasonable to expect different self-diffusion coefficients in different regions of an organ, for example, membrane diffusion, bulk fluid diffusion, intra- and extracellular diffusion, etc.

In the Carr–Purcell experiment, the echo amplitude following a 180° rf pulse at time 2τ decays as a function of the pulse spacing τ. The first echo amplitude is given by

$$E(2\tau) = \exp(-2\tau/T_2)\exp(-2\tau^3\gamma^2 G^2 D/3), \qquad (2.9)$$

where G is the applied field gradient. From this expression, the effective decay time attributable to diffusion above is

$$T_{2D} = 3/(\tau^2\gamma^2 G^2 D). \qquad (2.10)$$

In the diffusion experiment described, G is thought of as an externally applied uniform gradient. However, it is known from the early work of Senftle and Thorpe[87] in 1961 on bulk susceptibility of biological tissue, both normal and malignant, that in addition to the significant differences extant between normal tissue and its malignant counterpart, there are comparable differences among the various normal tissues. These bulk susceptibility differences will cause their own local gradients which may also introduce a T_{2D}, even in "uniform" fields.[86] However, we would not expect these effects to be very significant in fields of the order of 0.1 T. However, with the much higher fields contemplated for phosphorus imaging (see Chapter 7), these localized susceptibility effects could become significant.

[83] E. L. Hahn, *Phys. Rev.* **80**, 580 (1950).
[84] H. Y. Carr and E. M. Purcell, *Phys. Rev.* **94**, 630 (1954).
[85] E. O. Stejskal and J. E. Tanner, *J. Chem. Phys.* **42**, 288 (1965).
[86] K. J. Packer, *J. Magn. Reson.* **9**, 438 (1973).
[87] F. E. Senftle and A. Thorpe, *Nature (London)* **190**, 410 (1961).

TABLE 2.4

Self-Diffusion Coefficients for Pure Water and Water in Some Biological Tissues

Material	Diffusion coefficient D (m^2 sec^{-1})	Temp. (K)	Frequency (MHz)	Ref.
Striated muscle	1×10^{-9}	—	—	88
Porcine muscle	9.2×10^{-10}	300	10.7	57
Water	2.05×10^{-9}	293	—	89
Water	2×10^{-9}	300	—	83
Frog gastrocnemius muscle	1.56×10^{-9}	298	—	90
Frog liver	1.02×10^{-9}	298	—	90
Egg yolk	0.5×10^{-9}	295	30	90
Egg yolk	0.6×10^{-9}	295	30	90
Egg white	1.66×10^{-9}	295	30	90

Some diffusion coefficients have been reported in the literature[57,83,84,88] and in a few cases the temperature dependence of D has been measured and found to be characteristic of a thermally activated process.[57] Some experimental values of D are given in Table 2.4.

The question, still unanswered, is whether D will yield information in addition to that carried in T_1 and T_2. It should be remembered, of course, that T_1 and T_2 measurements carry information relating to both translational and rotational molecular motions, whereas D, of course, reflects translational molecular motions only. Indeed, for a classical liquid of spheres following the Stokes self-diffusion law, the contribution to the spin–lattice relaxation rate for translational diffusion is given by[52]

$$\left(\frac{1}{T_1}\right)_{\text{diff}} = \frac{\pi}{5}\frac{N\gamma^4\hbar^2}{aD} = \frac{6\pi^2}{5}\gamma^4\hbar^2\frac{N\eta}{kT}, \qquad (2.11)$$

where N is the density of spins and $2\pi\hbar$ is Planck's constant.

Our feeling is, therefore, that D will be a valuable adjunct which could be useful in connection with the other NMR parameters in the quest to characterize tissue. However, compared with T_1 and T_2 values, there seems to be a paucity of diffusion measurements in the range of tissues of interest.

[88] D. C. Chang, C. F. Hazlewood, B. L. Nichols, and H. E. Rorscharch, *Nature (London)* **235**, 170 (1972).

[89] R. Mills, *J. Chem. Phys.* **77**, 685 (1973).

[90] T. L. James, "Nuclear Magnetic Resonance in Biochemistry," p. 369. Academic Press, New York, 1975.

3. Basic Imaging Principles

3.1. Effects of Linear Gradients

3.1.1. Introduction

The fundamental relationship between the Larmor precessional angular frequency ω_0 of a nuclear spin I placed in a static magnetic field B_0 is given by

$$\omega_0 = \gamma B_0 \tag{3.1}$$

where γ is the magnetogyric ratio.[1] Because of this relationship, conventional NMR absorption techniques are inherently one dimensional in the sense that absorption is measured as a function of angular frequency ω, the only variable at our disposal.

In all imaging techniques, the general objectives are to measure a number of NMR parameters (e.g., spin density and spin–lattice relaxation time T_1), as function of their spatial coordinates. The assumption, implicit in many traditional NMR studies, that such parameters are independent of the spatial coordinates of the spin system is manifestly wrong, even in non-biological materials, where boundary effects can play an important role in relaxation mechanisms.[2] The question has been all along, how to measure such spatial variables? As we shall see, one answer lies in degrading the uniformity of the static magnetic field so that the magnetizations from different parts of the specimen, which now lie in slightly different static fields, precess at different frequencies. In other words, the spatial displacements are turned into frequency displacements. All the current imaging methods employ magnetic field gradients of one sort of another to achieve spatial differenti-

---

[1] C. P. Slichter, "Principles of Magnetic Resonance" (M. Cardona, P. Fulde, and H. J. Quessier, eds.), Vol. 1 in Springer Series in Solid State Sciences. Springer-Verlag, Berlin and New York, 1978.

[2] W. H. Tantilla and D. A. Jennings, *Bull. Am. Phys. Soc.* **3**, 145 (1958); **5**, 498 (1960).

ation. In most methods the gradients used have an approximately linear spatial dependence and are time modulated in some manner. In this chapter, therefore, we shall concentrate on the effects of linear magnetic field gradients.

3.1.2. LINEAR MAGNETIC FIELD GRADIENTS

A magnetic field gradient is, in general, a tensor $\mathscr{G}$ comprising nine components and is most easily written in Cartesian coordinates as a dyadic,[3]

$$\mathscr{G} = \begin{bmatrix} \mathbf{ii}\,\dfrac{\partial B_x}{\partial x} & \mathbf{ij}\,\dfrac{\partial B_x}{\partial y} & \mathbf{ik}\,\dfrac{\partial B_x}{\partial z} \\[2ex] \mathbf{ji}\,\dfrac{\partial B_y}{\partial x} & \mathbf{jj}\,\dfrac{\partial B_y}{\partial y} & \mathbf{jk}\,\dfrac{\partial B_y}{\partial z} \\[2ex] \mathbf{ki}\,\dfrac{\partial B_z}{\partial x} & \mathbf{kj}\,\dfrac{\partial B_z}{\partial y} & \mathbf{kk}\,\dfrac{\partial B_z}{\partial z} \end{bmatrix}. \tag{3.2}$$

The Hamiltonian for an isolated spin at position $\mathbf{r}$ in a magnetic field including a gradient is[4]

$$\mathscr{H} = \hbar\omega_0 I_z + \hbar\mathbf{I}\cdot\mathscr{G}\cdot\mathbf{r}. \tag{3.3}$$

The tensor–vector product ($\mathscr{G}\cdot\mathbf{r}$) can be contracted to a vector field $\Delta\mathbf{B}$ with components along the three principal axes. Since all three components are of similar magnitude, it is clear that when added to the main static field $\mathbf{B} = \mathbf{k}B_0$, the components $\Delta B_x\mathbf{i}$ and $\Delta B_y\mathbf{j}$ can be discarded provided that $B_0 \gg |\Delta\mathbf{B}|$. Upon examination of the gradient term in Eq. (3.2), we see that discarding these components of the contracted tensor–vector product is equivalent to retaining the bottom row only of $\mathscr{G}$ in Eq. (3.2).

Alternatively, the vector–tensor product $\mathbf{I}\cdot\mathscr{G}$ becomes, on discarding the same tensor components,

$$\mathbf{I}\cdot\mathscr{G} \simeq I_z\mathbf{G} \tag{3.4}$$

where the vector $\mathbf{G}$ has the components

$$G_x = \partial B_z/\partial x, \qquad G_y = \partial B_z/\partial y, \qquad G_z = \partial B_z/\partial z. \tag{3.5}$$

3.1.3. TRANSVERSE RESPONSE

The effect of applying a linear magnetic field gradient to a noninteracting spin system is to broaden the resonance absorption profile. In the time

---

[3] H. Goldstein, "Classical Mechanics." Addison-Wesley, Reading, Massachusetts, 1950.

[4] A. Abragam, "The Principles of Nuclear Magnetism." Oxford Univ. Press (Clarendon), London and New York, 1961.

domain, the normalized transverse response function or free induction decay (FID) following a β-degree rf pulse of duration t_w and with magnetic field $\mathbf{B}_1$ $(B_1 \cos \omega t, -B_1 \sin \omega t, 0)$ is given in the rotating frame by[5]

$$\langle I_+ \rangle = \sum_i \mathrm{Tr}\{\exp(-i\,\Delta\omega I_{z_i} t)\exp(-i\gamma\mathbf{G}\cdot\mathbf{r}_i I_{z_i} t)$$

$$\times R_x^\dagger(\beta)I_{z_i}R_x(\beta)\exp(i\gamma\mathbf{G}\cdot\mathbf{r}_i I_{z_i} t)\exp(i\,\Delta\omega I_{z_i} t)I_{+i}\}\Big/\sum_i \mathrm{Tr}\{I_{+i}^2\}, \quad (3.6)$$

where Tr is the trace or diagonal sum. The resonance offset angular frequency is denoted by $\Delta\omega = \omega_0 - \omega$, and I_{+i} is the ith spin raising operator. The rf pulse rotation operator $R_x(\beta) = e^{i\beta I_x}$ corresponds to a rotation of β about the x axis, where $\beta = \gamma B_1 t_w$. When $\beta = \pi/2$, for example, we see that $R_x^\dagger(\pi/2)I_z R_x(\pi/2) = I_y$, where the dagger (†) denotes the Hermitian adjoint and $I_z = \sum_i I_{z_i}$, etc.

Implicit in Eq. (3.6) is that the rf field $B_1 \gg \Delta\omega/\gamma, (\mathbf{G}\cdot\mathbf{r})$, which amounts to saying that the spectral distribution of the rf pulse is much wider than any offset or gradient shift from exact resonance. We emphasize this point, since it is in contrast to the case for selective rf pulses discussed later.

If the spins are distributed with a normalized density $\rho(\mathbf{r})$, then for $\beta = \pi/2$, Eq. (3.6) becomes[6]

$$\langle I_+ \rangle = S(t) = \int \rho(\mathbf{r})e^{i(\Delta\omega + \gamma\mathbf{G}\cdot\mathbf{r})t}\,dr^3. \quad (3.7)$$

Because of the linear relationship between frequency and displacement imposed by the field gradient, the inverse Fourier transform of $S(t)$ yields directly the distribution of spins which lie in isochromatic planes normal to $\mathbf{G}$. We illustrate this in Fig. 3.1 for a two-dimensional case in which $\mathbf{G} = \mathbf{i}G_x + \mathbf{j}G_y$. The locus of points for which $\gamma(xG_x + yG_y) = \omega_k$, where ω_k is a constant frequency (line through x_1y_1 and x_2y_2 with slope $-G_x/G_y$), lies normal to *the vector $\mathbf{G}$ (slope $+G_y/G_x$)*. Using this constraint in selecting the isochromatic planes, the integral, Eq. (3.7) at resonance, may be reordered to give for the two-dimensional case,

$$S(t) = \int P(\omega_k)e^{i\omega_k t}\,d\omega_k, \quad (3.8)$$

where

$$P(\omega_k) = \iint \rho_{\omega_k}(x, y)\,dx\,dy, \quad (3.9)$$

[5] P. Mansfield, "Progress in Nuclear Magnetic Resonance Spectroscopy" (J. W. Emsley, J. Feeney, and L. H. Sutcliffe, eds.), Vol. 8, p. 41. Pergamon, Oxford, 1971.
[6] P. Mansfield and P. K. Grannell, *Phys. Rev.* **12**, 3618 (1975).

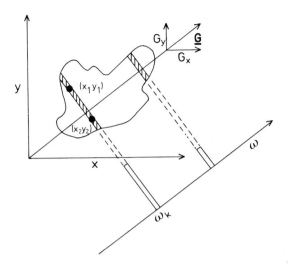

FIG. 3.1.   Sketch showing the integrated density projections of two narrow regions of a continuous arbitrary spin distribution in a gradient **G**. Note that the projection direction is orthogonal to **G** and the projection axis ω is parallel to **G**.

and $\rho_{\omega_k}(x, y)$ is the spin density subject to the constraint on ω_k. Thus, $P(\omega_k)$ is the total isochromatic spin contribution or spin density projection normal to the gradient **G**. The isochromatic constraint can be extended to three dimensions, in which case $\gamma(xG_x + yG_y + zG_z) = \omega_k$, thus allowing the three-dimensional form of Eq. (3.7) also to be reduced to a one-dimensional integral. This reduction of a three-dimensional Fourier transformation to one dimension has the unfortunate disadvantage that rather than full three-dimensional information, only the spin density projections are obtainable from a continuous spin distribution. We shall see in later sections how this difficulty may be circumvented.

3.1.4. PROJECTIONS

The idea of using linear magnetic field gradients in NMR is almost as old as NMR itself.[7-11] The early pioneers in the field[7-9] realized that gradients or inhomogeneities in the main field B_0 were often the limitation in obtaining

[7] R. Gabillard, *C.R. Acad. Sci. (Paris)* **232**, 1551 (1951).

[8] R. Gabillard, *Rev. Sci. Paris.* **90**, 307 (1952).

[9] H. Y. Carr and E. M. Purcell, *Phys. Rev.* **94**, 630 (1954).

[10] R. Bradford, C. Clay, and E. Strick, *Phys. Rev.* **84**, 157 (1954).

[11] E. R. Andrew, A. Finney, and P. Mansfield, *R. Radar Establishment Rep.* No. 167, 1970 (unpublished).

narrow absorption lines in liquids. In the measurement of self-diffusion constants,[12] large linear field gradients are required and these are often calibrated by observing the absorption profile of a regular-shaped homogeneous distribution of spins, such as a cylindrical volume of water.[10] Indeed, Gabillard[7,8] and Carr and Purcell[9] had calculated the FID shape for a homogeneous cylinder of spins in a field gradient G normal to the cylinder axis. They obtained the expression

$$S(t) = J_1(\gamma Gat)/\gamma Gat, \tag{3.10}$$

where J_1 is the first-order Bessel function and a is the cylinder radius.

These early workers also considered the FID resulting from other simple geometrical shapes containing homogeneously distributed material. Thus, although it could be said that they came close to the idea of spin imaging, none of the early workers seem to have considered using the FID to obtain structural information in a general inhomogeneously distributed spin system. This essential conceptual step in NMR imaging was taken much later in 1973 by Lauterbur[13] and independently by Mansfield and Grannell.[14]

As we have seen, two- and three-dimensional spin distributions can be transformed into one-dimensional projection distributions, but there still remain some crucial steps to be taken before an image of $\rho(\mathbf{r})$ can be obtained. There are now many different ways of obtaining $\rho(\mathbf{r})$ and these will be explained and discussed in later sections. It is nevertheless instructive to dwell a little longer on spin density projections. In the following we shall consider systems in which there is no variation in the spin distribution along one axis, so that our projections are in effect two-dimensional thin-slice transverse projections. We shall further assume the spins are uniformly distributed throughout a number of simple geometrical shapes. These shapes are listed in Fig. 3.2 together with their projection profiles in a linear gradient. The disk of Fig. 3.2a yields a semielliptical profile, which is independent of projection axis. It would not be difficult to devise a method of converting such a profile into a corresponding two-dimensional picture of the disk, with just one projection. Of course, we should need to be sure that the shape was indeed circular and not elliptical and this would require at least two projections to ascertain. Figure 3.2b is an annulus together with its corresponding projection profile. The square distribution of Fig. 3.2c gives two distinct profiles which would need to be recognized before we could deduce unequivocally that it was a square and not a rectangular or triangular distribution being shown. Variations on these shapes are given in Fig. 3.2d–f together with the expected profiles.

[12] E. O. Stejskal and J. E. Tanner, *J. Chem. Phys.* **42**, 288 (1965).
[13] P. C. Lauterbur, *Nature (London)* **242**, 190 (1973).
[14] P. Mansfield and P. K. Grannell, *J. Phys. C* **6**, L422 (1973).

Object   Projection

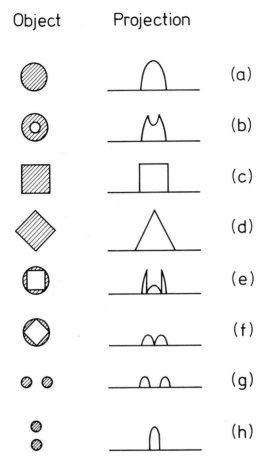

(a)

(b)

(c)

(d)

(e)

(f)

(g)

(h)

FIG. 3.2. Projections of a series of simple homogeneous geometric spin distributions in a uniform horizontal magnetic gradient.

With simple geometrical shapes and homogeneous spin distributions within these shapes, it is clear that just a few projection profiles are required to determine the shape of the object. In other words, the lower the symmetry, the more projections are required, in general, to ascertain the shape of the object uniquely.[15-20] Having said this, we should also add that from our

[15] R. N. Bracewell and A. C. Riddle, *Astrophys. J.* **150,** 427 (1967).
[16] R. Gordon, R. Bender, and G. T. Herman, *J. Theor. Biol.* **29,** 471 (1970).
[17] R. Gordon and G. T. Herman, *Commun. Assoc. Comput. Mach.* **14,** 759 (1971).
[18] B. K. Vainshtein, *Sov. Phys. Crystallogr.* **15,** 781 (1971).
[19] A. Klug and R. A. Crowther, *Nature (London)* **233,** 435 (1972).
[20] R. H. T. Bates and T. M. Peters, *N. Z. J. Sci.* **14,** 833 (1971).

list of shapes it is clear that if the object to be determined is always placed in the field gradient with a preferred orientation, for example, Fig. 3.2a,b, etc., then even though the more complicated shapes have lower symmetry, they can be uniquely differentiated. This leads one to enquire whether there are preferred orientations for more general shapes which allow a unique determination of shape. When both shape and density are varied, then there seems to be no substitute for looking at the object in as many directions as time and hardware permit. In fact, if we represent the object by an image field comprising a square array of m^2 density elements, then, roughly speaking, we must look at m projections of the object with equal angular displacements from $0°$ to $180°$ in order to completely determine the density array. This will be further clarified in Chapter 4. However, if density is kept constant, then it would appear that there is a class of shapes which can be uniquely determined in a single projection. This leads us into the area of character recognition which we shall touch upon only briefly.

Let us suppose that the object comprises a touching square matrix of 25 squares ($m = 5$) each of which may be filled or not with a uniform density of spins (Fig. 3.3a). The projection profile in G_x is shown in Fig. 3.3b, and it is clear that the same profile (Fig. 3.3b), can be obtained by filling box (1, 2) instead of box (1,1) (as drawn). Indeed, there are $^1C_5$ ways of filling one box in row 1, where

$$^nC_r = n!/[r!(n-r)!].$$

Similarly, there are $^2C_5$ ways of filling two boxes in row 2, $^3C_5$ ways of filling 3 boxes in row 3, $^4C_5$ ways of filling 4 boxes in row 4, and only one way of filling row 5 since all filled squares are indistinguishable. Thus the projection profile Fig. 3.3b is not unique, since there are 2500 ways of filling the squares, each version of which gives the same profile. Faced with one arbitrary profile it would be clearly impossible to determine the two-dimensional array shape.

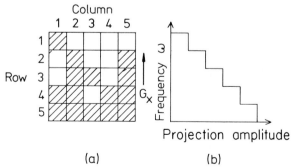

(a)　　　　　　　　(b)

Fig. 3.3.　Projection of a partially filled (shaded regions) square matrix in gradient G_x.

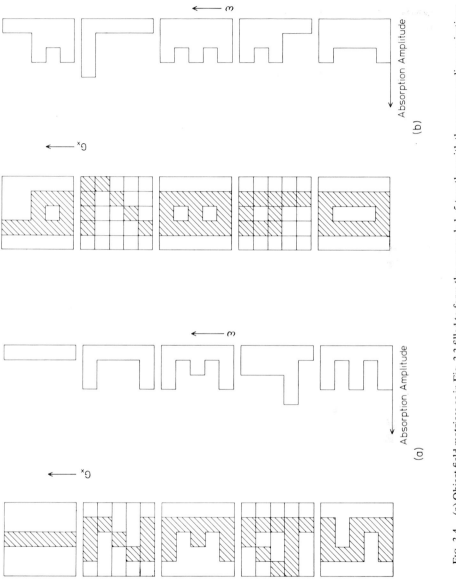

FIG. 3.4. (a) Object field matrices as in Fig. 3.3 filled to form the numerals 1–5, together with the corresponding projections in gradient G_x; (b) as in (a) but for numerals 6–0. Note in both (a) and (b), all projections are different and could thus be used for character recognition.

At this point it is worth enquiring how many *unique* profiles there are if we allow any number of boxes to be filled. This is clearly 5,[5] and although each unique profile may be produced by many different two-dimensional patterns, many of these patterns correspond to recognizable alphanumeric characters. Thus with suitable constraints placed on the object field, for example, that we are describing the numerals 0–9, we sketch in Fig. 3.4 the list of characters together with their "unique" one-dimensional profiles. The important point to note here is that two-dimensional character recognition is reduced to a set of distinguishable one-dimensional profiles.

All that has been discussed so far refers to continuously distributed material. We now turn briefly to the wider implications and possibilities of discretely distributed systems.

3.1.5. DISCRETE SPIN DENSITY DISTRIBUTIONS

When spins are discretely distributed, either artificially or naturally as in a crystal lattice, the FID from an orthorhombic three-dimensional distribution with lattice spacings a, b, c and lattice coordinates (la, mb, nc), becomes[6] from Eq. (3.7),

$$S(t) = \sum_{l,m,n} \rho_{lmn} \exp(i\,\Delta\omega t)\, \exp\big[i\gamma(laG_x + mbG_y + ncG_z)t\big], \quad (3.11)$$

where now ρ_{lmn} is the discrete spin density distribution. The same considerations regarding projections along the principal axes apply equally to discrete and continuous distributions as illustrated in Fig. 3.5 for a 5 × 5 point array of the letter P. In addition, however, because of the space between points in the array, there are certain special projection axes which allow all points in the two-dimensional distribution to be uniquely resolved as indicated by the projection in the special gradient $\mathbf{G}_s$. For the two-dimensional case illustrated, the condition to be met is

$$\mathbf{G}_s = G_x\mathbf{i} + G_y\mathbf{j} \qquad (3.12a)$$

with

$$G_x = 5G_y. \qquad (3.12b)$$

This means that one FID plus Fourier transformation contains all the information necessary to produce the original two-dimensional distribution.

Macroscopic spin distributions of interest do not occur naturally in discrete form so the ideas expressed above are not immediately applicable to spin imaging. However, we shall see later in Chapter 4 how a lattice structure may be superimposed on a continuous distribution and thus how a continuous spin distribution may be made to appear like a discrete lattice.

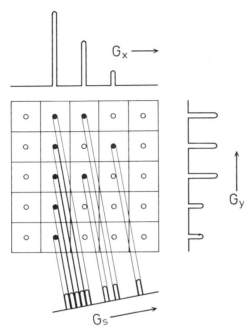

FIG. 3.5.   Projections of a discrete matrix filled to form the letter P. Projections in G_x or G_y alone do not contain sufficient information to uniquely determine the original object. Projection orthogonal to the special gradient direction G_s contains all information necessary to completely determine the original object distribution.

3.2. Equations of Motion of the Spin System

3.2.1. GENERAL FORMULATION

Although it is possible conceptually to perform many of the imaging experiments in solids, such experiments would have additional complications because of the necessity to artifically remove the large dipolar interactions which occur between static protons, for example. For this reason, we shall continue to restrict our discussion of all imaging techniques to noninteracting, or weakly interacting spin systems. In this case, and in the absence of electron coupled-exchange interactions, the spin magnetization behaves classically and is described by the phenomenological Bloch equations[4,21] in which relaxation effects are added empirically. The advantage of this approach is that it is mathematically more compact than the generalized density matrix method introduced briefly in Section 3.1.3. The method allows the

[21] E. R. Andrew, "Nuclear Magnetic Resonance." Univ. Press, Cambridge, 1956.

evolution of the total spin magnetization to be calculated either by treating it as a state vector, in which case the equation of motion is an eigenvalue equation controlled by the evolution operator, or by writing the magnetization as a column matrix, the evolution of which is controlled by the evolution matrix. However, before going on to the method of handling the equations of motion of the spin system, we first consider the vector form of the Bloch equations.[22,23] In this notation we shall write the total magnetization vector in the laboratory frame as

$$\mathcal{M} = \gamma \hbar \sum_i \mathbf{I}_i. \tag{3.13}$$

The Bloch equation in the laboratory reference frame becomes

$$\dot{\mathcal{M}} = \gamma \mathcal{M} \times \mathbf{B} - (\mathbf{i}\mathcal{M}_x + \mathbf{j}\mathcal{M}_y)/T_2 + \mathbf{k}(\mathcal{M}_0 - \mathcal{M}_z)/T_1, \tag{3.14}$$

where

$$\mathbf{B} = \mathbf{k}B_0' + \mathbf{B}_1(t), \tag{3.15}$$

in which the Zeeman term is

$$B_0' = B_0 + \Delta B + \mathbf{G} \cdot \mathbf{r}, \tag{3.16}$$

and ΔB is a field shift. $\mathcal{M}_0$ is the thermal equilibrium magnetization. The rf field $\mathbf{B}_1(t)$ rotates with angular frequency ω at or close to the spin Larmor frequency and has the form

$$\mathbf{B}_1(t) = \mathbf{i}B_1 \cos \omega t - \mathbf{j}B_1 \sin \omega t. \tag{3.17}$$

The terms T_1 and T_2 are the spin–lattice and spin–spin relaxation times, respectively.

As stated previously, the expanded form of Eq. (3.14) may be written compactly in matrix form as

$$d\mathcal{M}/dt = \mathcal{Q}\mathcal{M} + (\mathcal{M}_0/T_1), \tag{3.18}$$

where $\mathcal{M}$ is now a column matrix and the laboratory frame evolution matrix $\mathcal{Q} = \mathcal{A} + \mathcal{T}$ is given by

$$\mathcal{Q} = \begin{bmatrix} -1/T_2 & \gamma B_0' & \gamma B_1 \sin \omega t \\ -\gamma B_0' & -1/T_2 & \gamma B_1 \cos \omega t \\ -\gamma B_1 \sin \omega t & -\gamma B_1 \cos \omega t & -1/T_1 \end{bmatrix} \tag{3.19}$$

where $\mathcal{A}$ and $\mathcal{T}$ are the nondiagonal and diagonal parts of the matrix $\mathcal{Q}$, respectively.

[22] R. R. Ernst and W. A. Anderson, *Rev. Sci. Instrum.* **37**, 93 (1966).
[23] R. Freeman and H. D. W. Hill, *J. Magn. Reson.* **4**, 366 (1971).

We define a linear transformation U^{-1} to a new reference frame rotating about the z axis at angular frequency ω. Applying this to $\mathcal{M}$ we obtain

$$M = U^{-1}\mathcal{M}. \tag{3.20}$$

The equation of motion in the rotating reference frame becomes

$$dM/dt = QM + (M_0/T_1), \tag{3.21}$$

where

$$Q = U^{-1}\mathcal{A}U + \frac{\partial U^{-1}}{\partial t}U. \tag{3.22}$$

Since z components of magnetization are invariant to rotations about the z axis, $M_0 = \mathcal{M}_0$.

We emphasize that all spin rotations and matrix rotations will be described in terms of the Euler rotation angles α, β, γ, and we shall assume the convention adopted by Goldstein[3] in which all positive rotations are clockwise when looking along the positive vector axis. However, protons in a magnetic field precess anticlockwise about the magnetic vector and so correspond to negative Euler rotations.[1] This is no great problem in the rotating reference frame, since the rf field components can be chosen so as to give positive Euler angles.

The anticlockwise transformation matrix U^{-1} is therefore,

$$U^{-1} = \begin{bmatrix} \cos \omega t & -\sin \omega t & 0 \\ \sin \omega t & \cos \omega t & 0 \\ 0 & 0 & 1 \end{bmatrix} \tag{3.23}$$

Substituting Eqs. (3.19) and (3.23) into Eq. (3.22) we obtain

$$Q = \begin{bmatrix} -1/T_2 & \Delta\omega & 0 \\ -\Delta\omega & -1/T_2 & \omega_1 \\ 0 & -\omega_1 & -1/T_1 \end{bmatrix}, \tag{3.24}$$

where $\gamma B_0' = \omega_0'$, $\Delta\omega = \omega_0' - \omega$, and $\omega_1 = \gamma B_1$.

3.2.2. SOLUTIONS TO BLOCH EQUATIONS

It is helpful in considering the solutions of the Bloch equations to split Q into two parts, i.e.,

$$Q = A + T, \tag{3.25}$$

where the first term involves no relaxation terms and from Eq. (3.24) is given by

$$A = \begin{bmatrix} 0 & \Delta\omega & 0 \\ -\Delta\omega & 0 & \omega_1 \\ 0 & -\omega_1 & 0 \end{bmatrix} \qquad (3.26)$$

and the diagonal relaxation term is simply

$$T = \begin{bmatrix} -1/T_2 & & \\ & -1/T_2 & \\ & & -1/T_1 \end{bmatrix}. \qquad (3.27)$$

The solution of Eq. (3.21) is then

$$M(t) = e^{At}E(t)\chi(t)M(0) + M_0(1 - E_1), \qquad (3.28)$$

where $E(t)$ is the relaxation matrix given by

$$E(t) = e^{Tt} = \begin{bmatrix} E_2 & & \\ & E_2 & \\ & & E_1 \end{bmatrix}, \qquad (3.29a)$$

in which

$$E_1 = e^{-t/T_1}, \qquad (3.29b)$$

$$E_2 = e^{-t/T_2}, \qquad (3.29c)$$

and

$$\chi(t) = e^{-Tt}e^{-At}e^{Qt}, \qquad (3.30)$$

since A and T do not commute in general. The specific form of the propagator matrix e^{At} is discussed in detail below. The term $M(0)$ is the initial value of $M(t)$ at $t = 0$.

(1) *Intense β rf Pulses.* In the case of short, intense rf pulses, we assume that $\omega_1 \gg \Delta\omega$, in which case $\Delta\omega$ may be dropped from the matrix A [Eq. (3.26)]. Also, relaxation effects may be ignored, which corresponds to $E(t) = \chi(t) = 1$. Under these conditions the propagator matrix becomes the rf pulse rotation matrix $R_l(\beta)$, where l is the rotating frame axis along which the pulse is applied and β is the Euler nutation angle of the pulse generally given by $\beta = \gamma B_1 t_w$, where t_w is the rf pulse length.

Since we shall be requiring the various Euler rotation matrices, it is convenient to summarize them here. Positive rotations about the z axis (α and γ) take the form

$$R_z(\alpha) = \begin{bmatrix} \cos \alpha & -\sin \alpha & 0 \\ \sin \alpha & \cos \alpha & 0 \\ 0 & 0 & 1 \end{bmatrix}. \tag{3.31}$$

A positive rotation about the x axis is represented by

$$R_x(\beta) = \begin{bmatrix} 1 & 0 & 0 \\ 0 & \cos \beta & \sin \beta \\ 0 & -\sin \beta & \cos \beta \end{bmatrix}. \tag{3.32}$$

A non-Euler rotation through angle β about the y axis may be generated from Euler rotations, i.e.,

$$R_y(\beta) = R_z^{-1}(\pi/2)R_x(\beta)R_z(\pi/2) = \begin{bmatrix} \cos \beta & 0 & \sin \beta \\ 0 & 1 & 0 \\ -\sin \beta & 0 & \cos \beta \end{bmatrix}. \tag{3.33}$$

Returning to Eq. (3.28), we see that if

$$M(0) = \begin{pmatrix} 0 \\ 0 \\ M_0 \end{pmatrix}, \tag{3.34}$$

a $\pi/2$ rf pulse applied along the x axis gives from Eqs. (3.28), (3.32), and (3.34),

$$M(t_w) = \begin{pmatrix} 0 \\ M_0 \\ 0 \end{pmatrix} = M_x. \tag{3.35}$$

If at this point the rf pulse is switched off and the spin magnetization left to precess, the evolution is again described by Eq. (3.28), but in this case $\omega_1 = 0$, $\Delta\omega \neq 0$ in matrix A. Thus,

$$M(t + t_w) = e^{At}M(t_w) = R_{-z}(\Delta\omega t)M_x = \begin{bmatrix} M_0 \sin \Delta\omega t \\ M_0 \cos \Delta\omega t \\ 0 \end{bmatrix}. \tag{3.36a}$$

We note that the rotation matrices are unitary, i.e., $R_i(\alpha)R_i^{-1}(\alpha) = 1$. Also, in this notation, $R_i^{-1}(\alpha) \equiv R_{-i}(\alpha) \equiv R_i(-\alpha)$. Equation (3.36a) describes the FID of an isochromatic group of spins. For the FID of a sample experiencing a distribution of frequencies, as when a linear gradient is applied, Eq. (3.36a) would need to be summed over all isochromatic groups. When the relaxation matrix which now commutes with A is included we obtain from Eq. (3.28) with $\chi(t) = 1$ the well-known result for the $\pi/2$ pulse response,

$$M_x = M_0 e^{-t/T_2} \sin \Delta\omega t, \qquad M_y = M_0 e^{-t/T_2} \cos \Delta\omega t,$$

$$M_z = M_0(1 - e^{-t/T_1}). \tag{3.36b}$$

(2) *Spin Echoes.* If a second rf pulse $R_x(\beta)$ is applied at time τ following the initial $90°$ pulse $R_x(\pi/2)$, the resulting response may be calculated from a simple generalization of Eq. (3.28) In this case, neglecting the pulse lengths we obtain at time $t = \tau + t'$,

$$M(t) = e^{At'} E(t') R_x(\beta) \{ e^{A\tau} E(\tau) R_x(\pi/2) M(0) + M_0[1 - E_1(\tau)] \}$$
$$+ M_0[1 - E_1(t')]. \tag{3.37}$$

Straightforward multiplication of the matrices gives the following components of spin magnetization:

$$M_x = E_2(t)\{\cos \Delta\omega t' \sin \Delta\omega\tau + \sin \Delta\omega t' \cos \Delta\omega\tau \cos \beta\}M_0$$
$$+ \sin \Delta\omega t' \sin \beta \, E_2(t')M_0(1 - E_1(\tau)), \tag{3.37a}$$

$$M_y = E_2(t)\{\cos \Delta\omega t' \cos \Delta\omega\tau \cos \beta - \sin \Delta\omega t' \sin \Delta\omega\tau\}M_0$$
$$+ \cos \Delta\omega t' \sin \beta \, E_2(t')M_0(1 - E_1(\tau)), \tag{3.37b}$$

$$M_z = E_1(t')\{\cos \beta(1 - E_1(\tau)) - \sin \beta \cos \Delta\omega\tau \, E_2(\tau)\}M_0$$
$$+ M_0(1 - E_1(t')). \tag{3.37c}$$

For the particular case where $\beta = \pi$, the equations simplify considerably. In this case and for $t' = \tau$, M_x vanishes and $M_y(2\tau) = -E_2(2\tau)M_0$, independent of offset $\Delta\omega$. The spins in a specimen comprising a distribution of isochromatic regions would therefore all come into phase at this time to form a spin echo.[9] The peak amplitude of this echo decays with T_2. The z component is $M_z = M_0(1 - E_1(\tau))^2$. Spin echoes also occur for other angles and in particular for $\beta = \pi/2$, although in this case from Eq. (3.37a), $M_x = 0$ only for a symmetric isochromatic distribution.[24]

3.3. Selective Irradiation

3.3.1. GENERAL

In most conventional NMR experiments, the excitation spectrum of the applied rf pulses is many times broader than the spin absorption spectrum

[24] E. L. Hahn, *Phys. Rev.* **80**, 580 (1950).

and in this case, as we have discussed at some length in Section 3.2, the spin response to such pulses is straightforward to calculate. However, there is a class of experiments generally referred to as selective irradiative experiments, and originally introduced by Tomlinson and Hill,[25] in which the intrinsic absorption linewidth is much greater than the applied rf spectral width. Of course, the related ideas of selectively exciting particular transitions in both quadrupolar broadened resonances at high and low field[26,27] and multispin–species systems[28,29] has been used for many years to study relaxation mechanisms and spin dynamics. The selective irradiation experiments of Tomlinson and Hill[25] were motivated by the need to decouple complicated high-resolution spectra using variants of the "spin tickling" or steady-state double-resonance technique in order to obtain unambiguous chemical shift and exchange coupling assignments.[30]

3.3.2. ANALYSIS

In selective excitation experiments,[31–35] the spin absorption linewidth is arranged to be much wider than the rf spectral width by applying a magnetic field gradient during the rf irradiation period. Naturally, the formalism developed in Section 3.2 can be applied to selective pulses. We shall be particularly interested in situations where broadened spin systems are perturbed by the pulse to produce a nonequilibrium transverse response. In this analysis, therefore, we shall assume that rf pulses are always applied for times less than T_1 so that $E_1 = 1$ in the relaxation matrix [Eq. (3.29a)], but we shall retain the E_2 terms.

We consider a homogeneous spin distribution in the form of a cylinder (Fig. 3.6), the axis of which lies along the z direction. A linear static gradient G_z exists over the sample. We shall assume that in the initial equilibrium state, the spin magnetization is uniformly distributed along the cylinder axis with

[25] B. L. Tomlinson and H. D. W. Hill, *J. Chem. Phys.* **59**, 1775 (1973).

[26] R. E. Slusher and E. L. Hahn, *Phys. Rev.* **166**, 332 (1968).

[27] A. G. Redfield, *Phys. Rev.* **130**, 589 (1963).

[28] S. R. Hartmann and E. L. Hahn, *Phys. Rev.* **128**, 2042 (1962).

[29] F. M. Lurie and C. P. Slichter, *Phys. Rev.* **133**, A1108 (1964).

[30] W. A. Anderson, *Phys. Rev.* **102**, 151 (1956). See also Abragam.[4]

[31] A. N. Garroway, P. K. Grannell, and P. Mansfield, *J. Phys. C* **7**, L457 (1974).

[32] P. C. Lauterbur, C. S. Dulcey, C.-M. Lai, M. A. Feiler, W. V. House, D. Kramer, C.-N. Chen, and R. Dias, *Proc. Ampere Congr., 18th* **1**, 27 (1974).

[33] P. Mansfield, A. A. Maudsley, and T. Baines, *J. Phys. E* **9**, 271 (1976).

[34] D. I. Hoult, *J. Magn. Reson.* **26**, 165 (1977).

[35] P. Mansfield, A. A. Maudsley, P. G. Morris, and I. L. Pykett, *J. Magn. Reson.* **33**, 261 (1979).

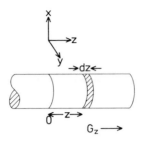

FIG. 3.6.   Slice selection from an extended object in a gradient G_z.

a magnetization per unit length $m_0(z)$ given by

$$m_0(z) = \int\int m(x, y, z) \, dx \, dy. \tag{3.38}$$

The total equilibrium spin magnetization M_0 is therefore,

$$M_0 = \int m_0(z) \, dz. \tag{3.39}$$

Initially, we shall apply an rf pulse of constant amplitude $\mathbf{i}B_1$ in a reference frame rotating with angular frequency ω and with the offset $\Delta B = 0$. In this case, $\omega = \gamma(B_0 + G_z z)$.

Spins which lie in the plane z are therefore at exact resonance and interact strongly with the rf pulse. Spins either side of this plane will be progressively less affected the further they lie from the isochromatic plane. The precise behavior of the spin system is best calculated for a particular pulse by considering spins displaced Δz from z which are then off resonance with respect to the pulse by

$$\Delta\omega_z = \gamma \, \Delta z G_z = \gamma b. \tag{3.40}$$

We consider the spins in the plane sheet of thickness dz at Δz. The equilibrium spin magnetization in the sheet is

$$\delta M_0 = m_0 \, dz. \tag{3.41a}$$

The motion of this magnetization

$$\delta M(t) = m(t) \, dz, \tag{3.41b}$$

in the tilted rotating frame follows a cone of precession about an effective field

$$\mathbf{B}_{\text{eff}} = B_1\mathbf{i} + b\mathbf{k}, \tag{3.42}$$

which makes a non-Euler angle $\theta = \tan^{-1}(B_1/b)$ with the z axis.

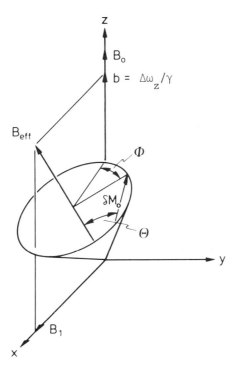

FIG. 3.7. Precession of elemental spin magnetization δM_0 in the tilted rotating reference frame. [From P. Mansfield, A. A. Maudsley, P. G. Morris, and I. L. Pykett, *J. Magn. Reson.* **33**, 261 (1979).]

The behavior of the elemental spin magnetization $\delta M(t)$ is shown in Fig. 3.7, and is described by a modification of Eq. (3.28), that is to say,

$$\delta M(t) = e^{Qt}\,\delta M(0), \tag{3.43}$$

where we set $1/T_1 = 0$ in the T matrix [Eq. (3.27)] throughout this section.

Ignoring T_2 effects for the moment, i.e., by replacing Q by A [Eq. (3.26)], we see that the propagator matrix e^{At} in Eq. (3.43) represents a negative Euler rotation about $\mathbf{B}_{\text{eff}}$. The tilted axis B_{eff} is itself a positive non-Euler rotation about y. Thus referred to the nontilted frame we see that

$$e^{At} = R_y(\theta)R_z^{-1}(\phi)R_y^{-1}(\theta), \tag{3.44}$$

or, alternatively,

$$R_z^{-1}(\phi) = \exp\{R_y^{-1}(\theta)AR_y(\theta)\}, \tag{3.45}$$

which, from Eqs. (3.26) and (3.33), yields an effective evolution matrix A' in the tilted reference frame of

$$A' = \begin{bmatrix} 0 & \omega_{\text{eff}} & 0 \\ -\omega_{\text{eff}} & 0 & 0 \\ 0 & 0 & 0 \end{bmatrix}. \tag{3.46}$$

If the selective pulse is applied for a time t_w, the total precessional Euler angle ϕ of the spins is, from Eq. (3.45) and (3.46), therefore

$$\phi = \omega_{\text{eff}} t_w, \tag{3.47a}$$

where

$$\omega_{\text{eff}}^2 = \omega_1^2 + \Delta\omega_z^2. \tag{3.47b}$$

If we now include the relaxation matrix $E'(t)$ in the tilted frame, with the plausible assumption that T_2' in this frame is offset independent, we obtain for the propagator matrix

$$e^{Qt} = R_y(\theta)R_z^{-1}(\phi)E'(t)R_y^{-1}(\theta). \tag{3.48a}$$

Using the rotation matrices, Eqs. (3.31) and (3.33) and Eq. (3.29a) and multiplying, we obtain

$$e^{Qt} = \begin{bmatrix} \cos^2\theta\cos\phi\, E_2' + \sin^2\theta & \cos\theta\sin\phi\, E_2' & \sin\theta\cos\phi\,[1 - E_2'\cos\phi] \\ -\sin\phi\cos\theta\, E_2' & \cos\phi\, E_2' & \sin\theta\sin\phi\, E_2' \\ \cos\theta\sin\theta\,(1 - E_2'\cos\phi) & -\sin\theta\sin\phi\, E_2' & \cos^2\theta + \sin^2\theta\cos\phi\, E_2' \end{bmatrix}. \tag{3.48b}$$

When $E' = 1$, this rotation matrix thus constitutes first a negative rotation about y through θ to bring the z' axis of B_{eff} along the z axis, an Euler rotation through $-\phi$ about z, and finally a positive rotation about y to take z back to z'.

Taking

$$\delta M(0) = \begin{bmatrix} 0 \\ 0 \\ \delta M_0 \end{bmatrix}, \tag{3.49}$$

and substituting Eqs. (3.48b) and (3.49) into Eq. (3.40), we obtain the following components of the elemental spin magnetization[35] $\delta M(t)$,

$$\delta M_x = m_x\, dz = \sin\theta\cos\theta\,(1 - E_2'\cos\phi)m_0\, dz, \tag{3.50a}$$

$$\delta M_y = m_y\, dz = \sin\theta\sin\phi\, E_2'm_0\, dz, \tag{3.50b}$$

$$\delta M_z = m_z\, dz = (\cos^2\theta + \sin^2\theta\cos\phi\, E_2')m_0\, dz. \tag{3.50c}$$

The magnitude of the total elemental magnetization δM_0 is, of course, constant when relaxation effects are excluded; that is, when $E_2' = 1$, namely,

$$\delta M_x{}^2 + \delta M_y{}^2 + \delta M_z{}^2 = \delta M_0{}^2 \qquad (3.50\text{d})$$

As with previous pulses, a 90° selective pulse corresponds to the condition $\omega_1 t_w = \pi/2$. The spin components, Eqs. (3.50a, b, and c), may be plotted as a function of offset b following a 90° (or other) selective pulse applied along x (see Fig. 3.7), and yield the nonequilibrium spin components existing at time t_w for a homogeneous spin system in a linear field gradient. The normalized components m_x/m_0, m_y/m_0, and m_z/m_0 are displayed graphically in Figs. 3.8a–c, respectively, for the case when $E_2 = 1$. These show a depletion of magnetization m_z/m_0 and a maximum in m_y/m_0 as expected on resonance when $b = 0$. An interesting point which we shall again refer to later is that

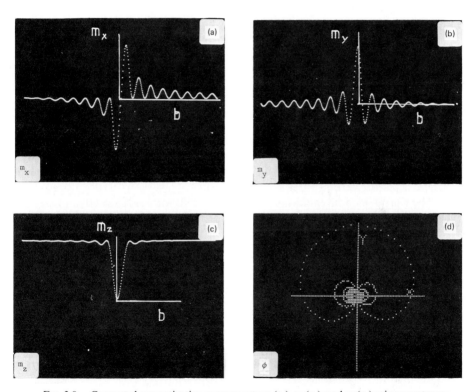

FIG. 3.8.   Computed magnetization components $m_x(t_w)$, $m_y(t_w)$, and $m_z(t_w)$ using a rectangular 90° selection pulse and a long homogeneous spin distribution (see Fig. 3.6); (a), (b), and (c) correspond to Eqs. (3.50a–c), respectively; (d) is the phase diagram of the spin distribution at time t_w in the x,y plane obtained by plotting m_x versus m_y. It demonstrates most clearly the poor directionality of the magnetization lobe.

m_y/m_0 and m_x/m_0 are reminiscent of the absorption and dispersion line shapes observed in normal NMR experiments. The actual distribution of magnetization existing in the x–y plane immediately following the selective pulse is shown graphically in Fig. 3.8d by plotting m_y/m_0 versus m_x/m_0 as a polar phase diagram. It is immediately obvious that although the phase lobe gives a net positive y magnetization and zero net x magnetization, a substantial fraction of the perturbed spin magnetization does not contribute to the observed FID and is therefore wasted.

The actual values of the total transverse magnetization components $M_x(t_w)$ and $M_y(t_w)$ immediately following the selective pulse at t_w may be obtained by integrating Eqs. (3.50a and b). Substituting for θ and ϕ in Eqs. (3.50a and b), we find the total x component is given by[35]

$$M_x(t_w) = \frac{m_0 B_1}{G_z} \int_{-\infty}^{\infty} db\, b\, \frac{1 - \cos \gamma t_w (B_1{}^2 + b^2)^{1/2}}{B_1{}^2 + b^2}. \tag{3.51}$$

We note from Fig. 3.8a and b that m_x/m_0 is an antisymmetric distribution which means that $M_x(t_w) = 0$ for all t_w. The y component, however, gives

$$M_y(t_w) = \frac{m_0}{G_z} \int_{-\infty}^{\infty} db\, \frac{B_1}{(B_1{}^2 + b^2)^{1/2}} \sin \gamma t_w (B_1{}^2 + b^2)^{1/2}, \tag{3.52}$$

which may be summed analytically to give,[35]

$$M_y(t_w) = \frac{\pi B_1 m_0}{G_z} J_0(\gamma B_1 t_w), \tag{3.53}$$

where J_0 is the zeroth-order Bessel function.

The normalized integrand m_y/m_0 of Eq. (3.52) for a $90°$ selective pulse is given by[35]

$$m_y/m_0 = (\pi/2)\, \mathrm{sinc}\, \gamma t_w (B_1{}^2 + b^2)^{1/2} \tag{3.54}$$

which approaches a normal sinc function when $B_1 \ll b$. We shall refer to this point later.

The evolution of the transverse y component of magnetization [Eq. (3.53)] is plotted in Fig. 3.9 for two values of B_1. This result shows that finite transverse magnetization exists very close to the time origin. This comes about because of the very broad frequency distribution of the spins assumed in the calculation. Short pulses are able to excite a large number of spins well off resonance, but through small nutation angles. However, as t_w increases, the m_y/m_0 distribution, Eq. (3.54), narrows, thus exciting fewer spins, but through larger nutation angles and in such a way as to maintain the total observed signal [Eq. (3.53)] at a high level. Thus a $90°$ selective pulse produces quite a large M_y component. As illustrated in Fig. 3.9, a $180°$ selective pulse can

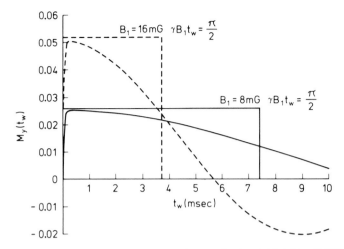

FIG. 3.9.   The total transverse y component of magnetization, $M_y(t_w)$ [Eq. (3.53)] as a function of the selective pulse length t_w. The parameters have been chosen to correspond to two
typical situations: $B_1 = 16$ mG (dotted) shows the 90° case when $t_w = 3.7$ msec, and magnetization reversal occurring when $t_w \geqslant 5.6$ msec, $B_1 = 8$ mG (solid line) shows the expected
magnetization around the 90° condition when $t_w = 7.4$ msec. [From P. Mansfield, A. A. Maudsley, P. G. Morris, and I. L. Pykett, *J. Magn. Reson.* **33**, 261 (1979).]

invert the magnetization, but since it is a nonlinear system, the inverted
signal magnitude is only about $\frac{2}{3}$ that following a 90° pulse.

3.3.3. Transverse Response in an Unswitched Field Gradient

The theory presented shows that following a single rf pulse, of length t_w,
applied to a homogeneous spin distribution of infinite extent residing in a
linear magnetic field gradient, an m_x and m_y spin distribution exists which
constitutes the initial conditions for free precession in the field gradient if left
switched on. The FID response $F(t)$ in the gradient may be calculated from the
complex Fourier transform of the two distributions using the expression

$$F(t) = rp \int e^{i\omega t}[M_y(t_w) + iM_x(t_w)](d\omega/G_z) = F_y(t) - F_x(t). \quad (3.55)$$

Figure 3.10 shows the separate Fourier-transformed components $F_y(t)$,
$F_x(t)$, and the difference signal $F(t)$ that one might expect to observe in an
actual experiment.

Since m_x is antisymmetric in the offset b, **gradient reversal immediately
following the pulse will cause $F_x(t)$ to add to F_y rather than subtract.**[35] We
have used the $F_y(t)$ and $F_x(t)$ expressions to **predict the effect of field reversal,**

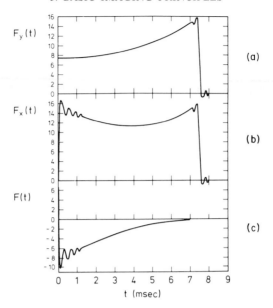

FIG. 3.10. Transient signals following a rectangular selective pulse calculated from Eq. (3.55): (a) $F_y(t)$, (b) $F_x(t)$, (c) $F(t)$. [From P. Mansfield, A. A. Maudsley, P. G. Morris, and I. L. Pykett, *J. Magn. Reson.* **33**, 261 (1979).]

or equivalently, the application of short 180° rf pulses, following an initial selection pulse of duration 7.4 msec and for B_1 values producing nutation angles of 90°, 60°, and 30°. The experimental results of Hoult[36] (Fig. 3.11a) confirm the predicted behavior shown in Fig. 3.11b, and also demonstrate another interesting feature of selective pulses, namely that for nutations less than 90°, the initial amplitude of the function $F(t)$ immediately following the first pulse rapidly decreases with decreasing nutation angle. This means, of course, that $F_y - F_x$ exactly cancels and the shapes of both functions approach rectangular temporal functions. When refocused, therefore, the echo–like signal is a very good rectangular approximation with amplitude double that of either F_y or F_x. The spin system behaves in an approximately linear manner and the signal shape, $F_y(t)$, approaches the shape of the rf excitation pulse, in this case a rectangular pulse.

Of course, $F_y(t)$ and $F_x(t)$ are not directly observable in these types of experiment, but are implied from their difference in $F(t)$ or their sum in an echo experiment.[36] However, it is possible to isolate $F_y(t)$ in an experiment involving both selective and nonselective pulses.[35] This is shown in Fig. 3.12. Figure 3.12a shows the Fourier transform of $F(t)$, the FID following a square

[36] D. I. Hoult, *J. Magn. Reson.* **35**, 69 (1979).

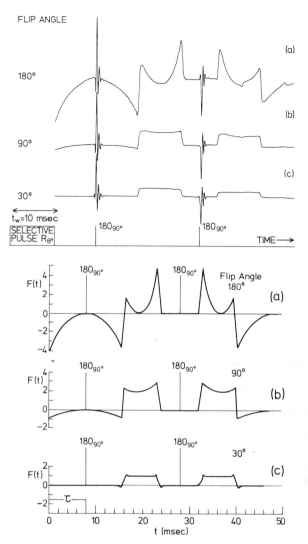

FIG. 3.11a. Response of homogeneous spin distribution in a linear gradient to various rectangular selective pulses followed by two nonselective 180° refocusing pulses. (a) The initial selection pulse nutation angle (flip angle) is 180°. Note the large initial signal and refocusing. The refocused signal corresponding to the spin evolution *during* the initial excitation is a grossly distorted square wave, both here and following the second 180° refocusing pulse. In (b) and (c), the initial selection pulse nutation angle is reduced to 90° and 30°, respectively. The initial FID following the selection pulse reduces considerably, but the refocused signals approach a rectangular shape corresponding to the limit of a 0° nutation to the rf excitation pulse shape. [From D. I. Hoult, *J. Magn. Reson.* **35**, 69 (1979).]

FIG. 3.11b. Theoretical responses to rectangular selective pulses calculated from Eqs. (3.51)–(3.53). The parameters have been chosen to correspond to the experimental results of Fig. 3.11a, above.

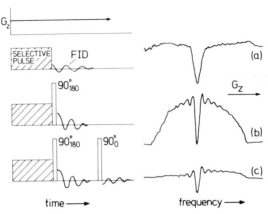

FIG. 3.12. Selective pulse experiments performed in a static field gradient. (a) Fourier-transformed FID following a rectangular selective pulse. (NB. The signal phase has been reversed.) (b) Combined effect of a selective 90° pulse and a short nonselective 90° pulse when applied to a cylindrical sample of water. The short nonselective pulse produces the overall projection profile. The chasm is produced by the selective pulse and in effect inspects the depletion of the localized z magnetization (see Fig. 3.8c). (c) Fourier transform of the transverse signal following a second nonselective 90° pulse as sketched. The transformed signal corresponds to the inverted m_y spin distribution of Eq. (3.50b) and Fig. 3.8b. [From P. Mansfield, A. A. Maudsley, P. G. Morris, and I. L. Pykett, *J. Magn. Reson.* **33**, 261 (1979).]

selective pulse observed in the same gradient. Figure 3.12b shows the frequency response of the same selective pulse immediately followed by a nonselective 90° rf pulse in which the rf phase has been reversed (denoted $90°_{-x}$). The effect of the nonselective pulse is to (1) return the selected magnetization along the z axis, and (2) produce a projection profile signal of the whole specimen (*sans* the selected region), which in this case was a cylinder of water. The chasm in the cylinder profile is thus m_z. If all transverse signal components are allowed to dephase following the first $90°_{-x}$ pulse, m_y is effectively stored along the z axis and may be inspected by further application of a second nonselective $90°_x$ pulse in phase with the selective pulse carrier. Figure 3.12c shows the experimentally observed m_y distribution of Eq. (3.50b) and Fig. 3.8b. The signal phases of all three results have been inverted for convenience.

Finally, we remark that for low B_1 fields, $F_y(t)$ being a good approximation to a rectangular function implies that m_y/m_0 is a sinc function, as suggested by Eq. (3.54) for $B_1 \ll b$. Thus, all theoretical predictions concerning a single square pulse irradiation are verified experimentally in the case when relaxation effects are ignored.

3.3.4. DISCRETE EXCITATION

So far we have described what happens when the selective excitation pulse is rectangular. In the original selective excitation method, as used for NMR imaging, in fact, the pulse comprised a number of discrete frequencies. These discrete frequencies arise naturally when dealing with Fourier transformations having a finite number of points.

The procedure is therefore to select the desired rf spectral distribution and set this up automatically in a particular block of computer memory. In our case, the block comprises 512 memory locations, and if interrogated on an oscilloscope display, would appear as in the sketch of Fig. 3.13. Here we illustrate a typical three- (even) point-high spectrum of constant height, and mean offset 4 $\Delta\omega$, where the angular frequency difference per point $\Delta\omega$ is

$$\Delta\omega = \pi/t_w. \qquad (3.56)$$

The angular frequency per point quoted is half the usual value $(2\pi/t_w)$ since for historical reasons, we use a nonsymmetric, real fast-Fourier transform based on the Bergland[37] modification of the Cooley–Tukey[38] algorithm. In this algorithm, $2N$ real points are transformed to N real points. In our case, $2N$ frequencies are transformed to N time domain points ($N = 256$). Because of this halving of the number of points, the time domain data contain "ghost" frequencies at $(2n + 1)\pi/t_w$ when n is an integer. This is an added complication which, although having some advantages, does not occur with $N \rightarrow N$ point complex transforms. However, if we ignore the ghost points by setting them equal to zero in the frequency domain (as in Fig. 3.13), then the output pulse contains frequencies at $2\pi n/t_w$, where t_w is the excitation pulse length. Thus

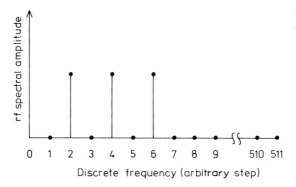

FIG. 3.13.   Discrete rf spectral distribution of a "three even-point-high" selective pulse.

[37] G. D. Bergland, *Commun. ACM* **11,** 703 (1968).
[38] J. W. Cooley and J. W. Tukey, *Math. Comput.* **19,** 296 (1965).

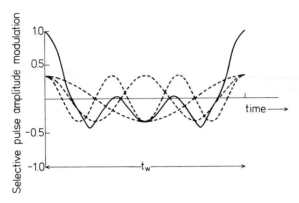

FIG. 3.14.   Amplitude modulation of a three even-point-high 90° selective pulse with spectral distribution as in Fig. 3.13. Broken lines show individual frequency components.

the desired spectral distribution is Fourier transformed and the result strobed out from the computer and used to amplitude modulate the fixed carrier frequency of the rf pulse using either a direct digital attenuator or a digital–analog converter (DAC) and linear gate.[33] Hardware details will be discussed in Chapter 8.

Figure 3.14 shows the expected rf envelopes for a three-point-high selective pulse with offset frequencies of $2\pi/t_w$, $4\pi/t_w$, and $6\pi/t_w$. Note that all rf signals rephase exactly at $t = t_w$. For a many-point-high spectrum, the rf envelope approaches a discrete sinc function which has the same periodic property that all rf signals rephase, in general, at times $t = mt_w$ (m is an integer). If we think of each discrete frequency as exciting its own independent nuclear signal, as in the previously discussed case for a one-point-high spectrum, then it is easy to see that the center frequencies of each adjacent excited line will all have relative phases differing by 2π. In other words, the separate rotating reference frames and, hence, signals of each excited line, although all different, will come into phase periodically and, in particular, when $t = t_w$. This result follows from the fact that the total accumulated phase of the nth offset point, when excited, is

$$\Phi = n\,\Delta\omega t_w = 2\pi n, \qquad (3.57)$$

since for n even,

$$\Delta\omega = 2\pi/t_w. \qquad (3.58)$$

Two important points emerge from this discussion. The first is that provided the individual NMR responses to the discrete frequencies do not overlap or interfere too much, separate and independent rotating frames may be defined. In fact, from Eq. (3.58) and the 90° pulse condition $\gamma B_1 t_w = \pi/2$, we see that the discrete lines are spaced $4B_1$ units apart which corresponds to the second zero crossing of m_y in Fig. 3.8b. So there will in general be some mild interference effect since m_y at this point has not decayed completely to zero. The second point is that for $\beta < \pi/2$, we have seen that the time response function $F_y(t)$ approximates closely to the rf pulse shape. Put together, these two points mean that any discrete selective excitation pulse shape will induce a nuclear signal component $F_y(t)$, which for low B_1 will follow the rf excitation envelope. To the extent that a continuous frequency selective pulse is the limit of touching discrete frequencies, these ideas carry over to analog-shaped selective pulses containing a continuum of frequencies. These ideas have recently been discussed by Hoult[36] and analyzed from a somewhat different point of view.

From the viewpoint developed here, we see that for more complicated selective pulses derived from given excitation spectra, the rf pulse envelope $B_1(t)$ becomes time dependent in the rotating reference frame defined by the carrier frequency. Each time domain point is strobed from the computer and applied via the DAC as a constant signal for a given time interval τ where $n\tau = t_w$. The manner in which this is achieved means that the spin magnetization for a *multiline* spectrum evolves exactly as in Eq. (3.43) but under a series of propagator matrices, each applied for time τ, and with matrix elements modulated according to the time data output. Neglecting the T operator, this becomes from Eq. (3.43),

$$\delta M(n\tau, \Delta\omega) = \hat{O} \prod_n e^{A_n\tau} \delta M(0), \tag{3.59}$$

where $\hat{O}$ is a product-ordering operator and

$$A_n = \begin{bmatrix} 0 & \Delta\omega & 0 \\ -\Delta\omega & 0 & \omega_1(n) \\ 0 & -\omega_1(n) & 0 \end{bmatrix}. \tag{3.60}$$

That is to say only $\omega_1(n)$ [or, alternatively, $B_1(t)$] varies, the offset $\Delta\omega$ remaining constant during the evolution of δM up to $n\tau = t_w$. This exact calculation gives one point of the distribution function $\delta M(\Delta\omega)$ at offset $\Delta\omega$ in the frequency domain and must therefore be repeated for all values of offset to obtain the full frequency distribution profiles. The importance of the ordering operator will become apparent when we discuss particular examples.

The simulated selective irradiation response for a four even-point-high excitation spectrum of constant amplitude and with zero ghost points is shown in Fig. 3.15. The m_x distribution (Fig. 3.15a) shows considerable oscillation well outside the four-point excitation region. The slight asymmetry of sidebands arises from the proximity of a higher frequency mirror

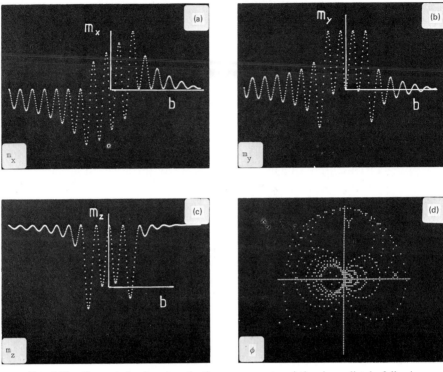

FIG. 3.15. Computed spin magnetization components existing immediately following a four even-point-high selective 90° pulse: (a) m_x, (b) m_y, (c) m_z, and (d) the x–y phase distribution of magnetization. Note the slight asymmetry in amplitude of all spectra caused by the relatively small frequency offset from the origin, in this case to the right of each spectrum.

spectrum, which exists in all our calculated spectra and which we have removed from the display. In a real experiment using single sideband modulation, the reflection spectrum, in this case above resonance, would not occur. Of course, with larger offset from resonance, the interference between the high and low sidebands which causes the asymmetry would automatically decrease. In fact, in imaging experiments using selective irradiation, the presence of the m_x distribution is not so critical as it might at first seem, provided that the total integral under the distribution is indeed zero. For one-point high this is the case and it is also true here provided the spin density over the excited range of frequencies is constant. From Fig. 3.15a it is clear that the constraint on spin density being constant over the excitation band becomes increasingly harder to justify for wide strips of excited material. Turning now to m_y (Fig. 3.15b) again we see extensive sideband modulation well outside the four central peaks which correspond to the four discrete frequencies actually excited. The integrated signal here, M_y, is certainly not zero, but the clear gaps between excited points means that useful spin magnetization is not contributing to the observed signal. The sideband oscillations are not too important in this type of excitation since their integrated contribution in any case approximates to zero. Thus, m_y is a reasonable approximation to a discrete rectangular excitation spectrum. An interesting feature of this result is that although m_y is constant over the four main central peaks, because of the substantial m_x variations, the depletion of the equilibrium z magnetization is not constant, as illustrated in Fig. 3.15c. (The slight asymmetry of the center peaks is, again, a manifestation of interference of the high-frequency mirror image.) The polar plot of magnetization in the x-y plane is shown in Fig. 3.15d and illustrates the y lobe consistent with a net positive M_y and a general symmetry about the y axis.

Although not shown here, the time domain pulse envelope looks similar to Fig. 3.14 and is therefore a symmetric sinc-type function exhibiting an initial decay up to $t_w/2$ and then a time-reversed growth fully rephasing at $t = t_w$. In this case, the ordering operator in Eq. (3.59) $\hat{\mathcal{O}} = 1$ for the whole pulse sequence. The missing signal components between the peaks in Fig. 3.15b may be made to contribute, thus improving the efficiency of excitation by employing the ghost points in the initial excitation spectrum. As pointed out in Eq. (3.56), the ghost points will introduce signal phases differing by π radians, so that, for example, if we set up a six-point-high excitation spectrum comprising three even and three odd points all of equal amplitude, the total signal amplitude at time t_w will be of necessity zero, since the in-phase even-point signals will exactly cancel the 180° phase-shifted signals from the odd points. Bearing in mind our previous discussion concerning the way in which the shape of $F_y(t)$ follows the rf excitation pulse envelope, the rf signal from such a six-point spectrum will start with all points in phase at $t = 0$, but each rf component will gradually dephase to give zero net signal at $t = t_w$. Here, therefore, we have a Fourier transform of the excitation spectrum which is *not* symmetric in the time interval t_w and so the ordering operator $\hat{\mathcal{O}} \neq 1$ and must be applied. In fact, the simple artifice of *reversing* the time ordering of the six-point-high excitation spectrum just described brings all rf and, hence, NMR signal components into phase at time t_w, even though the rf components start out in antiphase.

The simulated response to such a reverse-ordered three even-point-high three ghost-point-high selective pulse is shown in Fig. 3.16. The m_x component (Fig. 3.16a) shows much less sideband oscillation. However, although symmetric, so that M_x would vanish for a narrow excited strip, the magnitudes of the actual extrema of m_x are comparable to those of m_y (Fig. 3.16b). The m_y signal is much improved and approaches the ideal rectangular function. The m_z distribution still shows oscillation which in fact arises from

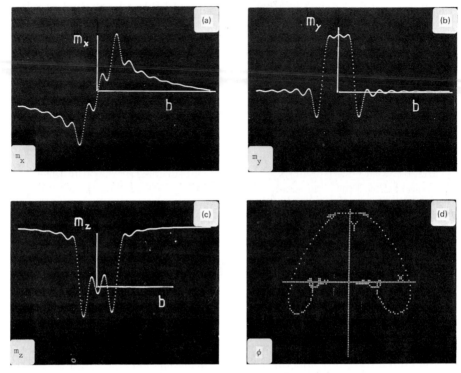

FIG. 3.16. Computed spin magnetization components existing immediately following a reverse-ordered three even, three odd-point 90° selective pulse: (a) m_x, (b) m_y, (c) m_z, and (d) the x–y phase distribution of magnetization. This indicates the somewhat improved directionality of the phaselobe.

partial inversion (low dips) rather than simple saturation of the localized magnetization. A summary of the total transverse magnetization in the x–y plane is shown in the interesting polar plot of Fig. 3.16d.

Returning to the point made earlier concerning the similarity of m_y and m_x with the absorption and dispersion spectra, respectively, it seems clear from Fig. 3.16 that these components are related. Indeed, m_y is very close to the cosine transform of the rf pulse envelope and m_x close to the sine transform. It would thus appear that in the limit of a well-chosen excitation spectrum, with perhaps more phase components in addition to the ghost points, the Kramers–Kronig relationship[1,4] prevents m_x from being zero. To reduce its value from that achieved so far we must do something extra.

Before turning to ways of reducing m_x, we should perhaps reiterate that for sufficiently narrow selected strips, it is actually unnecessary to be overly concerned about its size since as we shall see in more detail later, the gradient switch following selection in an imaging experiment effectively freezes in the m_x distribution existing at time t_w. This quenches any differential spin evolution between the positive and negative halves of the m_x distribution and so the *effective* distribution controlling the observed imaging signal is m_y as desired. Of course, if thicker strips are of interest, the spin distribution across the selected strip may be far from uniform. In this case the positive and negative components of m_x may not in general cancel and this could lead to signal phase shifts *along* a selected strip during the signal-read phase in an imaging experiment.

3.3.5. Focused Selective Excitation

We now turn to our final subject for discussion under the general selective irradiation heading, which we refer to as focused selective excitation. This method depends on a gradient reversal to produce a focused type of spin echo from an initially tailored excitation pulse. The idea of reversing field gradients to form echo-type signals in selective irradiation experiments was originally suggested by Hoult[34] to increase the signal obtained from rectangular excitation pulses. The idea has since been applied by Sutherland and Hutchison[39] to analog-shaped Gaussian pulses. Their pulse sequence and expected NMR signal is sketched in Fig. 3.17. Since the rf pulse shape is symmetric, starting and finishing with zero amplitude, one does not expect an m_y signal component at $t = t_w$. However, by reversing the gradient at t_w, a strong echolike signal is produced at $t = 3t_w/2$. At this time, the signal produced contains both m_x and m_y components, but the "dispersive" x-component extrema are somewhat smaller than m_y when a Gaussian profile is used. It would therefore seem that gradient reversal does have a useful effect in this type of selective irradiation, provided the spin evolution is interrupted at $t_w/2$ following the selective pulse.

In focused selective excitation, the pulse is chosen to spread the spin magnetization in the x–y plane. (Just the opposite of that desired in the previously described selective pulses, but in fact close to what is achieved experimentally.) Thus, although there is no signal expected at t_w, focusing after a further time $t_w/2$ in a negative gradient gives a very large m_y component with the maximum component of m_x reduced to about 20% that of m_y at its peak. A minor disadvantage of this method is that field-gradient reversal is required, imposing further technical problems.

Of course, focused selective pulses can be produced by discrete Fourier transformation of the appropriate spectral distribution. Indeed, the analysis for either the discrete or continuous case is exactly as already developed here. However, the discrete procedure which we favor, has the advantage that the selection pulse characteristics can be easily varied including the offset center

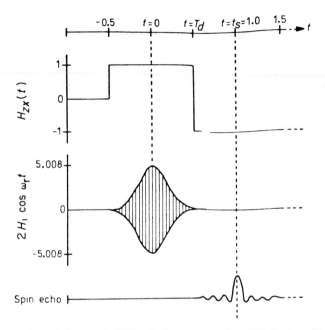

FIG. 3.17.   The scaled magnetic field excitation sequence for a 90° rf pulse, with a Gaussian envelope of amplitude $B_1(t) = 2.504 \exp[-\frac{1}{2}(0.4t)^2]$. [From R. J. Sutherland and J. M. S. Hutchison, *J. Phys. E Sci. Instrum.* **11**, 79 (1978).]

frequency, which obviates the necessity of using either a static field sweep, a gradient sweep, or an rf carrier sweep, when the selected line or plane of spins is required to be varied.

Figure 3.18 shows the simulated magnetization components for a discrete Gaussian-shaped excitation spectrum, which of course is also Gaussian in the time domain. The spectrum comprises discrete frequencies spaced at $2\pi/t_w$ with alternate positive and negative amplitudes. In this case it was not found necessary to employ "ghost" frequencies. The Gaussian-shape parameters used in our calculation were the same as those used by Sutherland and Hutchison.[39] Figures 3.18a and b show the m_x and m_y components, respectively, and demonstrate the absence of undesirable sidebands. Figure 3.18c shows the depletion of the equilibrium magnetization m_z, which is also a very smooth function. The efficacy of this focused selection procedure is summarized in the polar plot of the total x–y magnetization at time $t_w/2$ following the pulse. The directionality of the y lobe is a great improvement over nonfocused selection. Further developments of this work are in progress and will be reported elsewhere.[40]

[39] R. J. Sutherland and J. M. S. Hutchison, *J. Phys. E Sci. Instrum.* **11**, 79 (1978).
[40] P. Mansfield, R. Ordidge, and R. Rzedzian (to be published).

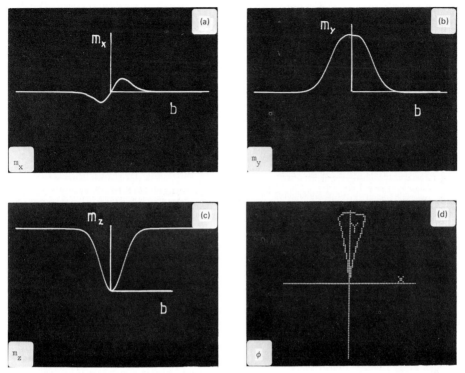

FIG. 3.18.   Computed spin magnetization components existing following a Gaussian selective rf pulse of length t_w and focusing period $t_w/2$ in a reversed gradient, as indicated in Fig. 3.17: (a) m_x, (b) m_y, (c) m_z, and (d) the $x-y$ phase distribution of magnetization. Note the much improved directionality of the phaselobe.

One final point is that selective excitation sequences which reduce m_x are valuable since they ameliorate the gradient switching problems from selection gradient to read gradient in practical imaging schemes.

3.4. Steady-State Free Precession

3.4.1. GENERAL DESCRIPTION

We have seen from Section 3.2 that the effect of a nonselective, nonresonant $90°$ rf pulse on an initially polarized spin system is to produce a transverse magnetization, the FID. But its effect is also to equilibrate the Zeeman spin energy levels so that M_z is initially zero and grows again with time constant T_1 to the full thermal equilibrium value M_0 if left long enough. However, if the spin system is again irradiated with a $90°$ pulse in a time $\tau \ll T_1$, it is easy to see that one cannot expect too much FID signal following the second pulse, or indeed, successive pulses if applied regularly. This procedure is in fact one way, used by some workers studying T_1 values in solids, of ensuring that the spin system is genuinely saturated following a comb of approximate $90°$ pulses spaced several T_2 apart, where $T_2/T_1 \ll 1$, so that recovery with T_1 really starts with zero M_z.

In inhomogeneously broadened liquidlike systems, and also solids, repetitive pulsing using $90°$ pulses is a standard method of measuring T_1 by measuring the quasi-equilibrium magnetization which obtains on resonance after the first pulse.[4] Here, $\tau \lesssim T_1$. As the pulse repetition spacing τ is increased, so more z magnetization is allowed to recover between pulses. In this simple example the quasi-equilibrium magnetization becomes [vide infra, Eq. (3.90b)]

$$M_y(\tau) = M_0(1 - e^{-\tau/T_1}),   (3.61)$$

when $T_2 \ll \tau$, which is a special case of Eq. (3.36b) in which M_z is interrogated. Because of this inequality, there is no quantal coherence in the transverse components of the spin system, and although it is a type of steady-state free precession (SSFP) experiment, the description SSFP is really reserved for the case where quantum coherence is preserved and sustained. This situation, as we shall see later, occurs when $\tau \lesssim T_2 \lesssim T_1$ and is not restricted to exact resonance. Indeed, we shall see that there are distinct advantages to working with the spin system *off* resonance.

The idea of SSFP was first described by Carr[41] and later applied particularly with regard to signal/noise enhancement in Fourier transform spectroscopy by Ernst and Anderson[22] and by Freeman and Hill.[23] However, it was Hinshaw[42] who first combined SSFP with NMR imaging to advantage in the sensitive point and multisensitive point imaging methods to be described later.

3.4.2. NAIVE APPROACH

Consider a spin system in a uniform polarizing field B_0. We shall not be too concerned (at least initially), about relaxation times. In the rotating reference frame, the FID following a nonresonant $90°$ rf pulse along the x axis comprises the full equilibrium magnetization M_0 which precesses coherently about the z axis at offset angular frequency $\Delta\omega = \gamma b$. Eventually, of course, relaxation processes will destroy the coherence and cause repolari-

[41] H. Y. Carr, *Phys. Rev.* **112,** 1693 (1958).
[42] W. S. Hinshaw, *J. Appl. Phys.* **47,** 3709 (1976).

zation along B_0. Consider, however, the situation where regular $90°$ pulses are applied along x every τ sec. We assume that the $90°$ pulses are pure rotations about the x axis even though the spin system is off resonance by b (i.e., $B_1 \gg b$, corresponding to short pulses). The pulse puts $M_z = M_0$ along the y axis (Fig. 3.19a). If b and τ are chosen such that

$$\gamma b \tau = (2n + 1)\pi, \tag{3.62}$$

where n is an integer, M_0 will precess in the interval τ to the $-y$ axis so that the second $90°$ pulse returns M_0 along the z axis, in which case there is no FID signal. The third pulse will, of course, bring M_0 back down along the y axis and so repeat the cycle. In fact, odd pulses will produce FID signals, whereas even pulses produce no signal. The important point is that τ may be arbitrarily short and yet it is clear that a sustained FID signal is produced. We shall see later how the simple picture is modified when relaxation times T_1 and T_2 are properly introduced. The particular experiment just described is not what is usually regarded as SSFP and is in some respects somewhat closer to the driven equilibrium Fourier transform (DEFT) sequence as discussed

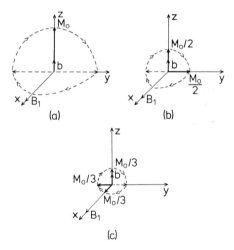

FIG. 3.19. Sketch showing the elementary principle of steady-state free precession (SSFP). (a) Case where full magnetization M_0 is first nutated to the y axis, allowed to precess in offset b to the $-y$ axis before being flipped back along z. (b) Case where it is assumed that magnetization $M_0/2$ exists initially along both the z and y axes. The y magnetization is again allowed to precess in b to $-y$. A $90°$ rf pulse along x then takes this magnetization $M_0/2$ back up along z, while at the same time establishing $M_0/2$ along y, thus maintaining a constant equilibrium magnetization $M_0/2$. (c) In this case, we assume that $M_0/3$ exists initially along the z, x, and $-y$ axes. In this case the magnetization between $90°$ pulses is allowed to precess through only $90°$. Note (a) corresponds to a practically realizable situation, whereas (b) and (c) violate spin conservation as drawn, excluding relaxation processes.

in the next section. However, it does serve to demonstrate how substantial magnetizations can be regularly and nondestructively interrogated by sequential preservation along the z axis, and in that sense it is the essence of the SSFP method.

Another related situation is sketched in Fig. 3.19b. In this case let us assume that *before* the first 90° pulse is applied, we have $M_0/2$ along the z axis and $M_0/2$ along the y axis. Again, we assume an offset field b and repetition period satisfying Eq. (3.62). Between pulses the magnetization $M_0/2$ will precess from the $+y$ to the $-y$ axis, creating the same cyclic initial condition for each pulse. In this case the FID following *all* pulses is $M_0/2$, and represents a dynamic equilibrium situation which again allows continuous interrogation of the spin system.

A third *Gedanke(n)* example is sketched in Fig. 3.19c where we assume that *before* the first 90° pulse we have $M_0/3$ along the x axis, $M_0/3$ along the $-y$ axis, and $M_0/3$ along the z axis. If the offset field and pulse spacing are chosen so that

$$\gamma b\tau = (2n + \tfrac{1}{2})\pi, \qquad (3.63)$$

a quasi-equilibrium situation obtains, then in each period τ spins along the x axis precess through 90° to $-y$, while spins placed along $+y$ following each pulse precess to the x axis, thus periodically creating the requisite initial conditions necessary to sustain the quasi-equilibrium state.

Because of the applied field gradients in imaging experiments, there is of, course, a distribution of precessional frequencies, lying between the discrete values mentioned, and as Hahn[24] demonstrated, such continuous inhomogeneous distributions can be rephased into spin echoes following successive 90° pulses. With this refocusing process plus the spin relaxation effects present in real systems, the simple view presented here becomes necessarily more complicated. We also point out that our simple examples Fig. 3.19b and c violate the conservation of spin magnetization in the absence of relaxation processes. For these reasons, therefore, we shall turn from pictorial models to an analysis based on solutions of the Bloch equations obtained by the matrix approach developed throughout this chapter.

3.4.3. SSFP ANALYSIS

As we have simply demonstrated, a sustained dynamic equilibrium (DE) magnetization may be produced by the SSFP method. Initially in our analysis we shall consider the steady-state case reached by a series of regularly spaced β pulses along the x axis, i.e., $R_{\pm x}(\beta)$. Although the situation for general angle pulses can be simply calculated, the central points of the method

and the major results and limitations may be brought out satisfactorily for
90° pulses, to which we restrict our discussion initially.

New results showing the approach to the dynamic equilibrium state are
also presented. The method of analysis is based on the Bloch equation
solution [Eq. (3.28)] developed in Section 3.2.2.

There are several ways of defining the steady-state response and these are
indicated in Fig. 3.20. In both Fig. 3.20a and b, a section of a stream of rf
pulses is shown, which we assume has been applied to the spin system long
enough to establish the dynamic equilibrium state. Between pulses there will
be, in general, a signal component which exists immediately following each
pulse, in addition to one immediately preceding each pulse. For this reason
we can define a spin magnetization evolution cycle as in Fig. 3.20a where the
signal is evaluated just before the rf pulse (denoted $-$) or just after the pulse
(denoted $+$). Incidentally, this iteration projection method can be performed
over many τ cycles, especially if the rf perturbing pulse is changing cyclically,
i.e., from $R_x(90)$ to $R_{-x}(90)$ (Fig. 3.20b), or in some other more general manner.

We shall first consider the cases corresponding to Fig. 3.20a. Let the
magnetization just prior to the nth rf pulse at time $t = n\tau$ be $M^-(n\tau)$. The
evolution between pulses for time t' is controlled by the relaxation and
propagation matrices so that at time $t + t'$ and with a simple generalization
of Eq. (3.28), it is given by

$$M^-(t + t') = e^{At'}E(t')R_\pm(\beta)M^-(t) + M_{eq}(1 - E_1(t')). \qquad (3.64)$$

For positive 90° pulses at $t' = \tau$, and if we assume that

$$M^-[(n + 1)\tau] = M^-(n\tau) = M_{DE}, \qquad (3.65)$$

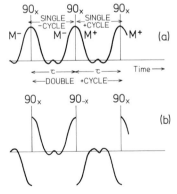

FIG. 3.20. (a) Dynamic-equilibrium response in an SSFP experiment using a train of
in-phase 90° rf pulses. The single positive cycle is defined between corresponding points on
M^+ over a period τ and likewise for the single negative cycle. Double positive or negative cycles
are defined over a 2τ period. (b) SSFP response to a phase-alternated 90° pulse sequence.

the dynamic equilibrium magnetization, Eqs. (3.64) and (3.65) may be rearranged to yield

$$M_{DE}^- = D_{-1}^{-1} M_{eq}(1 - E_1(\tau)),\qquad(3.66)$$

where the driving matrix is given by

$$D_{-1} = [1 - e^{A\tau}E(\tau)R_x(\pi/2)],\qquad(3.67)$$

and the thermal equilibrium matrix is given by

$$M_{eq} = \begin{bmatrix} 0 \\ 0 \\ M_0 \end{bmatrix}.\qquad(3.68)$$

Equation (3.66) may be evaluated straightforwardly by matrix methods[43] using the various forms of the operators already given in Section 3.2. In this case we find the following components of M_{DE}^-,

$$M_{xDE}^- = \{(1 - E_1)E_2 \sin \Delta\omega\tau\}M_0/\det D_{-1},\qquad(3.69a)$$

$$M_{yDE}^- = -(1 - E_1)\{E_2^2 \sin^2 \Delta\omega\tau$$
$$+ E_2 \cos \Delta\omega\tau E_2(1 - E_2 \cos \Delta\omega\tau)\}M_0/\det D_{-1}\qquad(3.69b)$$

$$M_{zDE}^- = \{(1 - E_1)(1 - E_2 \cos \Delta\omega\tau)\}M_0/\det D_{-1},\qquad(3.69c)$$

where the determinant of the matrix D_{-1} is given by

$$\det D_{-1} = (1 - E_2 \cos \Delta\omega\tau)(1 + E_1 E_2 \cos \Delta\omega\tau) + E_1 E_2^2 \sin^2 \Delta\omega\tau,\qquad(3.70)$$

and the exponentials E_1 and E_2 are evaluated at time τ.

In case (i), with $\Delta\omega\tau = 0$ and $\tau \leq T_1, T_2$, Eqs. (3.69) and (3.70) reduce to

$$M_{xDE}^- = 0,\qquad(3.71a)$$

$$M_{yDE}^- \simeq -\tau M_0/2T_1,\qquad(3.71b)$$

$$M_{zDE}^- \simeq \tau M_0/2T_1.\qquad(3.71c)$$

giving essentially negligible equilibrium signal.

In case (ii) with $\Delta\omega\tau = n\pi$ (n odd),

$$M_{xDE}^- = 0,\qquad(3.72a)$$

$$M_{yDE}^- \simeq [T_2/(T_1 + T_2)]M_0,\qquad(3.72b)$$

$$M_{zDE}^- \simeq [T_2/(T_1 + T_2)]M_0,\qquad(3.72c)$$

---

[43] G. Goertzl and N. Tralli, "Some Mathematical Methods of Physics." McGraw-Hill, New York, 1960.

whereas for even n, M_{yDE}^- changes sign in Eq. (3.69). When $T_1 = T_2$, Eq. (3.72b) predicts a transverse magnetization $M_0/2$ existing just *before* the application of the $(n + 1)$th rf pulse.

To examine what signal exists immediately *after* the pulses, we define the cycle as in the second case of Fig. 3.20b.

In this case the counterpart of Eq. (3.64) is

$$M^+[(n + 1)\tau] = R_x(\pi/2)e^{A\tau}E(\tau)M^+(n\tau) + R_x(\pi/2)M_{eq}(1 - E_1(\tau)). \quad (3.73)$$

Again, assuming that

$$M^+[(n + 1)\tau] = M^+(n\tau) = M_{DE}^+, \quad (3.74)$$

Eq. (3.73) may be rearranged to give

$$M_{DE}^+ = D_{+1}^{-1}R_x M_{eq}(1 - E_1(\tau)), \quad (3.75)$$

where now the driving matrix is given by

$$D_{+1} = 1 - R_x e^{A\tau}E(\tau), \quad (3.76)$$

and its determinant

$$\det D_{+1} = 1 - E_2(1 - E_1)\cos\Delta\omega\tau - E_1 E_2^2. \quad (3.77)$$

Evaluating Eq. (3.75) from Eq (3.76) and the relevant operators given in Section 3.2, we obtain the following dynamic equilibrium components following an rf pulse:

$$M_{xDE}^+ = -\{E_2(1 - E_1)\sin\Delta\omega\tau\}M_0/\det D_{+1}, \quad (3.78a)$$

$$M_{yDE}^+ = \{(1 - E_2\cos\Delta\omega\tau)(1 - E_1)\}M_0/\det D_{+1}, \quad (3.78b)$$

$$M_{zDE}^+ = \{E_2(E_2 - \cos\Delta\omega\tau)(1 - E_1)\}M_0/\det D_{+1}. \quad (3.78c)$$

As with M_{DE}^-, the case on resonance gives very small signals. However, for $\Delta\omega\tau = n\pi$, with n odd and $\tau \ll T_1, T_2$, Eqs. (3.78a, b, and c) reduce to

$$M_{xDE}^+ = 0, \quad (3.79a)$$

$$M_{yDE}^+ = [T_2/(T_1 + T_2)]M_0, \quad (3.79b)$$

$$M_{zDE}^+ = [T_2/(T_1 + T_2)]M_0, \quad (3.79c)$$

whereas for n even, we obtain

$$M_{xDE}^+ = 0, \quad (3.80a)$$

$$M_{yDE}^+ = (\tau/2T_1)M_0, \quad (3.80b)$$

$$M_{zDE}^+ = (\tau/2T_1)M_0. \quad (3.80c)$$

Equations (3.80a, b, and c) demonstrate that spins which are allowed to precess through 2π rad give very little signal. This state can be achieved as described here by doubling $\Delta\omega$ or τ. We shall shortly return to the implications of this result especially for the general case where the precessional phase $\Delta\omega\tau \neq n\pi$.

3.4.4. CONTINUOUS OFFSET DISTRIBUTION

In NMR imaging applications of the SSFP technique,[42,44] there will in general be a continuous distribution of offset frequencies present arising from the applied field gradients necessary for spatial discrimination. Our analysis above considers what happens particularly when $\Delta\omega\tau = n\pi$. There will, of course, always be parts of the specimen which satisfy this condition and these will give their full signal contribution at these particular frequencies. However, for the many spins which lie between these particular frequencies, the signals will be smaller. Figure 3.21 shows the magnitude of M_y^+ for fixed τ over the frequency range $\Delta\omega = 0$ to $4\pi/\tau$.

Regular sampling of the FID between pulses N times such that $N\,\Delta\tau = \tau$, where $\Delta\tau$ is the sampling period, will naturally correspond through the sampling theorem to N points in the frequency domain with equal spacing $\Delta\omega = 2\pi/\tau$, which, of course, exactly corresponds to the spacing between successive peaks in Fig. 3.21. To get the frequency sampling points correctly phased with the magnetization peaks in Fig. 3.21 requires a phase shift (or alternation) in the time sampling, a point we shall return to later. However, the broadening of the peaks in Fig. 3.21 implies a convolution corresponding to a damping function in the time domain.

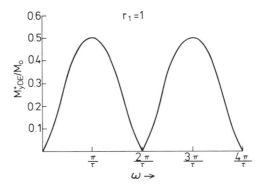

FIG. 3.21.   Normalized dynamic equilibrium magnetization $M_{y\mathrm{DE}}^+/M_0$ versus ω, for $r_1 = T_2/T_1 = 1$, showing the effective magnetization bunching produced by spins with a uniform frequency distribution.

[44] W. S. Hinshaw (to be published).

For a finite sampling time τ, the sampling theorem imposes a broadening sinc function on the otherwise discrete frequency spectrum. The broadening shown in Fig. 3.21 can be regarded as arising as a consequence of the sampling theorem. The deviation of the peaks in Fig. 3.21 from true, sinc functions could introduce a minor diminution of the observed signal in the frequency domain if Fourier transformation of the FID is used to generate the frequency spectrum. But our feeling is that this does not represent a very serious deviation from the theoretical maximum signal and, hence, from the efficiency obtainable by the method.

3.4.5. MULTICYCLE ITERATION

From our simple account of SSFP in Section 3.4.2 and the discussion of Fig. 3.19, it was clear that with no relaxation processes acting we could produce a cyclic situation where FID signals are produced after *pairs* of pulses. If the rf pulses are both 90^c and along the x axis, then by straightforward extensions of Eqs. (3.73)–(3.75) we get for the dynamic equilibrium magnetization just after the second pulse in the cycle, and for $\Delta\omega\tau = (2n + 1)\pi$,

$$M_{DE}^+ = D_{+2}^{-1}\{[R_x e^{A\tau}E(\tau) + 1]R_x M_{eq}\}(1 - E_1(\tau)), \qquad (3.81)$$

where now

$$D_{+2} = [1 - (R_x e^{A\tau}E(\tau))^2]. \qquad (3.82)$$

The new driving matrix D_{+2} can be expanded and we see immediately that Eq. (3.81) reduces identically to the single-cycle case. This implies that two-cycle iteration with relaxation *always* reduces to the case previously discussed. However, the approach to equilibrium cannot be instantaneous. As we have seen with $E(\tau) = 1$, the response after the first pulse gives M_0. When $E(\tau) \neq 1$, the response following this and successive even pulses must decay to the final equilibrium magnetization M_{DE}^+. The odd-pulse response must start from zero and *grow* to the equilibrium value. The odd/even pulse response evolution can be calculated exactly from Eq. (3.73) and we find for even pulses the following components of the magnetization $M^+(2n\tau)$ ($n = 0, 1, 2, 3, \ldots$),

$$M_x^+(2n\tau) = 0, \qquad (3.83a)$$

$$M_y^+(2n\tau) = E_1^n E_2^n M_0 + \frac{1 - E_1^n E_2^n}{1 - E_1 E_2}(1 - E_1)M_0, \qquad (3.83b)$$

$$M_z^+(2n\tau) = \frac{1 - E_1^n E_2^n}{1 - E_1 E_2}E_2(1 - E_1)M_0. \qquad (3.83c)$$

Again, from Eq. (3.73) we find by induction for odd pulses,

$$M_x^+[(2n + 1)\tau] = 0, \tag{3.84a}$$

$$M_y^+[(2n + 1)\tau] = \frac{1 - E_1^{n+1}E_2^{n+1}}{1 - E_1E_2}(1 - E_1)M_0, \tag{3.84b}$$

$$M_z^+[(2n + 1)\tau] = E_1^nE_2^{n+1}M_0 + \frac{1 - E_1^nE_2^n}{1 - E_1E_2}E_2(1 - E_1)M_0, \tag{3.84c}$$

where $n = 0, 1, 2, 3, \ldots$ Unless otherwise stated, E_1 and E_2 are evaluated at time τ.

Let $\tau/T_2 = r_0$ and $T_2/T_1 = r_1$ in Eqs. (3.83) and (3.84). The normalized curve M_y^+/M_0 is plotted in Fig. 3.22 and illustrates the approach of even and odd pulse responses to equilibrium $M_{DE} = r_1/(1 + r_1)$. Curve (a) corresponds to $r_0 = 1/30$, $r_1 = 1$, and in curve (b), $r_0 = 1/30$, $r_1 = 1/3$. Both sets of values of r_0 and r_1 correspond to plausible conditions that might be met in an SSFP experiment in biological systems, although we point out that r_1 values down to 1/10 are commonly observed in tissues (see Chapter 2). If r_0 is very small, the equilibration time can be quite long and therefore there could be value in sampling the response following just the even pulses over a sample time $2n\tau < T_2$ rather than waiting for the system to equilibrate. Of course, if odd pulse responses are also sampled, the advantage implied above is lost, since these signals are essentially zero and therefore add in unnecessary noise. As can be seen from Fig. 3.22, the advantage of even-pulse-response sampling can be quite striking since one is effectively sampling the full magnetization M_0 for half the time. In the most favorable case for maximum

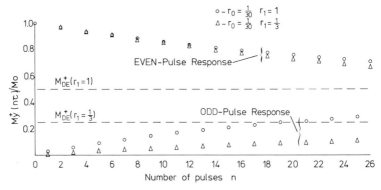

FIG. 3.22. Approach to dynamic equilibrium in an SSFP experiment. The even and odd pulse $M_y^+(n\tau)$ responses [Eqs. (3.83b) and (3.84b)] are plotted for two typical experimental situations, namely (a) $r_1 = T_2/T_1 = 1$ and (b) $r_1 = \frac{1}{3}$. In both cases we have taken $r_0 = \tau/T_2 = \frac{1}{30}$. The dynamic equilibrium values M_{DE}^+ are 0.5 and 0.25 for $r_1 = 1$ and $r_1 = \frac{1}{3}$, respectively.

equilibrium magnetization when $r_1 = 1$ and allowing for signal/noise considerations, one gains an initial improvement approaching $\sqrt{2}$ in sensitivity. For $r_1 < 1$, this sensitivity improvement is sustained for a greater number of pulses.

3.4.6. PHASE-ALTERNATED PULSES

In most applications of SSFP to imaging,[42,44] the phase of the rf pulses is alternated or modulated by 180°. That is to say, both $R_x(\beta)$ and $R_{-x}(\beta)$ pulses are used. Again, restricting our discussion to 90° pulses, we see that this particular experiment is a two-cycle iteration and the dynamic equilibrium magnetization is given by Eq. (3.81) suitably modified for the two types of 90° pulse present. The modification is in the driving operator D_{+2} which now becomes

$$D_{+2} = 1 - R_x e^{A\tau} E(\tau) R_{-x} e^{A\tau} E(\tau). \tag{3.85}$$

We examine the operator occurring in Eq. (3.82), i.e., $[R_x e^{A\tau} E(\tau)]^2$ when $\Delta\omega\tau = (2n + 1)\pi$, bearing in mind that $e^{A\tau} = R_{-z}(\Delta\omega\tau)$ [see Eqs. (3.26) and (3.31)]. Also by examining the operator part of D_{+2} [Eq. (3.85)] when $e^{A\tau} = 1$, i.e., when $\Delta\omega\tau = 2n\pi$, it follows that under these conditions,

$$(R_x e^{A\tau} E(\tau))^2 = R_x E(\tau) R_{-x} E(\tau) = \begin{bmatrix} E_2^2 & & \\ & E_1 E_2 & \\ & & E_1 E_2 \end{bmatrix}. \tag{3.86}$$

This result means, of course, that under the conditions specified, both the equilibrium value and the approach to equilibrium are the same for both in-phase and phase-alternated sequences. An important difference, and advantage, of the phase-alternated sequence is that the $M_y^+(\Delta\omega)$ distribution (Fig. 3.21) is shifted by π/τ from the origin, producing a signal maximum rather than zero as shown. This is helpful in the time domain since the sample points are regularly spaced in the period τ, the first point corresponding to the time origin. Phase alternation thus avoids the time-sampling difficulties referred to earlier when discussing the in-phase pulse sequence, though it should be emphasized that the received signal phases alternate in sign with phase-alternated pulses and this must be accounted for when sampling (see Fig. 3.20b).

3.4.7. GENERALIZED rf PULSES

Our discussion of SSFP so far has exclusively employed 90° rf pulses. This is satisfactory for a basic understanding of the technique. However, in the optimization of SSFP and related techniques dealt with later in this section and also in Chapter 6, it is useful to consider the general case when the rf pulses $R_{\pm x}(\beta)$ are allowed to have an arbitrary nutation angle β.

We reevaluate Eq. (3.75) for M_{DE}^+, using the expression for $R_x(\beta)$ [Eq. (3.32)]. The driving matrix D_{+1} [Eq. (3.76)] now has its determinant given by

$$\det D_{+1} = 1 + C(\tau)[E_1 - 1]E_2 - E_1E_2{}^2$$
$$+ \cos \beta [E_1 + C(\tau)E_2][C(\tau)E_2 - 1], \qquad (3.87)$$

where we introduce the shorthand notation $C(\tau) = \cos \Delta\omega\tau$, $S(\tau) = \sin \Delta\omega\tau$, and, again, E_1 and E_2 are evaluated at time τ. We thus obtain the following dynamic equilibrium magnetization components following a β rf pulse,

$$M_{xDE}^+ = -S(\tau)E_2 \sin \beta(1 - E_1)M_0/\det D_{+1}, \qquad (3.88a)$$

$$M_{yDE}^+ = (1 - E_2C(\tau)) \sin \beta(1 - E_1)M_0/\det D_{+1}, \qquad (3.88b)$$

$$M_{zDE}^+ = \{E_2(E_2 - C(\tau)) \sin^2 \beta + [1 - E_2C(\tau)(\cos \beta + 1)$$
$$+ E_2{}^2 \cos \beta(1 - C(\tau))S^2(\tau) \cos \beta] \cos \beta\}(1 - E_1)M_0/\det D_{+1}. \qquad (3.88c)$$

For our purpose, we focus attention on M_{yDE}^+. In SSFP experiments with $\Delta\omega\tau = n\pi$, and n an odd integer, $C(\tau) = -1$ and $S(\tau) = 0$. In this case Eq. (3.88b) reduces to

$$M_{yDE}^+ = \frac{(1 - E_1) \sin \beta M_0}{1 - E_1E_2 - \cos \beta(E_1 - E_2)}, \qquad (3.89a)$$

For the case when τ/T_1, $\tau/T_2 < 1$, Eq. (3.89a) reduces to

$$M_{yDE}^+ = \frac{r_1 \sin \beta M_0}{1 + r_1 - \cos \beta(1 - r_1)}, \qquad (3.89b)$$

where the ratio $r_1 = T_2/T_1$. For $r_1 < 1$, this expression is maximized approximately when $\cos \beta = r_1/(1 - r_1)$. To first order in r_1, $\sin \beta \simeq 1$. In this approximation $M_{yDE}^+ \rightarrow (T_2/T_1)M_0$ and $\beta \neq \pi/2$. However, if $\beta = \pi/2$, Eq. (3.89b) reduces to the previously derived result [Eq. (3.79b)], namely, $M_{yDE}^+ = M_0T_2/(T_1 + T_2)$.

Equation (3.88b) may also be used to evaluate the transverse equilibrium response in a repeated FID experiment. In this case, and at resonance where $C(\tau) = 1$ and $S(\tau) = 0$, the equilibrium magnetization M_{yDE}^+ becomes

$$M_{yDE}^+ = \frac{(1 - E_1) \sin \beta M_0}{1 + E_1E_2 - (E_1 + E_2) \cos \beta}. \qquad (3.90a)$$

Since a slowly repeated FID is really an incoherent SSFP experiment,

$\tau/T_2 \ll 1$. We may therefore approximate $E_2 \simeq 0$. Equation (3.90a) then reduces to

$$M_{y\mathrm{DE}}^+ = \frac{\left[1 - \exp(-\tau/T_1)\right] \sin \beta M_0}{1 - \cos \beta \exp(-\tau/T_1)}. \qquad (3.90b)$$

This result, derived by a different approach elsewhere,[47] is used to show that in a repeated FID experiment in the limiting case when both β and τ approach zero, the mean signal-to-noise power ratio $\rightarrow 0.5$. However, we point out that this limit can only be used when $T_1 \gg T_2$ as in many solid compounds. In biological materials where the ratio T_1/T_2 might be typically in the range 3–10, the approximation $E_2 \simeq 0$ in Eq. (3.90a) would no longer be valid.

3.5. Driven Equilibrium Fourier Transformation

3.5.1. BASIC PRINCIPLES

The driven equilibrium Fourier transform (DEFT) technique is discussed in this section not with a view to explaining imaging methods utilizing the technique, but rather as a potential alternative to SSFP. To our knowledge, no imaging results have been published so far using DEFT, although proposals employing the method have been made.

The nth cycle of a continuous sequence of pulse cycles is shown in Fig. 3.23. In the diagram, the assumption is made that the sequence has been running long enough to establish a dynamic equilibrium state when relaxation processes are present.

Let us assume from the start that we have a spin system comprising a distribution of isochromats. Following a short, nonselective 90° pulse in the $(n - 1)$th cycle, the FID is allowed to dephase in time τ, at which point a 180_{90}° pulse is applied. The subscripts denote rf carrier phase shifts of 0°, 90°, etc. This will produce a spin echo at a further time τ, at which point the transverse signal growth is at a maximum. At this time another 90_{180}° pulse is applied which returns the transverse magnetization along the z axis. Some

[45] See for example Bruker Physik Reports, (1968–1970).
[46] E. D. Becker, J. A. Ferretti, and J. C. Farrar, *J. Am. Chem. Soc.* **91,** 7784 (1969).
[47] J. S. Waugh, *J. Mol. Spectrosc.* **35,** 298 (1970).

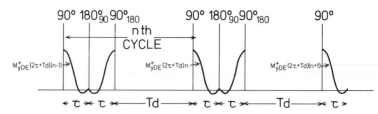

FIG. 3.23. Dynamic equilibrium response in a DEFT experiment. Note that when the delay period $T_d = 0$, the sequence reduces with suitable rf carrier phase changes of the 90° pulses to the standard Carr–Purcell spin–echo experiment. However, in this case the dynamic equilibrium magnetization $M_{yDE}^+(2n\tau)$ [Eq. (3.97)] approaches zero.

magnetization will, of course, be lost in the period 2τ. To prevent continuous degradation of the signal, the magnetization now along the z axis is allowed to recover partially in the delay period T_d, before applying the final 90°_0 read pulse. The cycle is thus defined to continuously interrogate the signal following the last pulse in each cycle, $M^+(2\tau + T_d)n$. Clearly, as with SSFP, the cycle could be defined differently so that the signal *preceding* the 90°_{180} pulse could be monitored.

3.5.2. ANALYSIS

Precisely the same methods of evaluating the dynamic equilibrium response are used here as in Section 3.4. Of course, the cycle is more complicated and therefore the expressions are correspondingly more unwieldy, but nevertheless not intractable. We obtain the following expression for the iterated response:

$$M^+(2\tau + T_d)n = O\{Pe^{A\tau}E(\tau)M^+(2\tau + T_d)(n - 1)$$
$$+ (P + 1)M_{eq}(1 - E_1(\tau))\}$$
$$+ R_x(\pi/2)M_{eq}(1 - E_1(T_d)), \qquad (3.91)$$

where the operators O and P are defined as follows:

$$O = R_x(\pi/2)e^{AT_d}E(T_d)R_{-x}(\pi/2), \qquad (3.91a)$$

$$P = e^{A\tau}E(\tau)R_y(\pi). \qquad (3.91b)$$

In the equilibrium state we assume that

$$M^+(n - 1)(2\tau + T_d) = M^+n(2\tau + T_d) = M_{DE}^+. \qquad (3.92)$$

Substituting Eq. (3.92) into Eq. (3.91) we obtain the following expression for the dynamic equilibrium:

$$M_{DE}^+ = D_{+3}^{-1}\{O[P + 1]M_{eq}(1 - E_1(\tau)) + R_x(\pi/2)M_{eq}(1 - E_1(T_d))\}, \qquad (3.93)$$

where the driving operator is given by

$$D_{+3} = [1 - OPe^{At}E(\tau)]. \qquad (3.93a)$$

The explicit forms for the operators O and P obtained from Eqs. (3.91a and b) and Eqs. (3.29a) and (3.31)–(3.33), are given by

$$O = \begin{bmatrix} E_2(T_d) \cos \Delta\omega T_d & 0 & E_2(T_d) \sin \Delta\omega T_d \\ 0 & E_1(T_d) & 0 \\ -E_2(T_d) \sin \Delta\omega T_d & 0 & E_2(T_d) \cos \Delta\omega T_d \end{bmatrix}, \qquad (3.93b)$$

$$P = \begin{bmatrix} -E_2(\tau) \cos \Delta\omega\tau & -E_2(\tau) \sin \Delta\omega\tau & 0 \\ -E_2(\tau) \sin \Delta\omega\tau & E_2(\tau) \cos \Delta\omega\tau & 0 \\ 0 & 0 & -E_1(\tau) \end{bmatrix}. \qquad (3.93c)$$

Evaluating Eq. (3.93) using the explicit matrix forms, we obtain the following components of M_{DE}^+:

$$
\begin{aligned}
M_{xDE}^+ = \{&(1 - E_1(T_d)E_2^2)(1 + C(T_d)E_2(T_d)E_1^2)S(T_d)E_2(T_d)(1 - E_1)^2 \\
&+ C(T_d)S(T_d)E_2(T_d)E_1^2(1 - E_1(T_d))E_2^2 E_2(T_d) \\
&\times (1 - E_2)^2\}M_0/\det D_{+3}, \qquad (3.94a)
\end{aligned}
$$

$$
\begin{aligned}
M_{yDE}^+ = \{&(1 + C(T_d)E_2(T_d)E_2^2)(1 + C(T_d)E_2(T_d)E_1^2) \\
&+ S^2(T_d)E_2^2(T_d)E_1^2 E_2^2\}(1 - E_1(T_d))M_0/\det D_{+3}, \qquad (3.94b)
\end{aligned}
$$

$$
\begin{aligned}
M_{zDE}^+ = \{&S^2(T_d)E_2^2(T_d)E_2^2(1 - E_1(T_d)E_2^2) \\
&+ (1 + C(T_d)E_2(T_d)E_2^2)(1 - E_1(T_d)E_2^2) \\
&\times C(T_d)E_2(T_d)\}(1 - E_1)^2 M_0/\det D_{+3}, \qquad (3.94c)
\end{aligned}
$$

where the determinant is given by

$$
\begin{aligned}
\det D_{+3} = \{&(1 + C(T_d)E_2(T_d)E_2^2)(1 + C(T_d)E_2(T_d)E_1^2) \\
&+ S^2(T_d)E_2^2(T_d)E_1^2 E_2^2\}[1 - E_1(T_d)E_2^2]. \qquad (3.95)
\end{aligned}
$$

In the above expressions we have used the shorthand notations $C(T_d) = \cos \Delta\omega T_d$, $S(T_d) = \sin \Delta\omega T_d$, and E_1, E_2 are evaluated at time τ unless explicitly indicated otherwise.

We note immediately that through cancellation, the y component of interest in these experiments reduces to

$$M_{yDE}^+ = \frac{[1 - E_1(T_d)]M_0}{1 - E_1(T_d)E_2^2}. \qquad (3.96)$$

For a symmetric distribution of isochromats, $M_{xDE}^+ \to 0$. In the case of interest, i.e., $T_d/T_1 \ll 1$ and $T_2/\tau \gg 1$, Eq. (3.96) reduces to

$$M_{yDE}^+ = \frac{M_0}{1 + (T_1/T_2)(2\tau/T_d)}. \tag{3.97}$$

This expression is directly comparable with Eq. (3.79b) for the SSFP experiment. The important difference is the factor $2\tau/T_d$ which can be made to balance the T_1/T_2 factor, even when the latter is greater than unity.

In our analysis of DEFT we have assumed implicitly that the pulse rotations are perfect. However, in experimental conditions this may be far from true. Pulse imperfections have led Waugh[47] to propose modified DEFT schemes in which, for example, the error in the $180_{90}°$ pulse is compensated by designing a longer cycle with more $180°$ pulses arranged in pairs. These z-re-stored spin echoes (ZRSE) amount to a truncation of a Carr–Purcell sequence[9] by re-storing transverse signal at the peak of the last echo. For one echo, ZRSE, of course, is equivalent to DEFT. However, the advantage of having pairs of $180°$ pulses suitably phased is that nutation errors in the $180°$ pulse may be automatically compensated. Two methods are proposed, the first is as in Fig. 3.23, but with the $180°$ pulse replaced by the cyclic sequence $[R_y(180) - 2\tau - R_y(180)]^n$. In the alternative $180°$ pulse compensation scheme, the first $180°$ pulse in Fig. 3.23 is substituted by $[R_x(180) - 2\tau - R_{-x}(180)]^n$. In both versions, the initial and terminating pulses are retained, that is to say, $R_x(90) - \tau - [\cdots]^n - \tau - R_{-x}(90)$, where n is the number of cycles in the sequence.

3.6. Comparison of Signal Responses

3.6.1. GENERAL

In this discussion we specifically exclude the question of sensitivity, since in practical imaging schemes this involves introducing noise, and, of course, is critically dependent on the imaging method as well as experimental parameters like probe design. These matters will be treated in later stages considering the signal and noise power, rather than simply the signal amplitude, as here.

The expressions derived for the transverse response of the SSFP, DEFT, and repeated FID experiments relate to the maximum signal observed at the beginning of each cycle. To these expressions, we can without further analysis add one more expression corresponding to a repeated Carr–Purcell[9] (CP) sequence. As discussed earlier, the FID following a $90°$ pulse can be sustained in a noncoherent manner when the repetition period or delay time T_d fulfills the condition $T_2 < T_d < T_1$ in which case, the magnetization in the dynamic equilibrium state is from Eq. (3.90b),

$$M_{DE} = \frac{M_0(1 - e^{-T_d/T_1}) \sin \beta}{(1 - \cos \beta \exp(-T_d/T_1))}, \tag{3.98}$$

and describes the initial value of the FID. Let the normalized FID shape be $G(t)$. The average signal value over the sampling period T_s is

$$\bar{G}(T_s) = (1/T_s) \int_0^{T_s} G(t)\, dt. \tag{3.99}$$

The average value of the dynamic equilibrium magnetization is therefore,

$$M_{DE} = M_0 F, \tag{3.100}$$

where for the repeated FID we obtain with $\beta = \pi/2$,

$$F_{FID} = \bar{G}(T_s)(1 - e^{-T_d/T_1})(T_s/T_d), \tag{3.101}$$

which represents the fraction of total possible signal continuously observed throughout the sampling period.

In the Carr–Purcell experiment, a train n of sustained spin echoes is produced, the envelope of which decays to zero with time constant T_2 in a total accumulation time $T_a = 2n\tau$. The average signal in such an echo sequence is the average over the decaying echoes of the FID average over period 2τ. If we assume that $\tau \ll T_2$, the FID average is $\bar{G}(\tau)$ [Eq. (3.99)], since we assume that $G(t)$ decays to zero, in this case in time τ. As the echoes decay to zero in T_a, the total average signal is simply $(T_2/T_a)\bar{G}(\tau)$. If the spin–echo sequence is repeated with delay time T_d, then the fraction F for the Carr–Purcell sequence becomes

$$F_{CP} = \bar{G}_{CP}(\tau)(1 - e^{-T_d/T_1})[T_2/(T_a + T_d)]. \tag{3.102}$$

The corresponding factor for SSFP becomes

$$F_{SSFP} = \bar{G}_{SSFP}(\tau)[T_2/(T_1 + T_2)], \tag{3.103}$$

and for the DEFT sequence,

$$F_{DEFT} = \bar{G}_{DEFT}(\tau)\left(1 + \frac{T_1}{T_2}\frac{2\tau}{T_d}\right)^{-1} \frac{2\tau}{2\tau + T_d}. \tag{3.104}$$

In the application of any of the above-mentioned data accumulation methods to NMR imaging, the FID shapes contain all the information necessary to reconstruct a picture point, line, or plane depending on the actual imaging method being used. For the same imaging technique, for example, the projection reconstruction method,[13] $\bar{G}(\tau)$ is the same for the last three data accumulation methods described and is determined by the shape of the object and the applied field gradients. For the FID case, however, we may choose $\bar{G}(T_s) = \bar{G}(T_2) = \bar{G}(\tau)$, assuming a reduced field gradient is applied, in which the shape function is still determined by the sample shape and gradient. In this case and as a preliminary guide to data accumulation efficiency we may directly compare Eqs. (3.101)–(3.104).

3.6.2. NORMALIZED COMPARISON

Our expressions, Eqs. (3.101)–(3.104) involve the parameters T_1, T_2, T_s, T_d, and $T_a = 2n\tau$. In the most favorable circumstances let us take $T_s = T_2$ and $T_d = T_1$. Bearing in mind the conditions on the FID and CP methods, namely, $\tau \ll T_2$ and $2n\tau > T_2$, let us, for example, take

$$T_d = 2n\tau = 3T_2. \tag{3.105}$$

For our comparison we shall also assume a typical biological $T_1 = 600$ msec, which, from Eq. (3.105), together with our other assumption and a value of $\tau = 3.0$ msec, gives $n = 100$. For further evaluation we reintroduce the ratio $r_1 = T_2/T_1$ and a new ratio $r_2 = 2\tau/T_d$ where T_d here is for the DEFT case and can be much shorter than T_1. Using these ratios we obtain the following approximate expressions corresponding to Eqs. (3.101)–(3.104),

$$F_{\text{FID}} \simeq \bar{G}(\tau)\tfrac{2}{3}r_1, \tag{3.106a}$$

$$F_{\text{CP}} = \bar{G}(\tau)\tfrac{2}{3}[r_1/(1 + 3r_1)], \tag{3.106b}$$

$$F_{\text{SSFP}} = \bar{G}(\tau)[r_1/(1 + r_1)], \tag{3.106c}$$

$$F_{\text{DEFT}} = \bar{G}(\tau)\,\frac{r_1 r_2}{(r_1 + r_2)(1 + r_2)}. \tag{3.106d}$$

Equations (3.106a–d) have been evaluated for two typical biological cases, $r_1 = 1$ and $r_1 = \tfrac{1}{3}$. The choice of r_2 is arbitrary in the DEFT experiment. The values chosen here are $r_2 = 1$ and $r_2 = \tfrac{1}{3}$. The values of F for the above-mentioned conditions are tabulated in Table 3.1.

This simple and restricted discussion brings out clearly several important points, namely: (1) Repetitive FID averaging can be as efficient or even better than the other three methods considered depending on circumstances; (2) The most efficient methods of averaging would appear to be SSFP and FID; (3) under typical conditions for biological materials of $r_1 = \tfrac{1}{3}$ taken with $T_d = T_1$ and $r_2 = \tfrac{1}{3}$, there is little to choose between the CP and DEFT signal-averaging methods. Indeed, it would appear that by using gradient reversals in the CP experiment instead of 180° pulses, simpler experimental conditions give essentially the same time averaged signal; (4) The advantage

TABLE 3.1

Comparison of the Normalized Signal Amplitudes $F/\bar{G}$
(Excluding Noise) for the Various Data-Acquisition
Techniques Referred to in the Table under
Typical Biological Conditions as Reflected
in the Values of r_1 and r_2

	$r_2 = 1$		$r_2 = 1/3$	
	$r_1 = 1$	$r_1 = 1/3$	$r_1 = 1$	$r_1 = 1/3$
FID	~2/3	~2/9	—	—
CP	1/6	1/9	—	—
SSFP	1/2	1/4	—	—
DEFT	1/4	1/8	3/16	1/8

in the time-averaged signal size of SSFP or FID over CP is about a factor two. However, in SSFP, one is restricted to using a train of rf pulses, which means that transmitter power requirements and rf deposition in a subject are unavoidably higher. Since $T_2 > \tau$, power deposition for the FID case is much less.

One final point is that all signal comparisons have been made with the assumption that it is necessary to repeat a cycle to obtain satisfactory S/N ratios. In the CP experiment, for example, this steady-state assumption loses us a factor $(1 - e^{-T_d/T_1})T_d/(T_a + T_d)$. In fact, if a sufficiently high signal–to–noise ratio could be achieved in a single CP sequence, then the comparison factor to be considered is simply $T_2/T_d = \frac{1}{3}$, using our previous assumption.

Of course, as we have already pointed out for SSFP experiments, under suitable conditions, even-integer pulse sampling only *before* the steady-state condition is achieved can also give signal improvements and is really analogous to the one-shot CP experiment.

4. Classification and Description of NMR Imaging Methods

4.1. Introduction

In this chapter we review all the imaging methods which have so far been proposed. Our treatment is largely descriptive with the critical comparison of techniques reserved for Chapter 5.

We begin by analyzing a number of methods which can be used to define an imaging plane. Most of these have in the past been associated with a particular imaging technique, for instance, oscillating gradients with the sensitive line or sensitive point methods, but many have a potentially wider application.

We then proceed to the detailed description of the various imaging methods which have been divided into three regimes, point, line, and plane, according to the manner in which the image data are acquired. Thus, in a point method, one would only receive information from a single pixel at any moment in time, whereas for a planar technique, information from a complete plane of spins is simultaneously collected. This classification is further clarified in Fig. 4.1, taken from the work of Brunner and Ernst.[1] Note that in our description we have made no distinction between plane and volume methods since the former are generally derivable from the latter by, for example, the addition of a selective pulse. Thus there is no fundamental difference in technique. However, we do indicate the modifications necessary and in Chapter 5, where performance comparisons are made, we treat the two cases separately.

---

[1] P. Brunner and R. R. Ernst, *J. Magn. Reson.* **33**, 83 (1979).

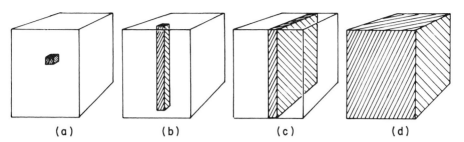

FIG. 4.1. Scheme of classification of the various imaging techniques: (a) sequential point measurement, (b) sequential line measurement, (c) sequential plane measurement, and (d) simultaneous volume measurement. [From P. Brunner and R. R. Ernst, *J. Magn. Reson.* **33,** 83 (1979).]

4.2. Slice Definition

Most methods of NMR imaging require the preselection of an imaging plane, either as an integral part of the technique or to achieve the reduction in signal amplitude which may be imposed by dynamic range considerations. [For example, a three-dimensional image consisting of a $100 \times 100 \times 100$ array of pixels, displayed on a 16-level gray scale, would require an accuracy of about 11 bits. This may prove difficult to handle with present systems, particularly with regard to data acquisition (see Section 8.5.6 for further discussion of dynamic range)].

Selection may be achieved by destroying all the sample magnetization apart from that existing in the desired spatial region (selective saturation). Alternatively, one may arrange to interact only with the required spins leaving the remaining population unaffected (selective excitation), or time averaging its effect to zero (e.g., sensitive point and sensitive line techniques).

4.2.1. rf FIELD SHAPING

The well-known response of the z component of magnetization to a continuous-wave rf field B_1 is

$$m_z(x, y, z) = m_0/(1 + \gamma^2 B_1{}^2 T_1 T_2). \tag{4.1}$$

As B_1 is increased, so m_z is progressively saturated. However, if B_1 is made spatially dependent, it becomes possible to selectively saturate a region of the sample.[2] Thus, if $B_1 = B_{13} y^3$, then

$$m_z(x, y, z) = m_0/(1 + \gamma^2 B_{13}{}^2 y^6 T_1 T_2), \tag{4.2}$$

[2] D. I. Hoult, *J. Magn. Reson.* **33,** 183 (1979).

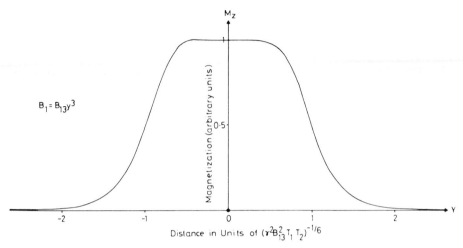

FIG. 4.2.   By tailoring the B_1 field, the degree of saturation in a sample may be varied along the length of the sample. The figure is a plot of magnetization versus length after saturation with a field having a cubic gradient. [From D. I. Hoult, *J. Magn. Reson.* **33**, 183 (1979).]

so that in the neighborhood of the origin $(y \sim 0)$, $m_z = m_0$, but as y increases, so m_z becomes increasingly saturated, falling off as y^{-6}. This response is illustrated in Fig. 4.2. (Many other responses are, of course, possible.) Thus, a plane of spins may be selected perpendicular to the y axis and centered about the origin. The width can be controlled via the constant B_{13} or, alternatively, by altering the functional dependence of B_1.

4.2.2. OSCILLATING FIELD GRADIENTS

4.2.2.1. *Introduction*

Oscillating field gradients were first introduced by Hinshaw[3] and find their greatest application in the sensitive point[3,4] and sensitive line[5] techniques to be described below. Their use in conjunction with projection reconstruction methods has also been recently demonstrated.[6]

In Section 4.2.2.2 we discuss the use of an oscillating field gradient with conventional Fourier transform NMR. The theory is extended to encompass steady-state free precession (SSFP) techniques in Section 4.2.2.3.

[3] W. S. Hinshaw, *Phys. Lett. A* **48**, 87 (1974).
[4] W. S. Hinshaw, *J. Appl. Phys.* **47**, 3709 (1976).
[5] E. R. Andrew, P. A. Bottomley, W. S. Hinshaw, G. N. Holland, W. S. Moore, and C. Simaroj, *Phys. Med. Biol.* **22**, 971 (1977).
[6] H. R. Brooker and W. S. Hinshaw, *J. Magn. Reson.* **30**, 129 (1978).

4.2.2.2. Fourier Transform NMR

The evolution of the x, y magnetization in an oscillating field gradient $\mathbf{G}(t) = G \cos(\Omega t)\mathbf{j}$ following a 90° rf pulse is given by the Bloch equation,

$$\partial m/\partial t = -m\{i(\omega_0 + \Delta\omega) + i\gamma Gy \cos(\Omega t) + [T_2(\mathbf{r})]^{-1}\}, \tag{4.3}$$

where $m = m(\mathbf{r}, t) = m_x(\mathbf{r}, t) + im_y(\mathbf{r}, t)$, and $\Delta\omega$ is the offset frequency. Then,

$$m = m_0 \exp\left\{-i(\omega_0 + \Delta\omega)t - i\gamma Gy \int_{t_1}^{t_1+t} \cos(\Omega t)\, dt - t/T_2(\mathbf{r})\right\}$$

$$= m_0 \exp\{-i(\omega_0 + \Delta\omega)t - (i\gamma Gy/\Omega)$$

$$\times\left[\sin(\Omega(t_1 + t)) - \sin(\Omega t_1)\right] - t/T_2(\mathbf{r})\}. \tag{4.4}$$

Making use of the identity,[7]

$$\exp(iA \sin B) = \sum_{n=-\infty}^{+\infty} J_n(A) \exp(inB), \tag{4.5}$$

where $J_n(A)$ are Bessel functions of the first kind of order n, we obtain,

$$m(\mathbf{r}, t) = m_0(\mathbf{r}) \exp\{-i(\omega_0 + \Delta\omega)t - t/T_2(\mathbf{r})\}$$

$$\times \sum_{k=-\infty}^{+\infty} \sum_{l=-\infty}^{+\infty} J_k(\gamma Gy/\Omega) J_l(\gamma Gy/\Omega)$$

$$\times \exp\{-i\Omega y[k(t_1 + t) - lt_1]\}. \tag{4.6}$$

We now consider the case in which the signal is phase sensitively detected at zero offset frequency. Equation (4.6) then reduces to

$$m(\mathbf{r}, t) = m_0(\mathbf{r}) \exp(-t/T_2(\mathbf{r})) \sum_k \sum_l J_k J_l \exp\{-i\Omega y[k(t_1 + lt) - lt_1]\}. \tag{4.7}$$

If this FID is recorded over an interval T and integrated over the same period, the resulting average signal is

$$s(\mathbf{r}) = \left(\frac{m_0(\mathbf{r})}{T}\right) \sum_k \sum_l J_k J_l \frac{\exp\{-i\Omega(k-l)t_1\}}{i\Omega k + 1/T_2(\mathbf{r})}$$

$$\times \left\{1 - \exp\left[-\frac{T}{T_2(\mathbf{r})} - i\Omega kT\right]\right\}. \tag{4.8}$$

If we further average over a number of FIDs which occur at random times t_1 with respect to the gradient cycle, this becomes

$$s(\mathbf{r}) = \left(\frac{m_0(\mathbf{r})}{T}\right) \sum_k J_k^2 \left(\frac{\gamma Gy}{\Omega}\right) \frac{1 - \exp\{-T/T_2(\mathbf{r}) - i\Omega kT\}}{i\Omega k + 1/T_2(\mathbf{r})}. \tag{4.9}$$

[7] A. S. Gradshteyn and I. M. Ryzhik, "Tables of Integrals, Series and Products." Academic Press, New York, 1965.

Taking the case for which $T \gg T_2(\mathbf{r})$ and considering the real component of the signal $s_R(\mathbf{r})$ we obtain finally

$$s_R(\mathbf{r}) = \left(\frac{m_0(\mathbf{r})}{T}\right) \sum_k J_k^2 \left(\frac{\gamma Gy}{\Omega}\right) \frac{T_2(\mathbf{r})}{1 + (\Omega k/T_2(\mathbf{r}))^2}. \tag{4.10}$$

Figure 4.3a illustrates the application of an oscillating field gradient to a sample. The spatial response, as given by Eq. (4.10) is plotted in Fig. 4.3b for the case in which $1/T_2(\mathbf{r})$ is 20 Hz; a variety of field gradient frequencies Ω are shown.

If $\Omega \gg 1/T_2(\mathbf{r})$, Eq. (4.10) can be approximated by

$$s_R(\mathbf{r}) \sim m_0(\mathbf{r})(T_2(\mathbf{r})/T)J_0^2(\gamma Gy/\Omega). \tag{4.11}$$

The time-averaged signal is thus spatially restricted along the y axis via the function $J_0^2(\gamma Gy/\Omega)$.[4,8] The half-width of its central maximum is $2.2\Omega/\gamma G$ and the first sidebands occur at $y = \pm 3.8\Omega/\gamma G$ with amplitudes 0.16 times that of the central peak. This response is also illustrated in Fig. 4.3b.

The spatial selectivity can be improved by using more general time-dependent functions. Thus, if we take a function $G(t)$, which is periodic modulo $1/\Omega$, then it can be expressed as a Fourier series

$$G(t) = \sum g_n \cos(n\Omega t), \tag{4.12}$$

where the Fourier coefficients are given by the usual inversion formula

$$g_n = (2\Omega/\pi) \int_0^{\pi/2} G(t) \cos(n\Omega t)\, dt. \tag{4.13}$$

The response of the spin system to $G(t)$ in the limit that $\Omega \gg 1/T_2(\mathbf{r})$ would thus be,[8]

$$s_R(\mathbf{r}) \sim m_0(\mathbf{r})(T_2(\mathbf{r})/T) \prod_{n=1}^{\infty} J_0^2(\gamma g_n y/n\Omega). \tag{4.14}$$

This is sharper than that of Eq. (4.11) and indicates the desirability of higher harmonics when greater selectivity is required.

The positioning of the slice can be achieved by moving the object relative to a fixed sensitive plane. This has the advantage of experimental simplicity. However, it is often necessary to leave the object in a fixed position and one then locates the sensitive plane by altering the current ratios in the gradient coil system.

4.2.2.3. The SSFP Technique

In most cases in which the oscillating field gradient method has found application it has been combined with the SSFP technique described in

---

[8] P. A. Bottomley, Ph.D. Thesis, Nottingham, 1978.

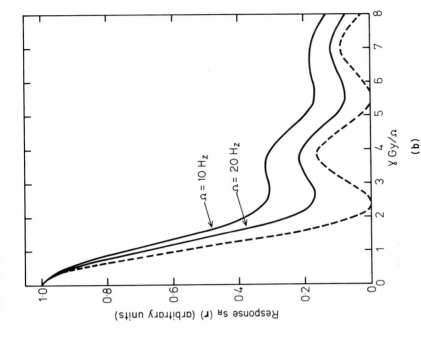

(a)

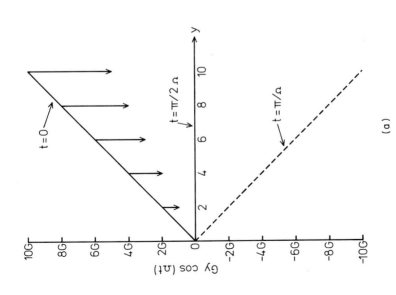

(b)

Fig. 4.3. The application of an oscillating field gradient $\mathbf{G}(t) = G\cos(\Omega t)\mathbf{j}$ to a sample: (a) spatial and time dependence of the magnetic field due to the gradient; (b) spatial response following a simple preparatory pulse. Gradient frequencies of 10 and 20 Hz are illustrated. The corresponding curves for 100 and 500 Hz are essentially indistinguishable from $J_0^2(\gamma G y/\Omega)$ shown as a broken line.

Section 3.4. We develop the relevant theory below. The nuclear signal obtained by the SSFP method with phase sensitive detection and removal of the sign alternation [see Eqs. (3.69a and b)] is

$$m(\mathbf{r}, t) = Q(\alpha) \sin \beta (1 + E_2 \exp(i\alpha))$$

$$\times \exp\left(i \Delta\omega t + i\gamma y \int_{t_1}^{t_1 + t} G(t)\, dt - t/T_2(\mathbf{r}) \right), \qquad (4.15)$$

where β is the pulse flip angle,

$$Q(\alpha) = \frac{m_0(\mathbf{r})(1 - E_1)}{(1 - E_1 \cos \beta)(1 + E_2 \cos \alpha) - (E_1 - \cos \beta)(E_2 + \cos \alpha)E_2}, \qquad (4.15a)$$

$$\alpha = \Delta\omega T + \gamma y \int_{t_1}^{t_1 + T} G(t)\, dt. \qquad (4.15b)$$

Since it is usual to operate with the pulse interval T short compared to the relaxation times T_2, T_1, we can make the following approximations

$$E_1 \simeq 1 - T/T_1, \qquad (4.16)$$

$$E_2 \simeq 1 - T/T_2. \qquad (4.17)$$

Then, for $\beta = 90°$, we have

$$Q(\alpha) \simeq \frac{m_0(\mathbf{r})}{(1 + 2T_1/T_2) + \cos \alpha}. \qquad (4.18)$$

When $\alpha \to 0$, this reduces to

$$Q(0) \simeq \frac{m_0(\mathbf{r})}{2(1 + T_1/T_2)}, \qquad (4.19)$$

and if, further, $T_1 = T_2$, then

$$Q(0) \simeq \tfrac{1}{4}[m_0(\mathbf{r})]. \qquad (4.20)$$

With the approximation of Eq. (4.19), $G(t) = G \cos \Omega t$, and taking the offset frequency $\Delta\omega$ as zero, we obtain for the real part of the signal during the nth pulse interval,

$$m_n(\mathbf{r}, t) = A_n(T)F_n(t) + B_n(T)G_n(t), \qquad (4.21)$$

where

$$A_n(T) = \frac{1 + \cos\{(\gamma G y/\Omega)[\sin(\Omega T + \phi_n) - \sin \phi_n]\}}{(1 + 2T_1/T_2) + \cos\{(\gamma G y/\Omega)[\sin(\Omega T + \phi_n) - \sin \phi_n]\}}, \qquad (4.22)$$

$$B_n(T) = \frac{-\sin\{(\gamma G y/\Omega)[\sin(\Omega T + \phi_n) - \sin \phi_n]\}}{(1 + 2T_1/T_2) + \cos\{(\gamma G y/\Omega)[\sin(\Omega T + \phi_n) - \sin \phi_n]\}}, \qquad (4.23)$$

$$F_n(t) = \cos\{(\gamma Gy/\Omega)[\sin(\Omega t + \phi_n) - \sin \phi_n]\} \exp(-t/T_2), \qquad (4.24)$$

$$G_n(t) = \sin\{(\gamma Gy/\Omega)[\sin(\Omega t + \phi_n) - \sin \phi_n]\} \exp(-t/T_2). \qquad (4.25)$$

If the gradient cycle is triggered at the start of the SSFP pulse sequence and evolves coherently during it for an integral number of periods, as illustrated in Fig. 4.4, then

$$\phi_n = n\Omega T, \qquad n = 0, 1, 2 \ldots . \qquad (4.26)$$

One way of achieving the required spatial resolution is to integrate the signal. The average response is then

$$s_R(\mathbf{r}) = \frac{1}{N}\left(\sum_n \left\{A_n(T) \int_0^T F_n(t)\,dt + B_n(T) \int_0^T G_n(t)\,dt\right\}\right), \qquad (4.27)$$

where N is the number of SSFP cycles over which we integrate. A typical response is shown in Fig. 4.5. It indicates that whereas the central slice definition is reasonable, the sidebands are by no means negligible; their envelope does not decay rapidly or even monotonically with distance (e.g., the sideband at $\gamma Gy/\Omega = 6.6$ is 25% of the central peak height). This particular implementation of the sensitive point technique cannot, therefore, be recommended as a particularly "clean" method for slice definition.

4.2.3. SELECTIVE EXCITATION METHODS

4.2.3.1. Introduction

By the application of magnetic field gradients and suitably shaped rf pulses, it is possible to restrict interaction to a selected portion of the total spin system. This technique is known as selective excitation or selective irradiation and was first proposed in 1974 by Garroway, Grannell, and Mansfield.[9] The general theory has already been developed in Section 3.3 so we restrict ourselves here to discussion of a number of experimental variants which may be useful in a variety of different contexts.

Each application of a selective irradiation (SI) cycle achieves spatial selectivity through a onefold reduction in dimensions of the sensitive region. We may therefore select a plane within a three-dimensional sample, a line within a plane, or a point along a line. Cycles can be applied consecutively to achieve further localization. Thus, for example, two such cycles would be needed to specify a line of spins within a three-dimensional object. This is the technique of line scanning by selective irradiation which is described in Section 4.4.2.

[9] A. N. Garroway, P. K. Grannell, and P. Mansfield, *J. Phys. C* **7**, L457 (1974).

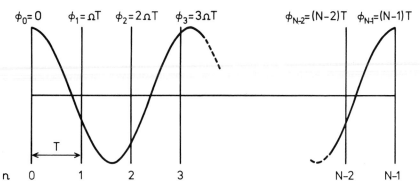

FIG. 4.4.   Gradient-cycle phase relationships during a SSFP cycle; ϕ_n denotes the phase at the start of the nth interpulse interval of duration T.

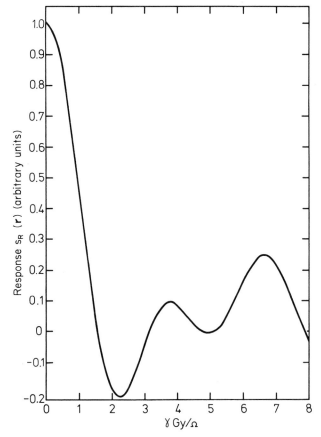

FIG. 4.5.   The spatial response [Eq. (4.27) with $N = 8$ and $T = 1$ msec] to an oscillating field gradient of frequency 250 Hz when SSFP is used.

We split the experiments into two categories:

(1)   those which destroy all magnetization, apart from that in the selected region where it is left unaffected (i.e., in its equilibrium state parallel to the static magnetic field);

(2)   those where the spins in the selected region are rotated (partially or wholly) into the x, y plane and the remainder are left unaffected.

In addition to selective irradiation in the laboratory frame, it has recently been demonstrated[2] that similar experiments are also possible in the rotating frame. In this case an rf field gradient is applied to the sample and a shaped orthogonal rf pulse is used to interact with a selected portion of the spin system. The nice feature of this method is that it can operate in the presence of a static field gradient and there is consequently no need for static gradient switching. It is further discussed in Section 4.5.2.

4.2.3.2. Type 1 Experiments

4.2.3.2.1.   *Selective Saturation.*   (a)   A linear field gradient G (along, say, the z direction) and an intense rf pulse with spectral components as shown in Fig. 4.6a are applied to the sample such that all spins except those in a narrow region of width $\Delta z = \Delta \omega / \gamma G$ are saturated. This leaves the spin system in the state shown in Fig. 4.6b and c for a time of the order of T_1. In practice, the pulse shapes are determined by Fourier transforming the required spectral distribution. Garroway *et al.*[9] originally made use of pulse-width modulation[10] techniques, but conventional amplitude modulation is now preferred.[11]

A major drawback with this technique is that it requires relatively high-power rf pulses with good dynamic range.[11] In addition, due to the non-uniformities of the transmitter coil, it is difficult to completely saturate all spins outside the required slice.[12] The small component of residual magnetization will then contribute to the signal following a 90° pulse and, in certain unfavorable circumstances, may even dominate it!

(b)   An alternative means of achieving a similar result would be to selectively saturate (by a similar process) those spins in the region Δz (see Fig. 4.7a). This would leave the z magnetization as shown in Fig. 4.7b with the x, y magnetization again zero. In order to achieve the final aim, an intense 180° pulse is applied to invert the entire system. Then after a time $t = T_1 \log_e 2$, the z magnetization in the slot will have grown to a value $m = m_0/2$, whereas

[10] B. L. Tomlinson and H. D. Hill, *J. Chem. Phys.* **59**, 1775 (1973).
[11] P. Mansfield, A. A. Maudsley, and T. Baines, *J. Phys. E* **9**, 271 (1976).
[12] A. A. Maudsley, *J. Magn. Reson.* **41**, 112 (1980).

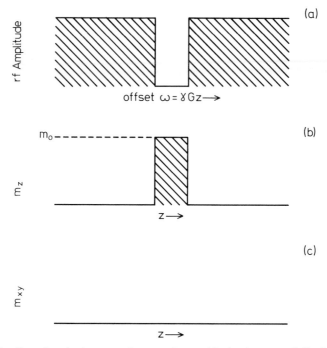

FIG. 4.6.   Type 1a selective saturation experiment: (a) rf pulse spectral distribution; (b) z component of magnetization following the rf pulse; and (c) x, y components of magnetization following the rf pulse.

everywhere else it will have decayed to zero (see Fig. 4.7c). The advantage of this over method (a) is that it does not require such intense saturating pulses. However, it suffers from the disadvantages that one only obtains half the equilibrium magnetization and, more seriously, that its efficiency is reduced if there is a distribution of T_1 values.

4.2.3.2.2.   *Selective Excitation.*   In this method one applies a linear field gradient and an rf pulse (Fig. 4.8a) such that all spins except those in a narrow slice are rotated through 90° into the x, y plane (Fig. 4.8b,c). After a time $t \geq T_2{}^*$, this magnetization dephases leaving just the z magnetization in the narrow slit. Clearly, this method is only practicable if $T_1 \gg T_2{}^*$, where $T_2{}^*$ is the transverse spin dephasing time in the field gradient.

4.2.3.3.   *Type 2 Experiments*

4.2.3.3.1.   *Selective Excitation.*   This is the most commonly applied version of the selective irradiation experiment, finding use as a means of

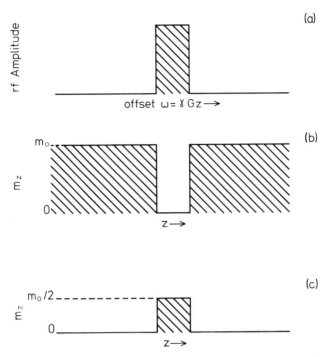

FIG. 4.7. Type 1b selective saturation experiment: (a) rf pulse spectral distribution; (b) z component of magnetization immediately following the rf pulse; and (c) after a further time $t = T_1 \log_e 2$.

plane definition with many of the planar techniques as well as being the basis of the selective irradiation line-scanning methods.

Again, a linear field gradient and a suitably tailored rf pulse are employed but, in this case, the narrow slot of spins receives a 90° pulse (see Fig. 4.9). The narrow region of x, y magnetization thus created is then immediately observable without interference from the remainder of the sample.

4.2.4. SPIN LOCKING

This method of achieving spatial resolution was proposed by Wind *et al.* in 1978.[13] To limit sensitivity to a plane normal to z, one applies a linear field gradient in this direction. A short 90° pulse in the x direction is then followed by a spin-locking pulse of amplitude ω_1 and length T. The y component of magnetization can then be expressed as the sum of two terms M_1 and $M_2(T)$

[13] R. A. Wind, J. H. N. Creyghton, D. J. Ligthelm, and J. Smidt, *J. Phys. C* **11**, L223 (1978).

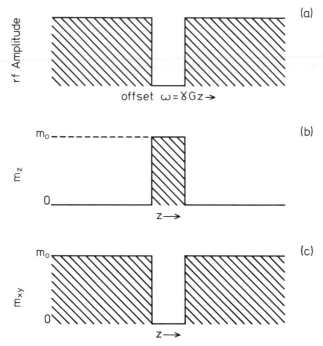

FIG. 4.8. Type 1 selective excitation experiment: (a) rf pulse spectral distribution; (b) z component of magnetization following rf pulse; and (c) x,y component of magnetization following rf pulse.

given by

$$M_1 = \int_{-\infty}^{+\infty} \rho(\Delta_z) V(\Delta_z) [\omega_1^2/(\omega_1^2 + \Delta_z^2)] \, d\Delta_z, \tag{4.28}$$

$$M_2 = \int_{-\infty}^{+\infty} \rho(\Delta_z) V(\Delta_z) [\Delta_z^2/(\omega_1^2 + \Delta_z^2)] \cos\{(\omega_1^2 + \Delta_z^2)^{1/2} T\} \, d\Delta_z. \tag{4.29}$$

Provided a sufficient time is allowed for the off-resonant spins to dephase (it is found experimentally that $T \geq 4/\omega_1$ is adequate), this spin-locking procedure selects those spins with offsets Δ_z in the range $-\omega_1 < \Delta_z < \omega_1$ [for an illustration see Fig. 4.10 taken from the work of Wind et al.[13]].

Further spatial selection (to a line or a point) can be achieved by switching the field gradient direction during the spin-lock pulse. Also, since the magnetization is destroyed in a coherent manner, it can be recalled by the use of a 180° pulse thus avoiding long delays between repeat experiments in the case of samples with long T_1's.

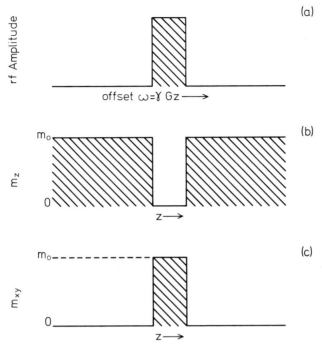

FIG. 4.9.   Type 2 selective excitation experiment: (a) rf pulse spectral distribution; (b) z component of magnetization following rf pulse; and (c) x,y components of magnetization following rf pulse.

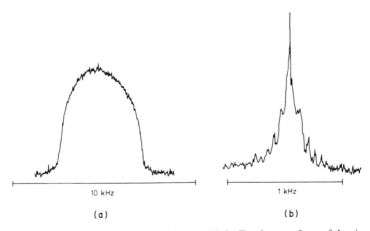

FIG. 4.10.   The selective spin-lock experiment: (a) the Fourier transform of the signal after a $\pi/2$ pulse of amplitude 120 kHz; (b) the Fourier transform of the signal obtained after 20-msec locking with an rf field having an amplitude of 80 Hz. [From R. A. Wind, J. H. N. Creyghton, D. J. Ligthelm, and J. Smidt, *J. Phys. C* **11**, L223 (1978).]

4.3. Point Methods

4.3.1. THE SENSITIVE-POINT METHOD

In 1974 Hinshaw[3] introduced a new approach to NMR imaging which he called the sensitive-point method.[4,14] This employs the SSFP technique with three mutually orthogonal oscillating field gradients in the manner described in Section 4.2.2. Only those spins in the region defined by the intersection of the three null planes are in a time-invariant magnetic field. The signal coming from this region, known as the sensitive point, can then be extracted with the aid of a simple narrow band filter.

The SSFP technique as we have seen (see Section 3.4) produces an alternation of signal sign in successive cycles which has to be removed as part of the detection procedure. However, this has the advantage that it gives the spectrometer the stability to baseline drift etc. of a chopper amplifier.

In the original implementation of the method the output of the spectrometer was connected through a 0.1–1.0 sec filter directly to a graph plotter. This demonstrates the relative ease of operation of the technique. There is, for instance, no requirement for computer control and the image is obtained directly without need for reconstruction or other computational procedures. The signal can therefore be monitored during acquisition. Many design problems that occur with line and planar techniques, namely bandwidth, dynamic range, gradient linearity, etc., are reduced to a minimum. However, against all these advantages must be weighed the one major drawback of long imaging times. For Hinshaw's early images which were of small objects about 1 cm in diameter (see, for example, the spring onion image described in Section 6.6.3), these times were of the order of a few hours. Motional artifacts are thus likely to present a major problem.

Finally, we mention that one can use the oscillating field gradient method to perform all conventional NMR experiments at a single point. Although such measurements are by no means restricted to this technique and can in fact be performed with almost any of the currently proposed imaging systems, the sensitive-point method may well be the easiest with which to implement them.

4.3.2. FONAR AND RELATED TECHNIQUES

FONAR or field-focused nuclear magnetic resonance is the name given by Damadian and his colleagues[15-19] to NMR imaging techniques which rely

---

[14] W. S. Hinshaw, *Proc. Ampere Congr., 18th, Nottingham* p. 433. North-Holland Publ., Amsterdam, 1974.

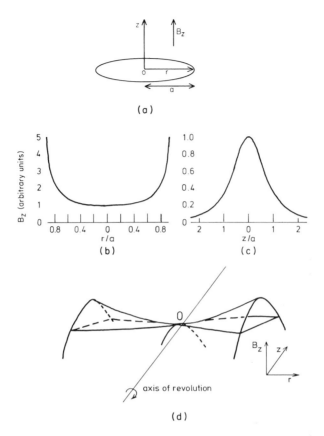

FIG. 4.11. The resonance aperture of an Amperian loop: (a) the Amperian loop; (b) magnetic field variation over the diameter and (c) along the axis of the loop; (d) schematic illustration of the field variation normal to the loop.

on the shaping of static and rf fields to isolate a signal-producing region or "resonance aperture" within a sample.

As an illustration, consider a simple circular coil or Amperian loop (Fig. 4.11a). The exact expressions for the fields across a diameter and along the axis of the coil are plotted in Fig. 4.11b and c and Fig. 4.11d indicates sche-

[15] R. Damadian, L. Minkoff, M. Goldsmith, M. Stanford, and J. Koutcher, *Physiol. Chem. Phys.* **8,** 61 (1976).

[16] R. Damadian, L. Minkoff, M. Goldsmith, M. Stanford, and J. Koutcher, *Science* **194,** 1430 (1976).

[17] R. Damadian, M. Goldsmith, and L. Minkoff, *Physiol. Chem. Phys.* **9,** 97 (1977).

[18] R. Damadian, L. Minkoff, M. Goldsmith, and J. A. Koutcher, *Naturwissenschaft* **65,** 250 (1978).

[19] R. Damadian, M. Goldsmith, and L. Minkoff, *Physiol. Chem. Phys.* **10,** 285 (1978).

matically the B_z field variation over a plane normal to the coil and containing its center. If the central field satisfies the Larmor condition, then those spins at or near resonance will lie predominantly within an elliptic region centered about the saddle point $\partial B_z/\partial x, \partial B_z/\partial z = 0$. There will also be some contribution from the lines of spins occurring at the intersection of the plane normal to B_z and containing the saddle point with the B_z surface. The full three-dimensional distribution is obtained by rotating the ellipse and lines of intersection about the z axis of the coil as generator. The result is a central ellipsoidal region surrounded by a ring of "V"-shaped cross section.

If the rf pulse is of sufficiently low power, i.e., long, then the range of frequencies it excites will be limited. The observed signal then arises almost exclusively from the ellipsoidal region since the width of "ring wall" excited is too small to provide any substantial contribution. The signal, in the form of the initial amplitude of the FID or some cumulative average, is stored and the resonance aperture scanned through the object to build up an image. (In practice, it may well be simpler to move the object relative to a fixed aperture.) It is, of course, possible to improve the response with the use of more sophisticated coil systems and by shaping the rf field, but we believe this to be the essence of the FONAR technique.

The first images using this method were of live normal and tumorous mice and were obtained by Damadian and colleagues in 1976.[15,16] The resonance aperture was adjusted to be approximately spherical with a diameter ~ 1–3 mm. The spectrometer operated at 10 MHz and the scan time was about 4 hr.

In 1977 this group extended the technique to human studies utilizing a superconductive magnet[20,21] designed and built by them to operate at 500 G (2.18 MHz for protons). The 90° rf pulses of 60 μsec duration were repeated every 800 μsec and the image points were video records of the maximum peak-to-peak amplitude of a constant 5-kHz off-resonance FID. Further details concerning the images are given in Chapter 6. Some medical evaluation of the FONAR technique and comparison with CT X-ray images has recently been reported by Partain et al.[21a]

A similar method has also been developed by Abe, Tanaka, et al.,[22a–c]

[20] L. Minkoff, R. Damadian, T. E. Thomas, N. Hu, M. Goldsmith, J. Koutcher, and M. Stanford, *Physiol. Chem. Phys.* **9**, 101 (1977).

[21] M. Goldsmith, R. Damadian, M. Stanford, and M. Lipkowitz, *Physiol. Chem. Phys.* **9**, 105 (1977).

[21a] C. L. Partain, A. E. James, J. T. Watson, R. R. Price, C. M. Coulam, and F. D. Rollo, *Radiology* **136**, 767 (1980).

[22a] Z. Abe, K. Tanaka, M. Hotta, and M. Imai, *in* "Biologic and Clinical Effects of Low-Frequency Magnetic and Electric Fields," pp. 295–317. Thomas, Springfield, Illinois, 1974.

[22b] K. Tanaka, Y. Yamada, T. Shimizu, F. Sano, and Z. Abe, *Biotelemetry* **1**, 337 (1974).

[22c] K. Tanaka, Y. Yamada, E. Yamamoto, and Z. Abe, *Proc. IEEE* **66**, 1582 (1978).

who achieve their localization by "degrading" the homogeneous magnetic field of a conventional spectrometer with one or more reversed Helmholtz coil pairs. Suitable filtering is used to ensure that the detected signal originates only from the resonance aperture.

It is also possible to define a resonance aperture by adding suitably shaped iron shims to the pole faces of a conventional electromagnet. Crooks *et al.*[23] have used this technique to image a number of phantoms and a rat at 15 MHz. Both the sample and coil system were mechanically moved through the resonance aperture taking about 6 hr to collect 200 pixels of estimated size: 0.6 cm FWHM (full-width half-maximum) in the image plane and 0.7 cm FWHM normal to it. (These figures were determined by moving a cubic water sample through the sensitive region.) Clearly, this method lacks the versatility of FONAR in that the shims have to be changed to alter the size of the resonance aperture.

4.4. Line Techniques

4.4.1. THE MULTIPLE SENSITIVE-POINT METHOD

4.4.1.1. *Introduction*

This method is a logical development of the sensitive-point technique described in Section 4.3.1 and uses one static and two oscillating field gradients. The latter serve to define a sensitive line within the sample, whereas the former allows spins to be differentiated along that line. This is shown schematically in Fig. 4.12.

Again, the pioneering work in this area was performed by Hinshaw and colleagues who obtained good quality, high-resolution images of objects up to 10 cm in size.[5,24-26] More recently, Moore, Holland, and Hawkes have constructed an imaging system, based on this principle, which is capable of scanning objects of human dimensions.[27,28] Some of the smaller scale images obtained with Hinshaw's original apparatus appear in Section 6.

[23] L. E. Crooks, T. P. Grover, L. Kaufman, and J. R. Singer, *Invest. Radiol.* **13,** 63 (1978).

[24] W. S. Hinshaw, P. A. Bottomley, and G. N. Holland, *Nature (London)* **270,** 722 (1977).

[25] W. S. Hinshaw, E. R. Andrew, P. A. Bottomley, G. N. Holland, W. S. Moore, and B. S. Worthington, *Br. J. Radiol.* **51,** 273 (1978).

[26] W. S. Hinshaw, E. R. Andrew, P. A. Bottomley, G. N. Holland, W. S. Moore, and B. S. Worthington, *Br. J. Radiol.* **52,** 36 (1979).

[27] W. S. Moore and G. N. Holland, *Phil. Trans. R. Soc. London Ser.* B **289,** 511 (1980).

[28] G. N. Holland, W. S. Moore, and R. C. Hawkes, *J. Comp. Assist. Tomog.* **4,** 1 (1980).

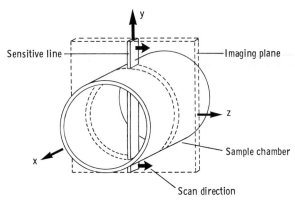

FIG. 4.12.   The multiple sensitive-point imaging method. Oscillating field gradients along the x and z directions generate a sensitive line parallel to y. The static gradient in this direction allows image points to be resolved along the sensitive line which can be scanned through the object by altering the current balance in the z-gradient coils. [From E. R. Andrew, P. A. Bottomley, W. S. Hinshaw, G. N. Holland, W. S. Moore, and C. Simaroj, *Phys. Med. Biol.* **22,** 971 (1977).]

4.4.1.2. *Description*

The experimental setup differs from the sensitive-point case in that one applies two rather than three oscillating nonconjugate field gradients. (These may be derived from separate supplies or else from the quadrature outputs of a single supply.) Furthermore, instead of integrating the phase sensitive detected nuclear signal, it is sampled for a discrete number of intervals (typically 128) during each SSFP cycle. Signals from successive cycles are cumulated after removal of the sign alternation (which again gives the system the stability to baseline drift, etc., of a chopper amplifier). This averaging serves both to improve the signal-to-noise ratio and to allow the oscillating gradients to achieve their spatial selectivity. The sets of averaged time data points (there will be two such sets if quadrature detection is employed) then undergo a discrete complex Fourier transformation which yields a line spin density profile—or image line. The z-gradient coil current ratio (see Fig. 4.12) is then altered to allow selection of a new line and the process is repeated to build up a full image array.

4.4.1.3. *Theory*

In this section we do not consider the manner in which the oscillating gradients produce a sensitive line, (this is a difficult problem which has been considered in some detail by Meiere and Thatcher[29]), but restrict ourselves

[29] F. T. Meiere and F. C. Thatcher, *J. Appl. Phys.* **50,** 4491 (1979).

to a discussion of the definition of the sensitive points along that line. One approach to this problem has been discussed by Hinshaw.[30] Here, however, we choose to follow the arguments developed in Section 4.2.2.3.

With the approximations that T_1, $T_2 \gg T$, the real and imaginary parts of the nuclear signal in a multiple sensitive point experiment are, respectively,

$$m_R(\mathbf{r}, t) = A(T)\cos(\gamma Gzt) - B(T)\sin(\gamma Gzt), \tag{4.30}$$

$$m_I(\mathbf{r}, t) = A(T)\sin(\gamma Gzt) + B(T)\cos(\gamma Gzt), \tag{4.31}$$

where

$$A(T) = \frac{1 + \cos(\gamma GzT)}{(1 + 2T_1/T_2) + \cos(\gamma GzT)} m_0(\mathbf{r}), \tag{4.32}$$

$$B(T) = \frac{\sin(\gamma GzT)}{(1 + 2T_1/T_2) + \cos(\gamma GzT)} m_0(\mathbf{r}), \tag{4.33}$$

where G is the static field gradient applied along the z direction and $m(\mathbf{r}, t)$ refer only to signals from the region close to the intersection of the null planes (the sensitive line) defined by the two orthogonal oscillating gradients.

Taking the complex Fourier transform of the signal over one SSFP cycle yields

$$R(z, \omega) = A(T)\left\{\int_0^T \cos(\gamma Gzt)\cos \omega t\, dt + \int_0^T \sin(\gamma Gzt)\sin \omega t\, dt\right\}$$

$$+ B(T)\left\{\int_0^T \cos(\gamma Gzt)\sin \omega t\, dt - \int_0^T \sin(\gamma Gzt)\cos \omega t\, dt\right\} \tag{4.34}$$

$$= A(T)\int_0^T \cos[(\gamma Gz - \omega)t]\, dt - B(T)\int_0^T \sin[(\gamma Gz - \omega)t]\, dt. \tag{4.35}$$

Using Eqs. (4.32) and (4.33) we obtain finally

$$R(z, \omega) = \frac{2m_0(\mathbf{r})\cos(\gamma GzT/2)}{T_1/T_2 + \cos^2(\gamma GzT/2)}$$

$$\times \frac{\cos(\gamma GzT - \omega T/2)\sin(\gamma GzT/2 - \omega T/2)}{\gamma Gz - \omega}. \tag{4.36}$$

We now ask what is the response $R(z, \omega)$ evoked by an isochromatic volume of spins at z. Taking first the case for which $z = 0$, Eq. (4.36) reduces to

$$R(0, \omega) = \frac{T}{(T_1/T_2 + 1)} \frac{\sin \omega T}{\omega T} m_0(\mathbf{r}). \tag{4.37}$$

---

[30] W. S. Hinshaw (to be published).

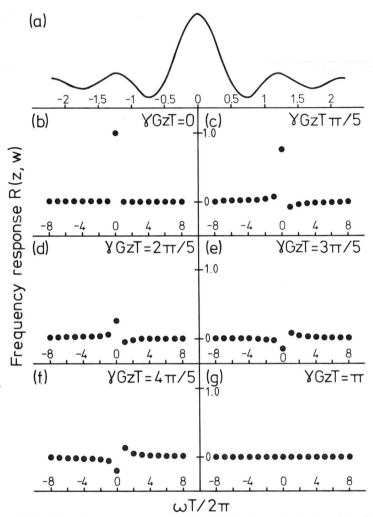

FIG. 4.13.   The multiple sensitive-point response to an isochromat of spins: (a) sinc function response $R(0,\omega)$ to an isochromat at $z = 0$; (b) digital response to an isochromat at $z = 0$; (c)–(g) digital response to isochromats at (c) $z = \pi/5\gamma GT$; (d) $z = 2\pi/5\gamma GT$; (e) $z = 3\pi/5\gamma GT$; (f) $z = 4\pi/5\gamma GT$; and (g) $z = \pi/\gamma GT$.

This is a sinc function centered at $\omega = 0$, with zeros at $\omega = n\pi/T$ and is illustrated in Fig. 4.13a. In an actual imaging experiment however, it is usual to perform a digital rather than a continuous Fourier transformation (see, for example, Bergland's[31] introduction to the fast Fourier transform). This, coupled with the fact that the nuclear signal is only observed for a finite

[31] G. D. Bergland, *IEEE Spectrum* **6,** 41 (1969).

period T (the interpulse interval), means that one samples only the frequency spectrum at a discrete set of points given by

$$\omega = 2n\pi/T, \qquad (4.38)$$

where n is a positive or negative integer. The response shown in Fig. 4.13a is therefore sampled only at these points which, apart from the central maximum at $z = 0$, all coincide with zeros of the sinc function (see Fig. 4.13b). The overall response to a δ function at $z = 0$ is therefore a corresponding δ function at $\omega = 0$. The response to an isochromatic volume of spins at $z = \pi/\gamma GT$ is readily seen to be zero.

For $z = 2n\pi/\gamma GT$ one obtains

$$R[(2n\pi/\gamma GT), \omega] = \frac{T}{T_1/T_2 + 1} \frac{\sin(2n\pi - \omega T)}{(2n\pi - \omega T)} m_0(\mathbf{r}), \qquad (4.39)$$

i.e., a set of sinc functions each centered at $\omega = 2n\pi$. When sampled discretely, we are again left only with a set of δ functions at frequencies $\omega = 2n\pi$. Similarly, the response to isochromats at $z = (2n + 1)\pi/\gamma GT$ is zero. In fact, one can show quite generally that

$$R(z + 2\pi/\gamma GT, \omega) = R(z, \omega - 2\pi/T), \qquad (4.40)$$

and that

$$R(-z, \omega) = R(z, -\omega). \qquad (4.41)$$

Thus, in order to determine $R(z, \omega)$ one needs only to calculate the responses for z in the range $0, \pi/\gamma GT$. A selection of these are illustrated in Fig. 4.13b–g. Note that, in general, the response is spread out over a considerable number of points. $R(z, \omega)$ tells us where in the frequency spectrum to find the information concerning a particular spin isochromat.

It is also of interest to know the total or summed response $S(z)$ for that isochromat. A measure of this is

$$S_t(z) = \int_{-\infty}^{+\infty} R(z, \omega)\, d\omega. \qquad (4.42)$$

This function is illustrated in Fig. 4.14a. However, as we have discussed at some length, the spectrum is sampled discretely so that a better measure of the total response is

$$S_d(z) = \sum_{n=-\infty}^{+\infty} R(z, 2n\pi/T). \qquad (4.43)$$

This function is illustrated in Fig. 4.14b. Its maxima, occurring at $z = 2n\pi/\gamma GT$ are the so-called sensitive points along the sensitive line. Note that there are

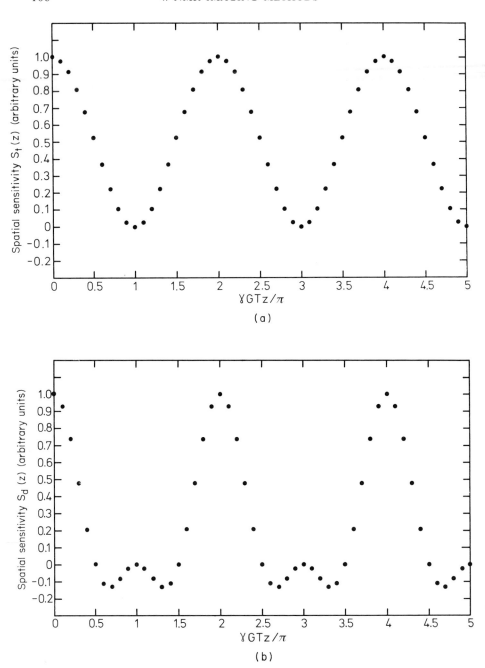

FIG. 4.14.   (a) The integrated multiple sensitive-point response $S_t(z)$ to an isochromat of spins. (b) The summed multiple sensitive-point response $S_d(z)$ to an isochromat of spins.

gaps between these points in which spins do not contribute to the nuclear signal at all.

4.4.1.4. *Tissue Contrast*

The amplitude of the response at a sensitive point is proportional to

$$\frac{m_0(\mathbf{r})}{(T_1/T_2 + 1)}, \tag{4.44}$$

where T_1, T_2 are also, in general, functions of $\mathbf{r}$ [see also Eq. (3.79)]. The image intensity is therefore not simply proportional to the spin density $m_0(\mathbf{r})$, but also reflects differences in T_1 and T_2. For instance, this method of imaging will discriminate very heavily against "solidlike" regions for which T_1/T_2 may be $\sim 10,000$, whereas for liquidlike areas in which $T_1 \simeq T_2$, the above factor assumes its maximum value of $m_0(\mathbf{r})/2$. As we have seen in Section 2.2, the ratio T_1/T_2 shows a strong frequency dispersion, but a "typical" value for biological tissue would lie in the range 3–10.

If $T_1 \gg T_2$ it is possible to achieve larger nuclear signals with the aid of small angle rf pulses (see Section 6.3.4). The simplification introduced in Section 4.2.2.3 is then no longer valid and we must use Eq. (4.15) to determine the new dependence on relaxation times. This raises another interesting point, namely, that besides altering the signal level (and hence the signal-to-noise ratio S/N) small-angle (or large-angle, $\beta > 90°$) pulses may affect the discrimination between regions of different T_1 and T_2. If such is the case, it may well be worth sacrificing overall S/N for increased tissue contrast [see Eq. (3.89a and b)].

4.4.1.5. *Signal Filtering*

The nuclear signal is normally integrated between sample points (e.g., by means of a simple integrating network with a time constant equal to the sampling interval). This is equivalent to filtering out all angular frequency components greater than $2\pi N/T$ (where N is the total number of sample points) and prevents frequency aliasing (the phenomenon whereby high frequencies are folded back into the lower end of the spectrum during the digital transformation). Such a procedure also, of course, improves the S/N through its band-limiting effect, and assists the oscillating field gradients in the averaging to zero of the "background" signal. Further discussion is given in Section 8.6.5.

4.4.1.6. *Choice of Gradients*

As we have seen, the sensitive points are separated in space by $2\pi/\gamma GT$. Thus, one should normally select the value of the linear gradient G such that

$$(2\pi/\gamma GT)N = L, \tag{4.45}$$

where L is the linear dimension of the object to be scanned. Taking as an example $L = 10$ cm and choosing typical values for N and T of 128 and 2.5 msec, respectively, we obtain $G \sim 0.8$ G cm^{-1}.

With regard to the oscillating gradient amplitudes, we can get an approximate idea of what is required from Fig. 4.3b. Thus for a spatial resolution of δL we have

$$(\gamma G_{osc} \, \delta L)/\Omega_{osc} \sim 5. \qquad (4.46)$$

One should, of course, choose the width of the sensitive line to be roughly equal to the resolution along it (namely, L/N). However, one may well require the plane thickness to be substantially greater in order to achieve an acceptable S/N. As an example, we take $\delta L = 2$ mm and $\Omega_{osc} = 50$ Hz, and obtain $G \sim 0.3$ G cm^{-1}.

4.4.2. SELECTIVE IRRADIATION LINE-SCANNING METHODS

4.4.2.1. *Early Experiments*

The first image to be produced by selective irradiation (SI) methods was reported by Garroway et al.[9] and used the technique of Section 4.2.3.2.1b to define a line within a two-dimensional sample. Resolution along the line was achieved by switching the gradient direction through 90° and recording the FID following a short 90° pulse. The full experiment is shown in Fig. 4.15a and the image obtained with it in Fig. 4.15b. As discussed in Section 4.2.3.2.1b this type of experiment will not work well in situations where there is a distribution of T_1's, and is in any case inefficient.

In 1974 Lauterbur et al.[32] also suggested a selective excitation method, which is illustrated in Fig. 4.16. Long, low-level rf pulses were used to define the sensitive region (though with less precision than a properly tailored pulse). Since these pulses were all resonant, the image was scanned by varying the field offset ΔB_0 in the presence of a field gradient G_1. As above, discrimination between the spins along the image line was achieved by recording the FID in a second gradient G_2. However, since in this case the selective excitation was of type 2, no further read pulse was required. No images obtained by this method were published. However, the possible use of the preparatory phase to define a plane for use with projection reconstruction techniques was realized.

[32] P. C. Lauterbur, C. S. Dulcey, Jr., C.-M. Lai, M. A. Feiler, W. V. House, Jr., D. M. Kramer, C.-N. Chen, and R. Dias, *Proc. Ampere Congr. 18th, Nottingham* p. 433. North-Holland Publ., Amsterdam, 1974.

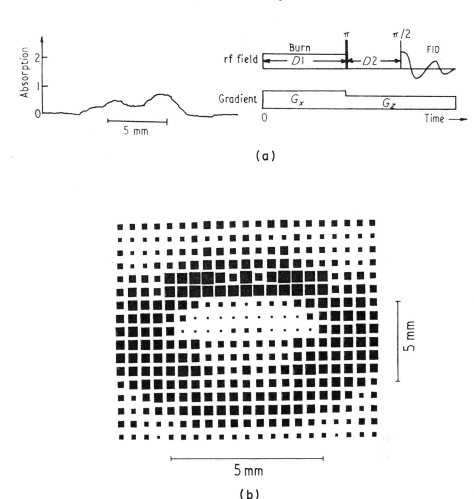

FIG. 4.15. (a) Proton absorption line shape obtained from a water annulus with the selective saturation sequence shown on the right. (b) Two-dimensional image of the water annulus. Each row of black dots corresponds to the spin density of a section through the annulus as in (a). The dot size is proportional to the spin density at that point. The whole picture is formed using ten gray levels or dot sizes. Elongation of the picture into an elliptical form is due to the inequality of the "burn" and "read" gradients. [From A. N. Garroway, P. K. Grannell, and P. Mansfield, *J. Phys. C* **7**, L457 (1974).]

4.4.2.2. *Chemical Shift Information*

One method of imaging the chemical shift distribution is to use small read gradients such that the width of a line profile is less than the separation of the various chemical shift contributions. Each spectral line is then broadened to yield the image line density of that particular component without interference

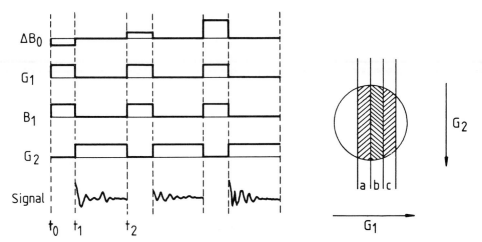

FIG. 4.16.   Selective excitation with rectangular pulses. By altering the offset ΔB_0 successive slices a, b, and c can be excited with an rf field B_1 in the presence of a gradient G_1. The gradient G_2 allows discrimination between spins in an orthogonal direction. [From Ref. 32.]

from neighboring spectral lines. Such a method is clearly possible only in situations in which there is a discrete number, rather than a continuous distribution of lines and will be particularly difficult for protons where the range of chemical shifts is of the order of only 10 ppm.

An alternative method was proposed by Lauterbur *et al.*[33] in 1975 and combines both selective irradiation and projection reconstruction techniques. One first selects a line within the (two-dimensional) sample by a Type 2 selective excitation method as described above. The FID is then recorded in the *absence* of a field gradient so that a high-resolution spectrum of the material lying along the line is obtained. This is then repeated for all lines across the sample (Figs. 4.17 and 4.18a). Finally, one alters the direction of the selection gradient and repeats the sequence (Fig. 4.18b). In this way, it is possible to acquire a sufficient number of projections to reconstruct an image of any region of the chemical shift spectrum (see, for example, Fig. 4.19; see also Section 4.5.2 for reference to another chemical shift measurement technique).

4.4.2.3. *Fast Scan Imaging*

In 1976 Mansfield, Maudsley, and Baines published details of a more efficient selective irradiation method[11] which was used extensively for small-

[33] P. C. Lauterbur, D. M. Kramer, W. V. House, Jr., and C.-N. Chen, *J. Am. Chem. Soc.* **97**, 6866 (1975).

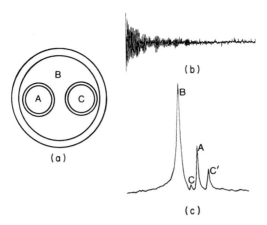

(b)

(a)

(c)

FIG. 4.17.   (a) A cross-sectional diagram of the composite sample. The outer 15-mm-o.d. tube contains sulfuric acid (B), one inner 5-mm-o.d. tube contains water (A), and the other contains $p-(CH_3)_3CC_6H_4NO_2$. (b) A 4-MHz FID of the proton resonance signals from the entire sample after a nonselective 5-msec 90° pulse in a homogeneous static magnetic field. The segment shown is 500 msec in length. (c) The Fourier transform power spectrum of (b). The peak labels correspond to those in (a), with C being the collapsed group of aromatic hydrogen peaks and C′ the tert-butyl peak. The chemical shift difference between B and C′ is about 10 ppm or 40 Hz. [From P. C. Lauterbur, D. M. Kramer, W. V. House, Jr., and C.-N. Chen, *J. Am. Chem. Soc.* **97,** 6866 (1975).]

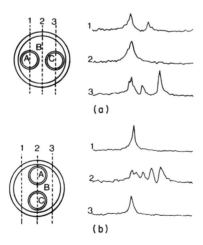

(a)

(b)

FIG. 4.18.   Spectra of the object in Fig. 4.17a produced by selective excitation in a field gradient: (a) excitation in an *x* gradient to give the spectra of compounds present near several planes of constant magnetic field perpendicular to the *x* axis; (b) excitation in an *x* gradient, after rotation of the object by 90°. [From P. C. Lauterbur, D. M. Kramer, W. V. House, Jr., and C.-N. Chen. *J. Am. Chem. Soc.* **97,** 6866 (1975).]

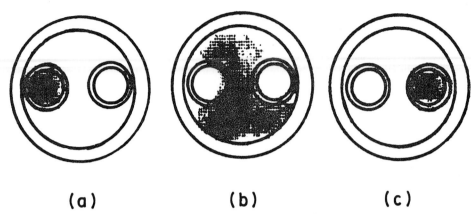

(a)   **(b)**   **(c)**

FIG. 4.19. Proton NMR zeugmatograms of the individual chemical species in the object shown in Fig. 4.17. Each was reconstructed by a multiplicative ART-type algorithm from six 32-point projections, at 30° intervals, each derived from 18 selective 7-msec half-width Gaussian pulse excitations in a 0.25-G cm^{-1} gradient. The images are displayed by a 16 gray-level over-printing program: (a) the distribution of water in one inner tube; (b) the distribution of sulfuric acid in the outer tube; (c) the distribution of p-tert-butylnitrobenzene in the other small tube, as derived from the intensities of the tert-butyl resonance. All data were obtained from the integrals of phase-corrected absorption mode transforms. [From P. C. Lauterbur, D. M. Kramer, W. V. House, Jr., and C.-N. Chen. *J. Am. Chem. Soc.* **97,** 6866 (1975).]

scale imaging[34-38] and, with minor modification, for the more recent human studies.[39-42]

The sequence is illustrated in Fig. 4.20 and commences with a type 1a plane selection (see Section 4.2.3.2.1) with the field gradient G_z along the z direction. A further stage of selection follows, this time type 2, with the field

[34] P. Mansfield and A. A. Maudsley, *Phys. Med. Biol.* **21,** 75 (1976).

[35] P. Mansfield and A. A. Maudsley, *Proc. Ampere Congr., 19th, Heidelberg* p. 247. Groupment Ampere, Heidelberg, 1976.

[36] P. Mansfield and A. A. Maudsley, *Br. J. Radiol.* **50,** 188 (1977).

[37] P. Mansfield and I. L. Pykett, *J. Magn. Reson.* **29,** 355 (1978).

[38] I. L. Pykett and P. Mansfield, *Phys. Med. Biol.* **23,** 961 (1978).

[39] P. Mansfield, I. L. Pykett, P. G. Morris, and R. E. Coupland, *Br. J. Radiol.* **51,** 921 (1978).

[40] P. Mansfield, P. G. Morris, R. J. Ordidge, R. E. Coupland, H. M. Bishop, and R. W. Blamey, *Br. J. Radiol.* **52,** 242 (1979).

[41] P. G. Morris, P. Mansfield, I. L. Pykett, R. J. Ordidge, and R. E. Coupland, *IEEE Trans.* **26,** 2817 (1979).

[41a] I. L. Pykett, P. Mansfield, P. G. Morris, R. J. Ordidge, and V. Bangert, *in* Lecture Notes in Physics 112, "Imaging Processes and Coherence in Physics," Proceedings, Les Houches 1979 (M. Schlenker, M. Fink, J. P. Goedgebuer, C. Malgrange, J. Ch. Viénot, and R. H. Wade, eds.). Springer-Verlag, Berlin and New York, 1980.

[42] P. Mansfield, P. G. Morris, R. J. Ordidge, I. L. Pykett, V. Bangert, and R. E. Coupland, *Phil. Trans. R. Soc. London Ser.* B **289,** 503 (1980).

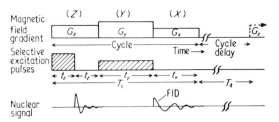

FIG. 4.20. Switching sequence showing applied field gradients, the selective excitation sequence and the nuclear free induction decay (FID) signals following the z and y tailored pulses. Note that only the FID in t_x is sampled in these experiments. [From P. Mansfield, A. A. Maudsley, and T. Baines, *J. Phys. E* **9**, 271 (1976).]

gradient G_y along y and selects a line of spins parallel to the x axis. The line density is finally obtained by Fourier transforming the FID recorded in a read gradient G_x.

The rf pulse shapes necessary for the two selective irradiation procedures are derived by digital Fourier transformation of the desired excitation spectra as discussed in Section 3.3.4. They are used to trigger a binary attenuation gate and thus modulate the low-level rf fed to the linear amplifier. Note that for single sideband irradiation at frequency $\omega_0 + \Delta\omega$, one requires two rf carriers in phase quadrature. However, provided no part of the sample is resonant at the image frequency $\omega_0 - \Delta\omega$, i.e., that the minimum value of $\Delta\omega$ is greater than the broadened profile half-width, a single carrier will suffice. The price one pays for experimental simplicity is the loss of half the rf power into the unwanted sideband.

In the experiments so far performed with this imaging system, it has been the custom to signal average each line prior to moving on to the next. One is thus obliged to allow a period T_d between shots in order for the z magnetization to recover. However, this can be turned to advantage; if the real spin density in the selected plane is $\rho(x, y)$ and the z magnetization recovers exponentially through a simple spin–lattice relaxation process in time $T_1(x, y)$, we can define[35] an effective spin density

$$\rho(x, y, \tau) = \rho(x, y)\{1 - \exp[-\tau/T_1(x, y)]\}, \qquad (4.47)$$

where $\tau = T_d + T_c$ is the repetition period of the experiment. The observed spin density $\rho(x, y, \tau)$ is thus dependent on the ratio $\tau/T_1(x, y)$ as well as on the actual spin density $\rho(x, y)$—an effect known as spin–lattice relaxation time discrimination. Much of the observed intensity variation between tissue types is attributable to this effect.

Faster scans can be obtained by incrementing the line offset after each shot. There is then no need for any delay since one is always irradiating a "fresh" part of the spin system, and a complete single-shot image is accumulated before any signal averaging takes place. The disadvantage of such a procedure is that one loses the spin–lattice relaxation time discrimination which contributes so much to tissue contrast. In addition the Fourier transforms must be performed rapidly or else computed prior to the experiment and stored.

The slower version of this experiment has been used to produce a number of small-scale images, some of which are illustrated in Section 6.6 and Chapter 7 (a human finger in Fig. 7.1 and an okra seed pod in Fig. 6.24). For these experiments on objects of approximately cylindrical symmetry, the first selection stage was omitted and the plane thickness was defined solely by the receiver-coil geometry. More recently a number of large-scale images have been produced and are discussed in Section 6.4.3 and Chapter 7.

4.4.2.4. *Focused Selective Excitation*

In 1976 Hutchison[43] proposed a selective irradiative line-scanning experiment using Gaussian-shaped rf pulses. This technique was developed theoretically with the aid of computer simulation[44] and in 1979 was used to image a human thoracic section.[45]

The experiment is split into two parts. In the first, the pulse and gradient sequence of Fig. 4.21 is applied to the sample. The initial Gaussian pulse combines with the G_x gradient to selectively nutate a plane of spins perpendicular to the x direction through 180°, the remainder of the sample being unaffected. The second pulse similarly combines with the G_y gradient to rotate a plane of spins perpendicular to the y axis through 90°. These spins thus end up in the x, y plane of the rotating frame. However, those which lie along the intersection of the two selected planes (the selected line) will have undergone both selection procedures and will consequently be 180° out of phase with the rest. Note that a spin refocusing procedure may be performed by reversing the direction of the y gradient following the 90° pulse.[44] This greatly improves the spatial selectivity, as discussed in Section 3.3.5. Discrimination between points along the selected line is achieved by recording the decay of the spin echo in a third linear gradient G_z directed along the line. This signal S_1 is, however, combined with that from the rest of the plane selected by the second selective pulse S_2 so that the total response S_a is given by

$$S_a = -S_1 + S_2. \tag{4.48}$$

[43] J. Hutchison, *in* Proc. 7th L. H. Gray Conf., Medical Images: Formation, Perception and Measurement (G. A. Hay ed.), p. 135. Wiley, New York, 1976

[44] R. J. Sutherland and J. M. S. Hutchison, *J. Phys. E Sci. Instrum.* **11**, 79 (1978).

[45] J. Mallard, J. M. S. Hutchison, W. A. Edelstein, C. R. Ling, M. A. Foster, and G. Johnson, *Phil. Trans. R. Soc. London Ser. B* **289**, 519 (1980).

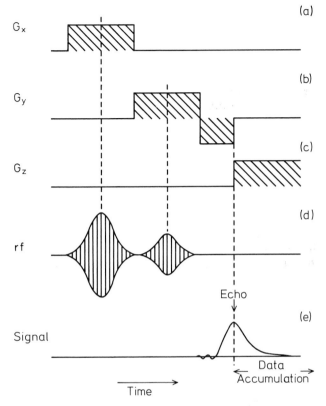

Fɪɢ. 4.21.   The focused selective excitation experiment: (a) G_x irradiation gradient; (b) G_y irradiation gradient; (c) G_z read gradient; (d) rf pulse sequence; and (e) nuclear signal following the rf pulse sequence.

In the second part of the experiment the preparatory 180° pulse is omitted. The response S_2 is unaffected but S_1 is reversed in sign so that the total response S_b is now given by

$$S_b = S_1 + S_2. \tag{4.49}$$

Thus, by subtracting the results of the two experiments one obtains the response of the selected line

$$2S_1 = S_b - S_a. \tag{4.50}$$

The procedure as discussed above can give rise to spurious signals arising from magnetization in the x, y plane following the imperfect selective 180° pulse. One way around this difficulty is to add a third part to the experiment in which the phase of this pulse differs from that of the subsequent 90° pulse by 180°. If we denote the signal due to this experiment by S_c, the required response S_1 is given by

$$4S_1 = 2S_b - (S_a + S_c). \qquad (4.51)$$

Alternatively, one may improve the selectivity of the 180° pulse by adding a refocusing gradient in this case as well.

The selected line can be scanned through the image either by stepping the static magnetic field or by altering the rf carrier frequency of the pulses. In practice, the latter method is preferred. Spin–lattice relaxation time discrimination can also be easily achieved by the introduction of appropriate delays in the timing sequence.

This method of imaging has recently been used with a whole-body imaging system and works well with phantom objects. However, the difference expression of Eq. (4.50) means that the technique is highly susceptible to any motion occurring between the first and second parts of the experiment. This is clearly evident in the human thoracic section shown in Fig. 6.18 and, whereas in this case it may be possible to obtain some improvement by suitable gating of the experiment from the ECG, the lesson to be learned is that "difference" techniques are generally undesirable in a practical imaging system.

4.4.2.5. *Other Techniques*

As discussed, the type 1a selective saturation used in the experiments of Mansfield *et al.* suffers from a number of experimental drawbacks. However, Maudsley[12] has recently proposed a selective irradiation imaging scheme which makes use of two type 2 excitation pulses in the manner of Fig. 4.22a. The effect of the first 90° pulse is to nutate only those spins within the dotted region of Fig. 4.22b. The 180° pulse then causes those spins in the shaded region or sensitive line to refocus after a delay t' equal to the selective pulse separation, and the line density can be recorded in the presence of the linear gradient G_y. This scheme is believed to be similar to that used by Crooks *et al.*[46] to obtain a number of good-quality images, including cross sections through live rats.

Maudsley[12] also outlined a scheme by which several lines could be imaged in each experiment. The idea is similar to that of planar imaging discussed in Section 4.5.3 but has the advantage that it does not suffer from the same loss of sensitivity. However, it does rely on a type of difference technique which, as we have seen, is strongly susceptible to motional artifact.

---

[46] L. Crooks, J. Hoenninger, M. Arakawa, L. Kaufman, R. McRee, J. Watts, and J. R. Singer, *Radiology* **136,** 701 (1980).

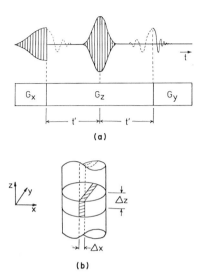

FIG. 4.22.   (a) The selective excitation pulse sequence for the observation of the spin density in the volume $y \, \Delta x \, \Delta z$ shown in (b). (From A. A. Maudsley, private communication, 1979.)

4.5. Planar and Three-Dimensional Techniques

4.5.1. FOURIER ZEUGMATOGRAPHY

4.5.1.1. Introduction

The method of Fourier zeugmatography was proposed in 1974 by Kumar, Welti, and Ernst[47] and is really an example of a broader class of NMR techniques known as two-dimensional Fourier transform spectroscopy,[48,48a] The idea is conceptually very simple and can be implemented in either the planar or full three-dimensional mode. However, the computational and storage problems associated with the latter are prohibitive so one may expect the initial applications of this technique to be limited to planar studies.

4.5.1.2. Mathematical Analysis

The description and mathematical analysis of these methods, which follow closely the original treatment of Kumar, Welti, and Ernst,[49] consider the most general three-dimensional case.

---

[47] A. Kumar, D. Welti, and R. R. Ernst, *Naturwissenschaft* **62,** 34 (1975).

[48] W. P. Aue, E. Bartholdi, and R. R. Ernst, *J. Chem. Phys.* **64,** 2229 (1976).

[48a] A. Wokaun and R. R. Ernst, "The Principles of NMR Fourier Spectroscopy in One and Two Dimensions," *in* Advances in Magnetic Resonance. Academic Press (in preparation).

[49] A. Kumar, D. Welti, and R. R. Ernst, *J. Magn. Reson.* **18,** 69 (1975).

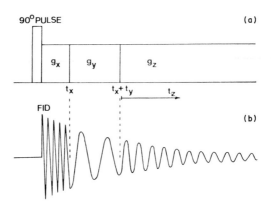

FIG. 4.23. (a) Diagram depicting the principle of the zeugmatographic method. (b) The FID is recorded during the third time interval as a function of t_z. For a three-dimensional zeugmatogram, usually N^2 such FID signals will be needed for a complete set of values for x and y gradients. N is the number of samples in an FID. [From A. Kumar, D. Welti, and R. R. Ernst, *J. Magn. Reson.* **18**, 69 (1975).]

The pulse and gradient sequence is illustrated in Fig. 4.23 and commences at $t = 0$ with a short nonselective rf pulse to nutate some or all of the equilibrium magnetization into the x, y plane of the rotating frame. This is then followed by periods t_x and t_y of variable length during which linear magnetic field gradients g_x, g_y are applied in the x and y directions, respectively. Finally, the FID is sampled in the presence of a z-gradient g_z for a time $t_z = t - (t_x + t_y)$.

The z component of the magnetic field is therefore given by

$$B_z(\mathbf{r}) = \begin{cases} B_0 + g_x x, & 0 < t < t_x, \\ B_0 + g_y y, & t_x < t < t_x + t_y, \\ B_0 + g_z z, & t_x + t_y < t, \end{cases} \qquad (4.52)$$

and the response of the spin system $S(t)$ will be a function of t_x and t_y, as well as t_z, i.e., $S(t) = S(t_x, t_y, t_z)$. The experiment is repeated for a full set of regularly stepped or incremented t_x and t_y values allowing a three-dimensional image of the object to be "reconstructed," as discussed below.

The observed signal $S(t)$ can be written as

$$S(t) = \iiint \rho(\mathbf{r}) s(\mathbf{r}, t) \, dv, \qquad (4.53)$$

where $s(\mathbf{r}, t) \, dv$ is the contribution from the volume element $dv = dx \, dy \, dz$ at position $\mathbf{r}$ and $\rho(\mathbf{r})$ is, as usual, the spin density.

We can denote the three-dimensional Fourier transform of $S(t)$ by

$S(\omega) = S(\omega_x, \omega_y, \omega_z)$ so that

$$S(\omega) = \int \int \int S(\mathbf{t}) \exp(-i\omega \cdot \mathbf{t}) \, dt_x \, dt_y \, dt_z. \qquad (4.54)$$

This, like its Fourier transform conjugate, can be expressed as an integral over all volume elements. Thus,

$$S(\omega) = \int \int \int \rho(\mathbf{r}) s(\mathbf{r}, \omega) \, dv, \qquad (4.55)$$

where $s(\mathbf{r}, \omega)$ is the Fourier transform of $s(\mathbf{r}, \mathbf{t})$. If we consider a single resonance, the cosine signal phase sensitively detected at frequency ω' is

$$s(\mathbf{r}, \mathbf{t}) = m_0 \cos\{(\Delta\omega + \eta_x x)t_x + (\Delta\omega + \eta_y y)t_y + (\Delta\omega + \eta_z z)t_z\}$$
$$\times \exp\{-(t_x + t_y + t_z)/T_2\}, \qquad (4.56)$$

where $\Delta\omega$ is the frequency offset $\gamma B_0 - \omega'$ and $\eta_k = -\gamma g_k$. $s(\mathbf{r}, \omega)$ is thus given by

$$s(\mathbf{r}, \omega) = \tfrac{1}{2}\{G(\Delta\omega + \eta_x x - \omega_x)G(\Delta\omega + \eta_y y - \omega_y)G(\Delta\omega + \eta_z z - \omega_z)$$
$$+ G(-\Delta\omega - \eta_x x - \omega_x)G(-\Delta\omega - \eta_y y - \omega_y)G(-\Delta\omega - \eta_z z - \omega_z)\},$$
$$(4.57)$$

where the complex line-shape function

$$G(\omega) = \frac{m_0/T_2}{(1/T_2)^2 + \omega^2} + \frac{im_0\omega}{(1/T_2)^2 + \omega^2}. \qquad (4.58)$$

The second term in Eq. (4.57) refers to the contribution of the resonance near $-\Delta\omega$ which can be neglected if $(1/T_2) \ll \Delta\omega$. Equation (4.57) shows that the following identity holds

$$s(\mathbf{r}, \omega) = s(\mathbf{0}, \omega - \eta\mathbf{r}), \qquad (4.59)$$

where η is a diagonal matrix

$$\begin{bmatrix} \eta_x & & \\ & \eta_y & \\ & & \eta_z \end{bmatrix}.$$

Thus, we can rewrite $S(\omega)$ as

$$S(\omega) = \int \int \int \rho(\mathbf{r}) s(\mathbf{0}, \omega - \eta\mathbf{r}) \, dv. \qquad (4.60)$$

If the frequency variable ω is now replaced by a spatial variable $\mathbf{r}'$ defined by

$$\omega = \Delta\omega\mathbf{1} + \eta\mathbf{r}', \qquad (4.61)$$

where $\mathbf{1}$ is the vector $(1, 1, 1)$ then neglecting the contribution of the resonance at $-\Delta\omega$,

$$S(\omega) = \bar\rho(\mathbf{r}')$$

$$= \frac{1}{2} \int\int\int \rho(\mathbf{r})G(\eta_x(x - x'))G(\eta_y(y - y'))G(\eta_z(z - z'))\,dv. \quad (4.62)$$

Thus the three-dimensional Fourier transform of the observed signal yields a spin density $\bar\rho(\mathbf{r})$ which is a convolution of the real spin density $\rho(\mathbf{r}')$ with the line-shape function. $\bar\rho(\mathbf{r})$ is a complex expression whose real and imaginary parts both contain absorptive and dispersive terms which may be positive or negative. It is therefore best to compute $|\bar\rho(\mathbf{r})|$ which, provided the gradients dominate the linewidth in the sense of Section 6.1.3, gives a good measure of $\rho(\mathbf{r})$.

Note that it is possible to use quadrature detection, in which case one obtains an additional signal $S'(\mathbf{t})$ given by Eq. (4.56) with the cosine replaced by sine. This allows complete separation of the resonance at $-\Delta\omega$ in exactly the same way that the use of two carrier phases in quadrature permits single sideband modulation experiments in selective excitation experiments (Section 4.4.2.3). Both $S(\mathbf{t})$ and $S'(\mathbf{t})$ still contain mixtures of absorptive and dispersive terms, however, so that it is still necessary to take the modulus of $\bar\rho(\mathbf{r})$. The net result is thus simply the expected $\sqrt 2$ improvement in sensitivity.

Since the absorption term is inherently narrower than the dispersive term it is generally desirable to form an image using it alone. A method for accomplishing this has been described by Kumar, Welti, and Ernst and we refer the reader to their paper for details.[49] We comment, however, that it involves the combination of the results from four separate experiments, a procedure which is susceptible to motional artefacts (see, for example, Section 4.4.2.4).

4.5.1.3. Illustrative Results

Kumar et al.[49] demonstrated the simpler version of their technique using a water-filled phantom (Fig. 4.24a). This was chosen to have cylindrical symmetry so that the second-gradient period t_y could be dispensed with, and the resulting image was a "shadowgraph" corresponding to a projection of the object onto the x, z plane. Figure 4.24b illustrates a number of typical FIDs; the dotted line indicates the boundary between periods t_x and t_y, and the corresponding Fourier transforms are shown on the right. Figure 4.24c

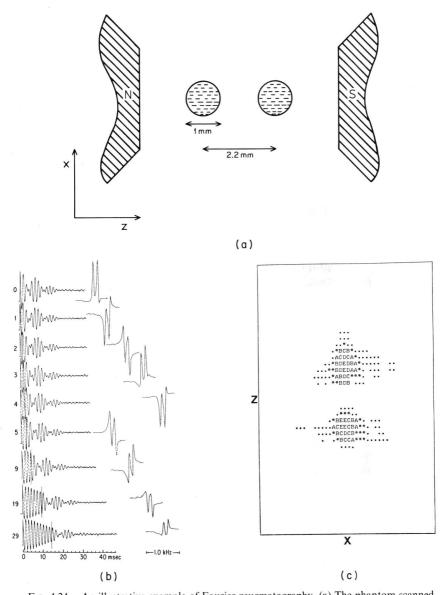

FIG. 4.24. An illustrative example of Fourier zeugmatography. (a) The phantom scanned consists of two water-filled glass capillaries of i.d. 1.0 mm, separated by 2.2 mm. (b) Nine typical FIDs selected out of a complete set of 64 signals obtained for pulsed linear field gradients along the x and z axes. The corresponding Fourier transforms are shown on the right-hand side. The numbers on the ordinate represent the time intervals in terms of sampling cycles during which the x gradient was on. The broken vertical lines in the FIDs indicate the point in time when the x gradient was stopped and the z gradient was switched on. At the same time, the recording of the FID was started. The sampling interval was 0.5 msec giving a total spectral width of 1 kHz. (c) The Fourier zeugmatogram obtained from the data partially shown in (b). The absolute value $|\bar{\rho}(\mathbf{r}')|$ of the spin density function is plotted as a function of x and z. Only the central section of the 64×64 zeugmatogram is shown. [From A. Kumar, D. Welti, and R. R. Ernst, *J. Magn. Reson.* **18**, 69 (1975).]

shows the image obtained by completing the two-dimensional Fourier transform. The intensity intervals are, in increasing order, , ., *, A, B, C, D, E.

4.5.1.4. *Experimental Variants*

In common with other planar techniques Fourier zeugmatography is susceptible to gradient nonlinearities and switching transients. In addition, care must be exercised in the choice of center frequency, gradient strengths, and sampling interval in order to avoid violating the sampling theorem and frequency foldover.

For the imaging of biological objects, it is necessary (unless a full three-dimensional experiment is envisaged) to define a slice. This is perhaps best achieved by using a selective excitation procedure (Section 4.2.3) involving the replacement of the nonselective preparatory pulse in Fig. 4.23 with a selective one and omitting one of the periods t_x or t_y. Alternatively, it is possible to invoke one of the other slice definition techniques discussed in Section 4.2. Even if one requires a three-dimensional image, it may still be more efficient to form the image a plane at a time in cases where the lattice relaxation time is long (see Chapter 5).

Quite recently, Edelstein *et al.*[49a] have used a simple modification of Fourier zeugmatography to produce 64^2 pixel whole-body images and, by extension in the usual manner, T_1 maps (see Chapter 7 for images). The variant relies simply on the fact that the spin phase accumulation $\int_0^{t_x} g_x(t)\, dt$ can be achieved either with the amplitude of g_x held constant and t_x varied, as in the original experiment, or by holding t_x constant at some convenient value such that $t_x < T_2$, and varying the amplitude of g_x.

4.5.2. ROTATING-FRAME FOURIER ZEUGMATOGRAPHY

4.5.2.1. *Introduction*

In 1979 Hoult[2] described and demonstrated a new imaging technique which he christened rotating-frame zeugmatography. It is essentially the rotating-frame analog of Kumar, Welti, and Ernst's method[47,49] and we therefore make use of their formalism in our analysis. As was the case with Fourier zeugmatography, the initial application is likely to be limited to planar rather than full three-dimensional studies. Nevertheless, we again preserve the generality of three dimensions in our mathematical description.

[49a] W. A. Edelstein, J. M. S. Hutchison, G. Johnson, and T. Redpath, *Phys. Med. Biol.* **25**, 751 (1980).

4.5.2.2. *Description and Mathematical Analysis*

The rf pulse and gradient timing sequence is illustrated in Fig. 4.25 and employs both static (G_z) and rf (B_{1x}, B_{1y}) field gradients. During the periods t_x, t_y a uniform rf field B_{10} is applied in addition to the respective rf field gradients B_{1x}, B_{1y}. Then, following t_x, a short nonselective 90° pulse, in phase quadrature with B_{10}, B_{1x}, B_{1y}, nutates the magnetization into the x, y plane of the rotating frame, allowing the FID to be recorded in the static field gradient G_z. Thus, the static and rf fields during the three periods $t < t_y$, $t_y < t < t_x + t_y$, $t_x + t_y < t$ are as given in Table 4.1.

If we now denote the rotating-frame axes by x', y', z' and take B_{10}, B_{1x}, and B_{1y} to be directed along y', the net effect of periods t_y and t_x is seen to be a rotation of the equilibrium magnetization (initially parallel to z, z') through an angle

$$\Theta = [\gamma B_{10}(t_x + t_y) + \gamma B_{11}(xt_x + yt_y)] \tag{4.63}$$

in the x', z' plane. (We have taken $B_{1x} = B_{1y} = B_{11}$. Note also that it is in fact possible to replace one of these rf field gradients by an oscillating z field as discussed in Hoult's paper.) The subsequent application of a short 90° pulse along the x' direction flips this magnetization into the x', y' plane such that the phase relative to the x' axis is $\Theta + 90°$. The FID recorded in the gradient G_z is thus given by

$$S(t) = \iiint \rho(\mathbf{r})s(\mathbf{r}, t)\, dv, \tag{4.64}$$

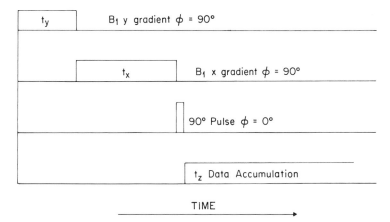

FIG. 4.25. A timing diagram showing the pulse sequence required to obtain three-dimensional information. The times t_x, t_y, and t_z are the independent variables and normally, t_x, $t_y \ll T_2$. [From D. I. Hoult, *J. Magn. Reson.* **33**, 183 (1979).]

TABLE 4.1

RADIO FREQUENCY AND STATIC FIELD REQUIREMENTS
DURING SPECIFIED TIME INTERVALS FOR ROTATING-
FRAME FOURIER ZEUGMATOGRAPHY EXPERIMENTS

Period	Static field	rf field
$0 < t < t_y$	B_0	$B_{10} + B_{1y}y$
$t_y < t < t_x + t_y$	B_0	$B_{10} + B_{1x}x$
$t_x + t_y < t$	$B_0 + G_z z$	0

with

$$s(\mathbf{r}, t) = \exp\{i[\pi/2 + \gamma B_{10}(t_x + t_y) + \gamma B_{11}(xt_x + yt_y)$$
$$- \gamma G_z z t_z] - t_z/T_2\}. \tag{4.65}$$

In similar manner to the case of Fourier zeugmatography, the response is dependent on t_x and t_y as well as on t_z, i.e., $S(\mathbf{t}) = S(t_x, t_y, t_z)$. A three-dimensional image can thus be obtained in like manner by repeating the experiment for a series of equally spaced values of t_x and t_y and performing the three-dimensional Fourier transform to convert a dependence on t_x, t_y, and t_z to a dependence on the conjugate variables x, y, and z. As before we define

$$S(\boldsymbol{\omega}) = \int\int\int S(\mathbf{t}) \exp(-i\boldsymbol{\omega} \cdot \mathbf{t}) \, dt_x \, dt_y \, dt_z, \tag{4.66}$$

which can again be expressed as an integral over all volume elements

$$S(\boldsymbol{\omega}) = \int\int\int \rho(\mathbf{r}) s(\mathbf{r}, \boldsymbol{\omega}) \, dv, \tag{4.67}$$

where $s(\mathbf{r}, \boldsymbol{\omega})$ is the Fourier transform of $s(\mathbf{r}, \mathbf{t})$ defined in analogous manner to Eq. (4.66). Thus,

$$s(\mathbf{r}, \boldsymbol{\omega}) = \frac{\exp\{iT_x[-\omega_x + \gamma(B_{10} + B_{11}x)]\} - 1}{\gamma(B_{10} + B_{11}x) - \omega_x}$$
$$\times \frac{\exp\{iT_y[-\omega_y + \gamma(B_{10} + B_{11}y)]\} - 1}{\gamma(B_{10} + B_{11}y) - \omega_y} \frac{T_2}{i - (\gamma G_z z + \omega_z)T_2}, \tag{4.68}$$

where T_x and T_y are the maximum values for t_x and t_y, respectively and lead to truncation effects in the corresponding Fourier transforms. Using a similar procedure to that used in Section 4.5.1.2, one can show that

$$S(\boldsymbol{\omega}) = \rho'(\mathbf{r}') = \int\int\int \rho(\mathbf{r}) T(x - x') T(y - y') \frac{T_2}{1 + i\gamma G_z T_2(z - z')} \, dv, \tag{4.69}$$

where

$$T(x) = \frac{\exp(i\gamma B_{11} T_x x) - 1}{\gamma B_{11} x}. \tag{4.70}$$

Thus, the calculated spin density $\rho'(\mathbf{r}')$ is related to the real spin density via a convolution with the usual line-shape function along z and truncation functions $T(x)$, $T(y)$, along x and y. The result should be compared with Eq. (4.62), which is the analogous expression derived for Fourier zeugmatography. The differences result from the absence in the present case of T_2 effects in the periods t_x and t_y (at least to a first approximation). Note, however, that the truncation effects described above must also be present in the Fourier zeugmatographic experiments, so that the functions $G(\eta_x(x - x'))$ and $G(\eta_y(y - y'))$ should really be regarded as convolutions of the line-shape expression with truncation functions. If the experiment is properly designed this effect should make little difference to the observed resolution. Figure 4.26 illustrates the point-spread function derived from the product of the

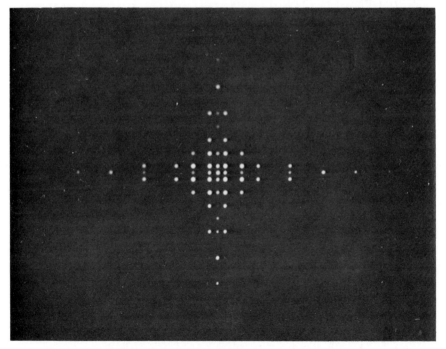

FIG. 4.26. When the transient is restricted in both x and y time domains and negligible decay occurs, two-dimensional transformation generates a characteristic diffraction pattern, the like of which may be found in any standard optics text. Filtering with, for example, exponentials in the time domains, attenuates the patterns but does not remove the characteristic star shape. [From D. I. Hoult, *J. Magn. Reson.* **33**, 183 (1979).]

two truncation functions. It is characteristically star shaped and was obtained by deliberately shortening the periods t_x and t_y.

As shown previously, the final expression for $\rho'(\mathbf{r})$ is a complex function whose real and imaginary parts both contain mixtures of absorption and dispersive terms. This means again that one either has to work with the power spectrum $|\rho'(\mathbf{r})|$ or else devise a series of imaging experiments to allow separation of the absorption term. This has been discussed by Hoult for the two-dimensional rotating-frame zeugmatography case.

4.5.2.3. Separation of the Absorption Mode— a Two-Dimensional Example

Inspection of the expression for the FID [Eq. (4.65)] in the above experiment indicates that the effect of each of the periods t_x, t_y, and t_z is to generate signal phase variation between different regions of the sample. Herein lies the problem.

If we now consider a two-dimensional example, we can overcome this difficulty by the removal of the short 90° pulse. (The timing diagram is now as shown in Fig. 4.27.) During the period t_x the magnetization is nutated through an angle

$$\Theta = \gamma(B_{10} + B_{11}x)t_x, \tag{4.71}$$

in the x', z' plane. In the absence of the 90° pulse the component of magnetization in the x', y' plane lies along x' and has an amplitude given by $\sin \Theta$. The ensuing FID is thus proportional to

$$s(\mathbf{r}, t) = \sin\{\gamma(B_{10} + B_{11}x)t_x\} \exp\{[-1/T_2 - i\gamma G_z z]t_z\}, \tag{4.72}$$

and we now see that the rf field gradient gives rise only to amplitude variations. This allows the phase variations, still produced by the z magnetic field gradient, to be separated in a conventional manner.

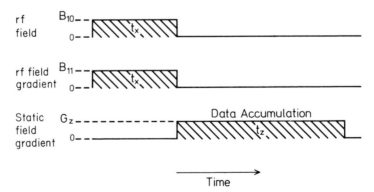

FIG. 4.27.   The timing sequence for the two-dimensional amplitude variation experiment.

The expressions for $s(\mathbf{r}, \omega)$ and $S(\omega)$ are

$$s(\mathbf{r}, \omega) = \frac{\sin\{[\omega_x - \gamma(B_{10} + B_{11}x)]T_x\}}{2[\omega_x - \gamma(B_{10} + B_{11}x)]} \frac{1/T_2}{(1/T_2)^2 + (\omega_z + \gamma G_z z)^2}, \quad (4.73)$$

$$= (T_x/2) \operatorname{sinc}\{[\omega_x - \gamma(B_{10} + B_{11}x)]T_x\}$$

$$\times \frac{1/T_2}{(1/T_2)^2 + (\omega_z + \gamma G_z z)^2}, \quad (4.74)$$

$$S(\omega) = \rho'(\mathbf{r}) = \int\int\int \rho(\mathbf{r})(T_x/2) \operatorname{sinc}\{\gamma B_{11} T_x(x - x')\}$$

$$\times \frac{1/T_2}{(1/T_2)^2 + [\gamma G_z(z - z')]^2} \, dv, \quad (4.75)$$

where we have selected the absorption mode.

The price paid for the use of amplitude variation is a loss of $\sqrt{2}$ in sensitivity arising from the "wasted" z component of magnetization. This z component also causes further difficulties; since it is proportional to $\cos \Theta$ and therefore varies across the sample, it is essential to wait $\sim 5T_1$ before repeating the experiment. This leads to a further loss in sensitivity of about 40%. These points notwithstanding, the technique is very efficient and promises to be of major future importance.

Figure 4.28a and b are examples of two-dimensional ^{31}P images of simple phantoms using the amplitude variation method described above. Further details of the experimental technique and a full discussion of the resolution may be found in Hoult's original paper.

Finally, we mention that the two-dimensional amplitude variation method can be extended for use with three-dimensional objects by suitably preparing a plane of spins. This can be achieved, for example, with simple saturation or selective excitation techniques. Modifications to measure chemical shift have also recently been proposed.[49b]

4.5.3. PLANAR SPIN IMAGING

4.5.3.1. *Principles*

Planar spin imaging is a logical extension of the selective irradiation line-scanning method. It was proposed by Mansfield and Maudsley in 1976[50] and was motivated by the desire for faster imaging speeds.

Suppose that by some means or other we can arrange to interact only with those spins lying on a two-dimensional grid or three-dimensional lattice. This

[49b] S. F. J. Cox and P. Styles, *J. Magn. Reson.* **40**, 209 (1980).
[50] P. Mansfield and A. A. Maudsley, *J. Phys. C* **9**, L409 (1976).

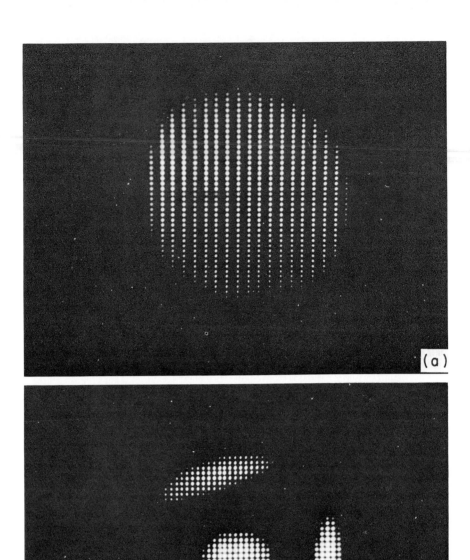

FIG. 4.28. Two-dimensional ^{32}P images formed by the amplitude-variation method. (a) A cross section 5 mm thick of a 6-mm tube containing 85% phosphoric acid. The data were accumulated in about 3 min. The resolution is about 0.4 mm. (b) A phantom was created by milling 0.75-mm-deep flats on the side of a Teflon cylinder which slid into a 6-mm tube. In addition, a hole was drilled through the center of the phantom and its length restricted to 5 mm. The resolution is about 0.2 mm and the data were accumulated in about 3 min. [From D. I. Hoult, *J. Magn. Reson.* **33**, 183 (1979).]

need not be regular, but for simplicity, we shall assume an orthorhombic lattice with "cell constants" a, b, and c. We shall then be interested solely in the spin density $\rho(x, y, z)$ at the lattice points $x = la$, $y = mb$, $z = nc$ (l, m, and n integral). The FID from such a system to which linear orthogonal field gradients G_x, G_y, G_z are applied would thus be

$$S = \sum_{lmn} \rho(la, mb, nc) \cos\{[l(G_x a) + m(G_y b) + n(G_z c)]t\} \, \Delta v, \qquad (4.76)$$

where Δv is the size of a volume element centered about the lattice point. Now, provided the gradients obey the conditions

$$N(G_z c) \leqslant (G_y b) \leqslant (G_x a)/M, \qquad (4.77)$$

where M, N are the largest values of m, n, respectively, each lattice point is uniquely defined in frequency. Thus, Fourier transforming Eq. (4.76) yields $\rho(la, mb, nc)$ directly. The problem of forming a complete image in a single experiment thus reduces to one of imposing a discrete structure on an otherwise continuous spin distribution. This can be achieved by selective excitation methods as discussed below.

4.5.3.2. *Means of Operation*

Rather than consider the full three-dimensional case which has been discussed in detail by Mansfield and Maudsley,[51] we restrict our attention to the two-dimensional situation.

Selection to a plane can, as usual, be achieved by a type 1 selective excitation method (Section 4.2.3.2) or, if coarser resolution is acceptable, by relying on the receiver-coil geometry. This is the A phase in the timing sequence shown in Fig. 4.29a. The B phase consists of a multiline excitation using a linear field gradient in the y direction (G_y) and a tailored pulse with spectral components as shown in Fig. 4.29b. The result of this step is to nutate the magnetization occurring in a series of narrow strips into the rotating frame (see Fig. 4.30).

In phase C the FID from these strips is recorded in the presence of the read gradient G_z and the preparation gradient G_y. The use of both gradients, in contrast to the single gradient required for simple line scanning, allows discrimination between the strips as well as between the spins along a strip. Fourier transformation followed by appropriate data ordering allows the full image to be constructed.

4.5.3.3. *Finite Selection Width*

In our discussion of the principles of planar imaging we assumed a point lattice. However, practical considerations require that we work with finite

[51] P. Mansfield and A. A. Maudsley, *J. Magn. Reson.* **27**, 101 (1977).

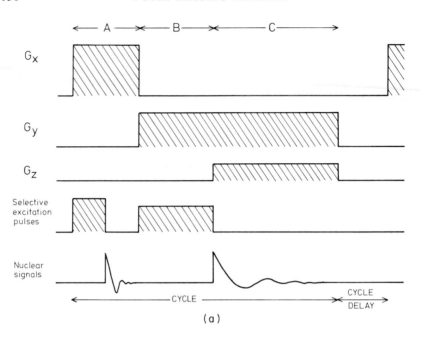

(a)

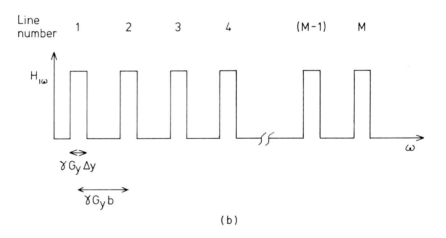

(b)

FIG. 4.29.   Planar spin imaging in two dimensions: (a) the timing sequence; (b) the discrete multiple rectangular spectral distribution for the tailored rf selective excitation of phase B.

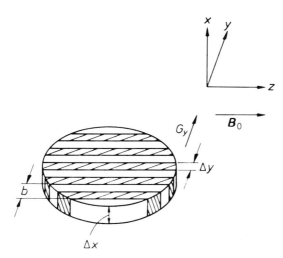

FIG. 4.30.   Sketch showing multistrip or grid selection in G_y within one layer of magnetization of thickness Δx. The shading corresponds to the perturbed spin regions. [From P. Mansfield and A. A. Maudsley, *J. Phys. C* **9**, L 409 (1976).]

volume elements. (This is also desirable from the S/N viewpoint!) The revised gradient requirements are then

$$G_x \Delta x + G_y \Delta y + NG_z c \leqslant G_y b \leqslant (1/M)[G_x a - G_x \Delta x - G_y \Delta y], \quad (4.78)$$

where Δx, Δy are the plane thickness and linewidth, respectively. The gradients G_x, G_y thus give rise to an additional broadening which, however, may be removed by a deconvolution procedure as discussed by Mansfield and Maudsley.[51]

As we shall see in Chapter 5, there are several limits on the volume element size imposed by resolution requirements. This means in practice that although the technique is capable of rapid imaging, the S/N per unit time is poor.

4.5.3.4. *Results*

Figure 4.31 shows the response profiles from a mineral oil annulus. Figure 4.31a was obtained with only the preparation gradient present in the read phase of the experiment. Each of the thirteen peaks thus represents the projection of the spin density along a single image line. In Fig. 4.31b both preparation and read gradients were present causing the projections to broaden into line densities. Figure 4.32 illustrates the image obtained by appropriate stacking of these line-density profiles. For finer line resolution (not shown), a fourfold interlace procedure was employed in which four consecutive planar imaging experiments were performed in order to fill in

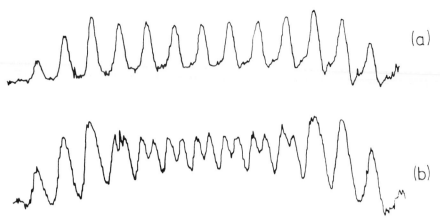

FIG. 4.31. (a) Discrete projection spectrum of a mineral oil annulus comprising ($M =$) 13 lines. This was obtained by irradiating and reading in G_y only. (b) Response profile as in (a) but broadened in the read phase by an additional gradient G_z. [From P. Mansfield and A. A. Maudsley, *J. Magn. Reson.* **27**, 101 (1977).]

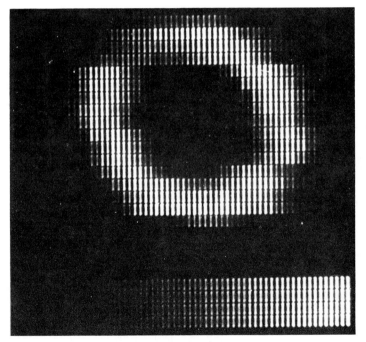

FIG. 4.32. A cross-sectional proton spin density image of a mineral oil annulus produced by planar spin imaging. An intensity scale corresponding to a 16-level linear density wedge is also included across the bottom of the picture. [From P. Mansfield and A. A. Maudsley, *J. Magn. Reson.* **27**, 101 (1977).]

the missing lines. There is, of course, no necessity for any delay between these strips, since each one excites "fresh" spins.

A number of variations to the basic scheme which allow imaging of relaxation times and improvements of the overall efficiency have been proposed. We refer the reader to the original papers of Mansfield and Maudsley for details.

4.5.4. Projection Reconstruction

4.5.4.1. *Introduction*

We have seen in Chapter 3 that the absorption line shape in a linear magnetic field gradient is the projection of the spin density in a direction normal to that gradient. The problem of how to reconstruct an image, given a sufficient number of such projections, is the subject of this section.

This was tackled by Lauterbur in 1973[52] who devised his own reconstruction algorithm to produce the first ever NMR image illustrated in Fig. 4.33. He christened his imaging technique zeugmatography from the Greek $\zeta \epsilon \nu \gamma \mu \alpha$, "that which is used for joining."

The problem of reconstruction from projections, however, is a general one and has been investigated by workers in a diversity of research fields. The mathematical foundation was laid by Radon in 1917[53] and the first application was to radio astronomy by Bracewell in 1956.[54] Since then the technique has been "rediscovered" a number of times and applied, for instance, to electron microscopy and optics. Unquestionably, the most important development, however, was the announcement in 1972 of the EMI X-ray CT (computerized tomographic) scanner.[55] This has revolutionized modern radiology and has given tremendous impetus to the development of improved reconstruction algorithms. Although we must refer the reader to the literature for a full review of reconstruction methods,[56] we nevertheless feel it worthwhile to discuss some of the techniques which have potential application to NMR imaging. We restrict ourselves initially to the two-dimensional experiment and base our treatment on the Brooks and Di Chiro review article.[56] Three-dimensional experiments are briefly discussed in Section 4.5.4.7.

4.5.4.2. *Reconstruction Methods: Notation*

The notation to be used in the subsequent analysis is defined in Fig. 4.34. We assume that we wish to reconstruct on an $n \times n$ grid with cell width w as

[52] P. C. Lauterbur, *Nature (London)* **242,** 190 (1973).

[53] J. Radon, *Ber. Verh. Saechs. Akad. Wiss.* **69,** 262 (1917).

[54] R. N. Bracewell, *Aust. J. Phys.* **9,** 198 (1956).

[55] G. N. Hounsfield, *Br. J. Radiol.* **46,** 1016 (1973).

[56] R. A. Brooks and G. Di Chiro, *Phys. Med. Biol.* **21,** 689 (1976).

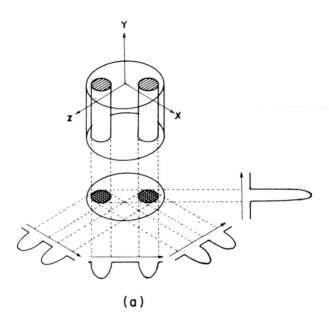

(a)

(b)

FIG. 4.33. (a) Relationship between a three-dimensional object, its two-dimensional projection along the y axis, and four one-dimensional projections at 45° intervals in the x,z plane. The arrows indicate the gradient directions. (b) Proton NMR zeugmatogram of two 1-mm-i.d. water-filled glass capillaries attached to the inside wall of a 4.2-mm-i.d. glass tube. Reconstruction was from four projections as diagrammed in (a). [From P. C. Lauterbur, *Nature* (*London*) **242**, 190 (1973).]

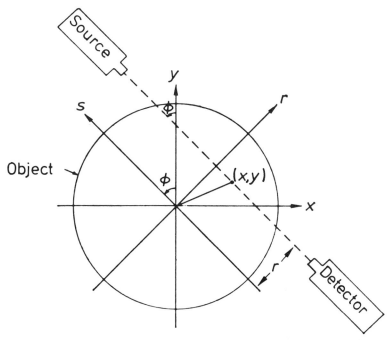

FIG. 4.34.   Coordinate systems. Points within the object are described by a fixed (x, y) coordinate system. Rays (dashed line) are specified by their angle with respect to the y axis, ϕ, and their distance from the origin, r. The s coordinate denotes distance along the ray. [From R. A. Brooks and G. Di Chiro, *Phys. Med. Biol.* **21**, 689 (1976).]

shown in Fig. 4.35. We should then expect that about n projections comprising n rays each would be required to accurately determine the image. Then,

$$r = x \cos \phi + y \sin \phi, \tag{4.79}$$

and, if the two dimensional spin density at the point (x, y) is denoted by $\rho(x, y)$, the projection $P(r, \phi)$ is defined by

$$P(r, \phi) = \int_{r, \phi} \rho(x, y) \, ds. \tag{4.80}$$

4.5.4.3. *Back Projection*

This is the simplest of all reconstruction methods and was widely used in the early days of reconstructive tomography. We illustrate the procedure for the case of a simple rectangular object in Fig. 4.36. Each profile is back projected in a direction orthogonal to the applied field gradient with which it

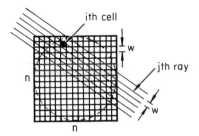

FIG. 4.35. Cellular array used for reconstruction. The object is bounded by the dashed circle, which contains n cells along a major diameter. The rays have a finite width, here taken to be equal to the cell width w. The contribution of the ith cell to the jth ray (heavy outline) is the weighting factor w_{ij}. [From R. A. Brooks and G. Di Chiro, *Phys. Med. Biol.* **21**, 689 (1976).]

was produced. This leads to a central region of reinforcement corresponding to the original rectangle. (Only two projections have been illustrated; in practice many more would be used.) For the general case we may express this mathematically as

$$\bar{\rho}(x, y) = \sum_{j=1}^{m} P(x \cos \phi_j + y \sin \phi_j, \phi_j)\, \Delta\phi, \qquad (4.81)$$

where $\bar{\rho}(x, y)$ is the calculated spin density. The sum extends over all m projections whose angular separation is $\Delta\phi$.

Note that the procedure is by no means exact. For instance, in Fig. 4.36 regions lying outside the object have been assigned a nonzero spin density. As the number of projections increases this effect gives rise to the familiar "star artifact" around regions of high density. For this reason back projection is no longer used per se. Nevertheless, as we shall see below, it is possible to modify or filter the projections prior to back projection in such a way as to make the method analytic.

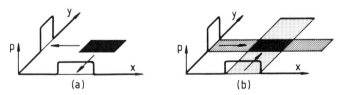

FIG. 4.36. Back projection: (a) two profiles of a rectangular object are shown graphically; (b) these profiles are back projected onto the image plane and superimposed to form an approximation to the original object. [From R. A. Brooks and G. Di Chiro, *Phys. Med. Biol.* **21**, 689 (1976).]

4.5.4.4. *Iterative Reconstruction*

Iterative reconstruction involves the choice of starting values (usually uniform intensity) for the spin densities in each of the n^2 pixels. One then compares the projections from this configuration with the measured ones and applies an appropriate correction to obtain an improved estimate of the spin density. The process is repeated and (hopefully) converges to a faithful representation of the actual spin density after a small number (e.g., five or six) of iterations.

There is a whole range of such techniques differing in the way in which the correction is applied, e.g., additive and multiplicative algebraic reconstruction techniques (ART). Although such methods are now used predominantly with γ-ray imaging systems, an iterative procedure was used for the early NMR imaging experiments.[52,57–59] It has the advantage that any corrections required to remove rf penetration effects, for example, can be more easily implemented than is the case with the analytic techniques to be discussed below. Nevertheless, the latter are much faster and are preferred in situations where the imaging speed is dominated by the reconstruction time.

4.5.4.5. *Analytic Methods: Fourier Reconstruction*

Let us denote the two-dimensional Fourier transform of $\rho(x, y)$ by $F(k_x, k_y)$ where k_x, k_y are the wave numbers in the x and y directions, respectively. Then,

$$F(k_x, k_y) = \int_{-\infty}^{\infty} \int_{-\infty}^{\infty} \rho(x, y) \exp\left[-2\pi i(k_x x + k_y y)\right] dx \, dy. \quad (4.82)$$

Now, if we express this in terms of the r, s coordinate system (see Fig. 4.34), we obtain

$$F(k_x, k_y) = \int_{-\infty}^{\infty} \int_{-\infty}^{\infty} \rho(x, y) \exp(-2\pi i k r) \, dr \, ds, \quad (4.83)$$

where

$$k = (k_x^2 + k_y^2)^{1/2}, \quad (4.84)$$

and the angle of rotation ϕ relating (x, y), (r, s) is given by

$$\phi = \tan^{-1}(k_y/k_x). \quad (4.85)$$

[57] P. C. Lauterbur, *Proc. Int. Conf., 1st, Stable Isot. Chem. Biol. Med., May 9–11* AEC, CONF-730525 (1973).

[58] P. C. Lauterbur, *Pure Appl. Chem.* **40**, 149 (1974).

[59] P. Mansfield, P. K. Grannell, and A. A. Maudsley, *Proc. Ampere Congr., 18th, Nottingham* p. 431. North-Holland Publ., Amsterdam, 1974.

The integral of $\rho(x, y)$ over s is just the projection $P(r, \phi)$ (see Eq. 4.80) so that

$$F(k_x, k_y) = \int_{-\infty}^{\infty} P(r, \phi) \exp(-2\pi i k r) \, dr \qquad (4.86a)$$

$$= F(k, \phi), \qquad (4.86b)$$

where $F(k, \phi)$ is the Fourier transform of $\rho(r, \phi)$ with respect to r and Eq. (4.86) is of fundamental importance and is sometimes known as the central slice theorem. It states that the Fourier coefficient of the spin density $F(k_x, k_y)$ is equal to the Fourier transform of the projection taken in the direction of the Fourier wave defined by k_x, k_y. It also forms the basis for a reconstruction method: One obtains the Fourier transforms of the projections, interpolates to give a square array of coefficients $F(k_x, k_y)$, and then performs a reverse two-dimensional transform to obtain $\rho(x, y)$. The time-consuming step is the two-dimensional interpolation, which has tended to restrict the application of the method.

4.5.4.6. Relationship of Back Projection to Filtered Back Projection

If we write the expression for back projection [Eq. (4.81)] in its integral form we have

$$\bar{\rho}(x, y) = \int_0^{\pi} P(x \cos \phi + y \sin \phi, \phi) \, d\phi, \qquad (4.87)$$

and replacing $\rho(x \cos \phi + y \sin \phi, \phi)$ by its Fourier representation as in Eq. (4.86), we obtain

$$\bar{\rho}(x, y) = \int_0^{\pi} \int_{-\infty}^{\infty} (F(k, \phi)/|k|) \exp[2\pi i k(x \cos \phi + y \sin \phi)]|k| \, dk \, d\phi. \qquad (4.88)$$

If we now take the Fourier transform of $\bar{\rho}(x, y)$, then

$$\bar{F}(k_x, k_y) = F(k, \phi)/|k| = F(k_x, k_y)/|k|. \qquad (4.89)$$

Thus we see that the back-projected image is equivalent to the true image, except that the Fourier coefficients have been scaled by the wave-vector amplitude $|k|$. This knowledge allows us to correct or filter the back-projection method to make it exact. The procedure is as follows: Take the Fourier transform of each projection, multiply by $|k|$, perform the inverse transform, and then back project onto the image plane with interpolation. There are a number of problems associated with the procedure and, in practice, alternative methods are preferred. They are related to the simple method described above via the convolution theorem. (Multiplication by $|k|$ in the frequency domain is equivalent to a convolution with the Fourier transform of $|k|$ in the spatial domain.) One version is known as Radon filtering where the filtered projection P^* is given by

$$P^*(r, \phi) = (1/2\pi^2) \int_{-\infty}^{\infty} \frac{\partial P(r', \phi)/\partial r'}{r - r'} \, dr'. \tag{4.90}$$

Another version is known as convolution filtering in which

$$P^*(r, \phi) = k_m P(r, \phi) - \int_{-\infty}^{\infty} P(r', \phi) \frac{\sin^2[\pi k_m(r - r')]}{\pi^2(r - r')^2} \, dr', \tag{4.91}$$

where k_m is the maximum spatial frequency present ($=\frac{1}{2}\omega$). This method is widely used by manufacturers of X-ray CT machines and is becoming increasingly popular with NMR imagers, too. A simple FORTRAN coding is given in Ref. 60. Other problems associated with the band limiting may be corrected for by additional filtering of the projections (see, for example, Shepp and Logan[61]). It is also possible, by appropriate choice of the filter function, to trade off spatial resolution against tissue contrast. For example, if one were interested only in abnormalities of size 1 cm or greater, then one could observe much smaller variations in water content or relaxation time than would be the case for a resolution of 1 mm.

We now clarify the filtered back-projection technique procedure as applied to NMR imaging experiments. The NMR conjugate variables are of course ω and t. In X-ray experiments these correspond to x and k, respectively. Thus the first step in the X-ray procedure of taking the FT of the projection is unnecessary since this is already available experimentally in the form of the FID. The FID for a given gradient projection is then multiplied by $|t|$. This product is then inversely transformed to give one filtered projection. The back-projection and interpolation procedures necessary to co-add the projection set to form the final image are then the same as those used in CT imaging.[56]

4.5.4.7. *Experimental Considerations and Three-Dimensional Imaging*

The early experiments of Lauterbur *et al.* used CW NMR techniques.[52,57,58] However, Fourier transform methods (first suggested in connection with projection reconstruction by Kumar, Welti, and Ernst[49]) offer a number of significant advantages and are now universally used.[43,62,63] In addition to the well-known improvement in sensitivity, many of the

[60] R. A. Brooks and G. Di Chiro, *Radiology* **117**, 561 (1975).
[61] L. A. Shepp and B. F. Logan, *IEEE Trans.* **21**, 21 (1974).
[62] J. M. S. Hutchison, J. R. Mallard, and C. C. Goll, *Proc. Ampere Congr., 18th, Nottingham* p. 283. North-Holland Publ., Amsterdam, 1974.
[63] P. C. Lauterbur, *in* "NMR in Biology" (R. A. Dwek, I. D. Campbell, R. E. Richards, and R. J. P. Williams, eds.), p. 323. Academic Press, New York, 1977.

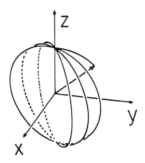

FIG. 4.37. The gradient scanning path for a three-dimensional projection-reconstruction experiment. [From C.-M. Lai and P. C. Lauterbur, *J. Phys. E* **13**, 747 (1980).]

analytic reconstruction methods require the Fourier transforms of the projections to be calculated as intermediates and these are directly available as FIDs in the case of transient techniques.[49] Hutchison *et al.* have exploited this property in their Fourier transform spin-echo method.[43,62] It is also simpler and more reliable to use Fourier methods if T_1 discrimination is required. However, CW saturation methods have been employed[64] for this purpose.

A number of planar images obtained by projection reconstruction methods are presented and discussed in Chapters 6 and 7. We also remind the reader (see Section 4.4.2.2) that, provided the applied field gradients are sufficiently small, it is possible to image the distribution of chemical shifts within a sample while retaining all the sensitivity advantages (see Chapter 5) of the projection reconstruction method.[65] If the range of the chemical shifts present does not allow this direct approach then a combination of projection reconstruction and selective excitation methods may be appropriate.[33]

Undoubtedly the most exciting recent advance in projection reconstruction imaging has been the development by Lauterbur and colleagues[66,67] of the full three-dimensional technique. One way in which this can be achieved is to direct the field gradient around the path illustrated in Fig. 4.37 so as to produce a series of two-dimensional projections by reconstruction in the conventional manner.[68] These projections can then be combined using the same algorithm to yield a three-dimensional image as shown schematically in Fig. 4.38. Figure 4.39 shows a set of 16 two-dimensional projections ob-

[64] P. C. Lauterbur, C.-M. Lai, J. A. Frank, and C. S. Dulcey, Jr., *Phys. Can.* **32**, Special July Issue: *Dig. Int. Conf., 4th Med. Phys.* Abstract 33. 11 (1976).

[65] P. Bendel, C.-M. Lai, and P. C. Lauterbur, *J. Magn. Reson.* **38**, 343 (1980).

[66] C.-M. Lai, *J. Appl. Phys.* **52**, 1141 (1981).

[67] P. C. Lauterbur and C.-M. Lai, *IEEE Trans.* **27**, 1227 (1980).

[68] C.-M. Lai and P. C. Lauterbur, *J. Phys. E* **13**, 747 (1980).

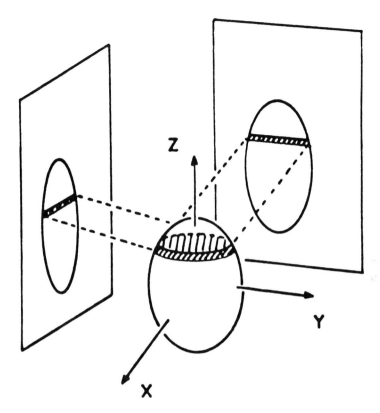

FIG. 4.38.   The geometry of a two-stage reconstruction method. [From C.-M. Lai and P. C. Lauterbur, *J. Phys. E* **13**, 747 (1980).]

tained from a phantom consisting of a series of glass tubes (56-mm long × 13-mm diameter) filled with water doped to have a T_1 of about 50 msec. The experiment was performed at 4 MHz with a gradient strength of 4.7 μT cm^{-1} and the reconstruction onto a 33 × 33 array was by filtered back projection of 30 one-dimensional projections distributed at 12° intervals over 360°. The 30 two-dimensional projections were also distributed at 12° intervals over 360° so they could be similarly reconstructed to yield a three-dimensional image, which is shown in Fig. 4.40. The full experiment consisted of 30 × 30 = 900 FIDs (no signal averaging) and took 12 min to complete. The resolution is relatively low (~3 mm) but nevertheless some fine detail, such as the constrictions at the ends of the tubes, is visible.

The advantage of three-dimensional reconstruction is its great information-gathering potential resulting in a high S/N per unit time. When it comes to image presentation, however, a full three-dimensional array constitutes

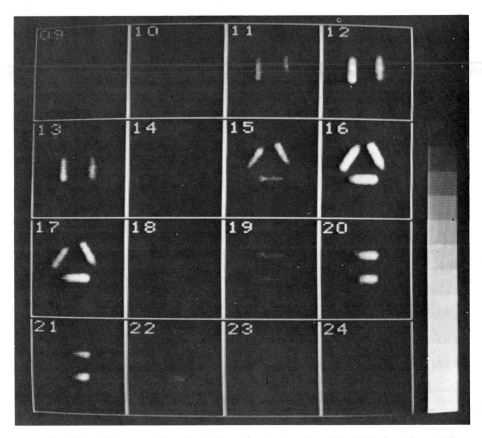

FIG. 4.39.   Sixteen two-dimensional views of a phantom, from 6° to 198° of rotation at 12° intervals, each reconstructed from 30 one-dimensional projections measured over 360°. [From P. C. Lauterbur and C.-M. Lai, *IEEE Trans.* **27,** 1227 (1980).]

something of an embarrassment of riches. Nevertheless, one then has the facility to select for display any image plane regardless of orientation.

The next stage of development will involve the increase in matrix size and, hence, in resolution. This will probably involve some difficulties with signal handling and in particular with dynamic range. Nevertheless, the present results are most encouraging. (See Notes Added in Proof, Note 4.1.)

We mention finally that it is not necessary to perform the three-dimensional reconstruction in two stages and a number of direct schemes have also been proposed.[68,69] It is as yet not clear which method will be the most efficient.

[69] L. A. Shepp, *J. Comput. Assist. Tomogr.* **4,** 94 (1980).

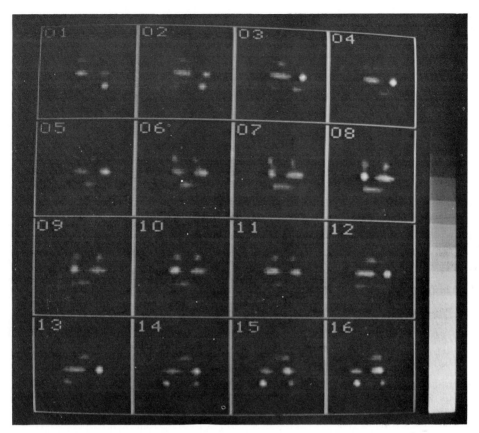

FIG. 4.40.   Tomographic images from a three-dimensional array reconstructed from a set of two-dimensional views some of which are shown in Fig. 4.39. [From P. C. Lauterbur and C.-M. Lai, *IEEE Trans.* **27**, 1227 (1980).]

4.5.5. ECHO-PLANAR IMAGING

4.5.5.1. *Introduction*

We discussed at some length in Chapter 3 how, given a discrete lattice of spins, it was possible to simultaneously observe and differentiate the signals from *all* nuclei. The problem of how one imposes a discreteness on what is in reality a continuous spin distribution was first considered in connection with planar spin imaging (Section 4.5.3) where the approach was to restrict observation to narrow strips or lines of spins singled out by a selective irradiative procedure. This process as we shall see in Chapter 5 is inefficient, since in any one experiment the majority of spins do not contribute to the signal.

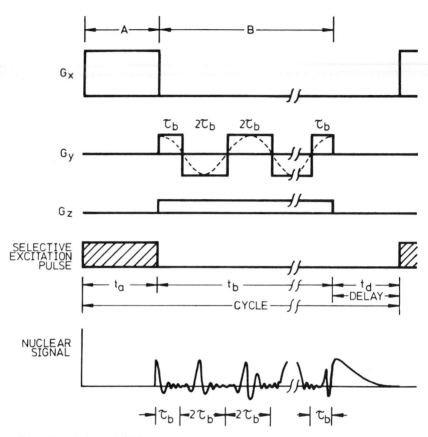

FIG. 4.41. Pulse and field gradient timing diagram for a two-dimensional echo-planar imaging experiment. [From P. Mansfield and I. L. Pykett, *J. Magn. Reson.* **29,** 355 (1978).]

However, in 1977 Mansfield[37,70] devised a new scheme, now known as echo-planar imaging, which is free from this disadvantage. As usual, the technique exists in full three-dimensional or planar forms, but we restrict the preliminary discussion to the latter case.

4.5.5.2. *Echo-Planar Imaging in Two Dimensions*

The gradient and rf pulse-timing sequence is illustrated in Fig. 4.41 and commences (phase A) with the definition of a plane by a type 2 selective excitation procedure (90° or θ° selective pulse in a gradient G_x). Immediately following this preparation, the FID is observed in the presence of an oscil-

[70] P. Mansfield, *J. Phys. C* **10,** L55 (1977).

lating gradient $G_y(t)$ and a linear static gradient G_z of much smaller magnitude. The effect of the oscillating gradient is perhaps best understood by considering the Fourier transform of the periodic function illustrated in Fig. 4.42a and described mathematically by

$$f(t) = h(t) \sum_{n=0}^{\infty} g(t - 2n\tau). \qquad (4.92)$$

Such an echo train can be produced in an NMR experiment if the FID in a single gradient G_y is recalled with $180°$ rf pulses at times $(2n + 1)\tau$ following an initial $90°$ pulse. As an alternative to $180°$ pulses, one may periodically reverse the gradient G_y in the manner of Fig. 4.41, where $\tau = \tau_b$. In either case, the echo amplitude will decay by a relaxation or diffusion process which we describe by a function $h(t)$.

Now the Fourier transform of $f(t)$ is given by

$$F(\Omega) = (\pi/\tau) \sum_{n=-\infty}^{+\infty} \{G(n\pi/\tau)H(\Omega - n\pi/\tau)\}, \qquad (4.93)$$

where $G(\Omega)$ and $H(\Omega)$ are the respective Fourier transforms of $g(t)$ and $h(t)$, and $H(\Omega)$ is normalized to unity. $F(\Omega)$, $G(\Omega)$, and $H(\Omega)$ are illustrated in Fig. 4.42b. If the echo train does not decay significantly, i.e., $h(t) = 1$, then $H(\Omega - n\pi/\tau)$ becomes the Dirac delta function $\delta(\Omega - (n\pi/\tau))$ and $F(\Omega)$ reduces to a series of "spikes" regularly spaced at intervals of $\Delta\omega = \pi/\tau$. Thus the formation of an echo train imposes a discreteness on the Fourier-transformed projection profile.

In the full two-dimensional planar experiment the additional gradient G_z broadens the individual discrete lines or spikes to yield a complete set of cross-sectional profiles. This is illustrated for the case of an annulus in Fig. 4.43 where the upper diagram shows the discrete lines obtained by modulating with G_y in the absence of G_z and the lower diagram shows the full

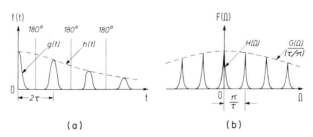

FIG. 4.42.   (a) Periodic function showing an FID followed by a series of damped spin echoes. (b) Fourier transform of the function illustrated in (a). [From P. Mansfield and I. L. Pykett, *J. Magn. Reson.* **29**, 355 (1978).]

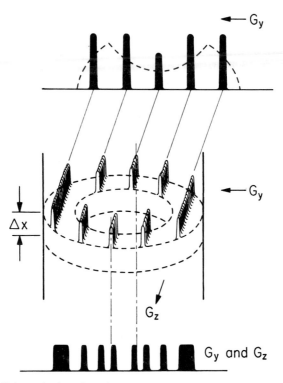

FIG. 4.43.   Spin projections for a homogeneous annulus. The upper projection (dotted) is the projection in G_y alone. This turns into the discrete profile if G_y is modulated. The lower projection corresponds to the application of a modulated G_y gradient together with G_z on continuously. [From P. Mansfield and I. L. Pykett, *J. Magn. Reson.* **29**, 355 (1978).]

cross-sectional profiles obtained in the presence of both G_y and G_z. It then only remains for these to be appropriately ordered and displayed.

The use of 180° pulses in place of gradient reversal allows the relaxation due to static field inhomogeneity to be eliminated but unfortunately requires the use of relatively high rf power and necessitates the periodic reversal of G_z. Since G_z is much smaller than G_y, however, the latter requirement is not hard to meet.

4.5.5.3. *Echo-Planar Imaging in Three Dimensions*

The planar experiment can be extended to three dimensions by the application of a second modulated field gradient G_x and the replacement of the selective pulse by a nonselective 90° (or $\beta°$) pulse. The FID in the rotating frame at time t following this pulse is then[37]

$$S(t) = \mathrm{Re} \iiint \rho(x, y, z)$$

$$\times \exp\left[i\gamma \int_0^t (xG_x(t') + yG_y(t') + zG_z) \, dt' \right] dx \, dy \, dz. \quad (4.94)$$

Taking the integral over x first we have

$$f(y, z, t) = \int dx \, \rho(x, y, z) \exp\left[i\gamma \int_0^t xG_x(t') \, dt' \right] \quad (4.95a)$$

$$= \sum_{P=0}^{N} g(y, z, (t - 2P\tau_a)), \quad (4.95b)$$

where $G_x(t)$ is on for $N/2$ cycles of period $4\tau_a$ and we have assumed $h(t) = 1$. When all three gradients are applied we may write $\rho(x, y, z)$ as a function of angular frequency and, for large even N, the Fourier transform of $f(y, z, t)$ becomes

$$F(\omega_x, \omega_y, \omega_z) = \sum_{l=-\infty}^{+\infty} 2\pi a \rho(\omega_x, \omega_y, \omega_z) \, \delta(\omega_x - l\Delta\omega_x). \quad (4.96)$$

Integration over y introduces a similar Dirac delta function $\delta(\omega_y - m\,\Delta\omega_y)$ where l, m are integers and the angular frequency intervals are

$$\Delta\omega_x = \pi/\tau_a = \gamma a G_x, \quad (4.97)$$

$$\Delta\omega_y = \pi/\tau_b = \gamma b G_y. \quad (4.98)$$

If we now take the inverse Fourier transform of Eq. (4.96) and substitute it back in Eq. (4.94) together with the corresponding expression for the y integral, we obtain

$$S(t) = (ab/\gamma G_z) \sum_{l,m} \int d\Omega(l, m) \, \rho[\Omega(l, m)] \cos \Omega(l, m)t, \quad (4.99)$$

where the angular frequency $\Omega(l, m)$ is given by

$$\Omega(l, m) = l \, \Delta\omega_x + m \, \Delta\omega_y + \omega_z. \quad (4.100)$$

If the signal is sampled digitally for a total period τ_c, then the corresponding frequency interval is given by

$$\Delta\omega_z = 2\pi/\tau_c = \gamma c G_z. \quad (4.101)$$

Thus, Eq. (4.99) may be written as a discrete sum

$$S(t) = \sum \rho_{lmn} \cos\{t[l \, \Delta\omega_x + m \, \Delta\omega_y + n \, \Delta\omega_z]\} \, \Delta V_{lmn}, \quad (4.102)$$

where $\Delta V_{lmn} = abc$ is the unit cell volume. To the approximation that we can replace $H(\omega_x - l \Delta\omega_x)$ by $\delta(\omega_x - l \Delta\omega_x)$ and $H(\omega_y - m \Delta\omega_y)$ by $\delta(\omega_y - m \Delta\omega_y)$, therefore, all spins in each unit contribute to the appropriate lattice point. Further, provided that

$$\Delta\omega_x/M = \Delta\omega_y = N \Delta\omega_z, \qquad (4.103)$$

where M, N are the maximum values of m and n, respectively, then all values of ρ_{lmn} are uniquely defined in the frequency domain. Thus, Fourier transformation of Eq. (4.102) followed by appropriate reordering yields in one step the complete three-dimensional spin density distribution ρ_{lmn}. As was the case with multiplanar imaging (Section 4.5.3), therefore, a three-dimensional function has been obtained with a one-dimensional transform. This effect is referred to as Fourier transform nesting.[37]

4.5.5.4. *The Spatial Response Function*

There is a minor problem associated with image reconstruction in the presence of a periodically switched gradient which can be understood by focusing our attention on a point (y_0, z_0) within a planar sample. During the even cycles of Fig. 4.41, these spins evolve in an off-resonance field of $[G_y y_0 + G_z z_0]$, whereas during the odd cycles they evolve in a field of $[-G_y y_0 + G_z z_0]$. Thus, unless the sample is symmetrically distributed about the y axis, one should use only all even or all odd echoes in the image reconstruction process. Of course, once reconstruction is complete both positive and negative field images can be added to regain the $\sqrt{2}$ loss in S/N.

It is of interest to enquire what the response is of a region of isochromatic spins. For simplicity, we consider only a single switched gradient G_y and take the static gradient $G_z = 0$. Figure 4.44a and b illustrate the even and odd echoes, and Fig. 4.44c and d indicate the response of spin isochromats at points chosen so that their frequencies are, respectively integral (S_I) and nonintegral (S_{NI}) multiples of the gradient switching rate ($f_g = 1/2\tau$). We may consider the response of a particular isochromat as the product of a cosine wave of appropriate frequency multiplied by a "top-hat" function of width 2τ and convoluted with a comb of delta functions of separation 2τ. This is illustrated in the upper part of Fig. 4.45. In order to determine where our isochromat of spins will appear in the final image, we take the Fourier transform of the response. The result, as illustrated in the lower part of Fig. 4.45, is a sinc function centered at the isochromat frequency and sampled by the comb of deltas of angular frequency separation $\Delta\omega = \pi/\tau$. This is similar to the result obtained for the sensitive-point method and illustrated in Fig. 4.13. Thus, for example, the isochromat shown in Fig. 4.44c would contribute only to the appropriate delta or line projection, whereas an isochromat of frequency $(2n + 1)f_g/2$ would not contribute at all, and an

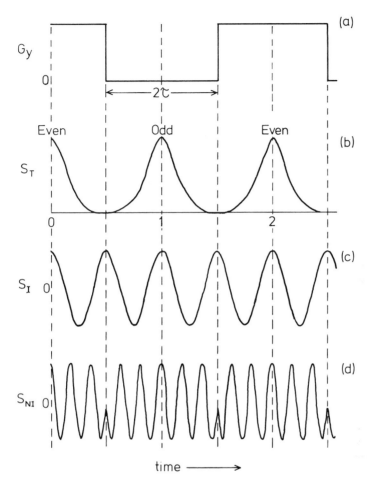

FIG. 4.44.   The NMR response to a switched field gradient: (a) the field gradient $G_y(t)$; (b) response envelope or spin echo train due to the switched gradient; (c) the response of a spin isochromat whose frequency is an integral multiple of the gradient repetition rate; and (d) the response of a spin isochromat whose frequency is a nonintegral multiple of the gradient repetition rate.

isochromat of intermediate frequency would contribute to a number of line projections, albeit predominantly to those which lie immediately adjacent to it. The signals obtained from the latter regions represent the great S/N advantage which echo-planar imaging has over the planar and multiplanar techniques.

We now ask what the total response function is. This can be obtained by summing for each isochromat the contributions to all line projections. The

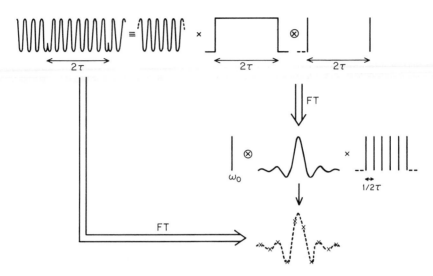

FIG. 4.45.   The determination of the Fourier transform (FT) of the response of a spin iso-chromat; × denotes multiplication and ⊗ denotes convolution.

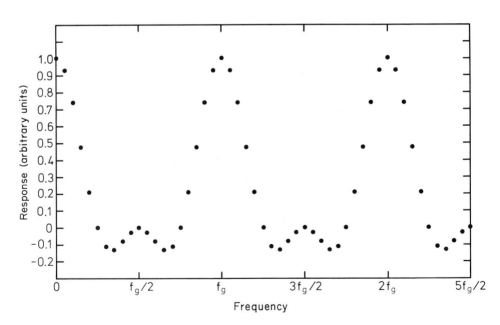

FIG. 4.46.   The total response function.

result is shown in Fig. 4.46 where the zero response for the $(2n + 1)f_g/2$ condition is again apparent. This should not, however, be construed as a loss in sensitivity since the same principle applies to the simple digitally sampled Fourier transform experiment.

4.5.5.5. *Experimental*

Figure 4.47 illustrates some early results obtained at 15 MHz from a homogeneous mineral oil cylinder of diameter 13.8 mm. Figure 4.47a is a

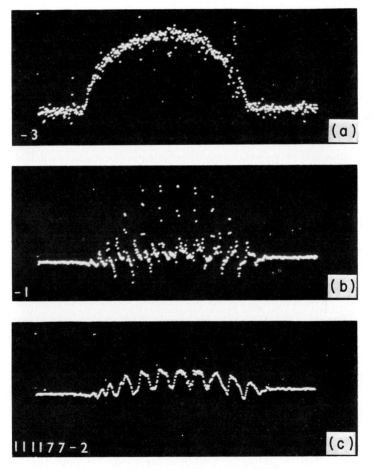

FIG. 4.47.   Experimental spin projections obtained at 15.0 MHz for a homogeneous cylinder of mineral oil: (a) continuous spin projection with steady gradient G_y; (b) discrete projection where G_y is modulated; (c) broadened discrete projection obtained as in (b) above, but with the addition of a steady gradient G_z. [From P. Mansfield, P. G. Morris, R. J. Ordidge, I. L. Pykett, V. Bangert, and R. E. Coupland, *Phil. Trans. R. Soc. London Ser. B* **289**, 503 (1980).]

conventional projection through the cylinder with $G_y = 1.26$ G cm^{-1}. Figure 4.47b shows the set of line profiles obtained by switching G_y with a period of 1.28 msec in the absence of G_z ($t_a = 7.4$ msec, $t_b = 10.24$ msec, and the sampling interval $= 20$ μsec). Note the great improvement in S/N over Fig. 4.47a. Finally, Fig. 4.47c shows the projections broadened in a G_z gradient of 0.23 G cm^{-1} to yield a full set of profiles. These should be rectangular but depart somewhat from this shape due to a slight imbalance in the positive and negative amplitudes of G_y. The effects of this also manifest themselves in the undershoot of the discrete projections in Fig. 4.47b.

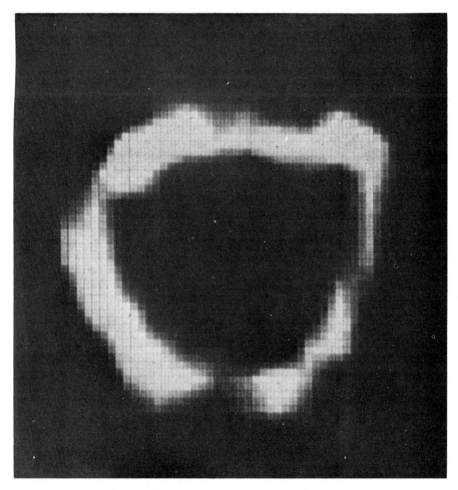

FIG. 4.48.   The first two-dimensional echo-planar image of an oil-filled annulus showing strong image distortion. [From P. Mansfield and I. L. Pykett, *J. Magn. Reson.* **29,** 355 (1978).]

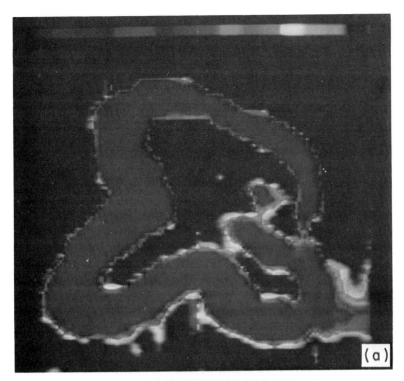

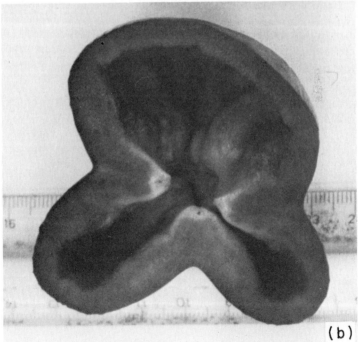

FIG. 4.49. (a) Thin-slice echo-planar image through an intact red pepper, together with (b) a photograph of the actual section obtained after the experiment: slice thickness 1.0 cm; imaging time 5.0 sec. [From P. Mansfield and R. J. Ordidge (to be published).]

Figure 4.48 illustrates the first image to be formed by an echo-planar method. It is of a mineral oil annulus (o.d. 9.3 mm and i.d. 3.8 mm) corresponding to the schematic sketch of Fig. 4.43 and obtained under similar conditions to those described above. It is an average of 16 shots taking 1.9 sec to acquire and a further 3 sec to transform. The matrix size was originally 32 × 16, but has been interpolated to 64 × 64.

More recently, trial echo-planar experiments have been performed with a $\frac{1}{3}$ human-body-sized probe at 4 MHz.[42] Recent results, now free of the early distortions, are extremely encouraging. For example, Fig. 4.49a and b compare a 1.0-cm-thick slice echo-planar image through an intact red pepper with the corresponding section obtained subsequently by cutting through the vegetable. This image was obtained in 5.0 sec.

With regard to the future, the single-shot imaging time of ∼10–100 msec and high efficiency of the method suggests that it will find application in real-time cardiac studies of moderate resolution. For high resolution, it may be advantageous to combine the method with one of the steady-state techniques along the lines of DEFT or SSFP. However, the broad-bandwidth and large-gradient requirements present a major challenge which is likely to limit the matrix size to ∼64 × 64. (See Notes Added in Proof, Note 4.2.)

5. Comparison of Imaging Methods

5.1. Introduction

The detailed evaluation and comparison of imaging sensitivities is a complex, often emotive, subject involving not just an appraisal of imaging methods but also a discussion of the nature of the sample in terms of its electromagnetic and NMR properties. We have therefore divided the problem into two parts.

In this chapter we carry out a quantitative comparison of the various methods discussed in Chapter 4 leading us to the concept of an optimum imaging system. In Chapter 6 we then calculate the S/N to be expected from such a system in a variety of different imaging regimes ranging from human whole-body studies to "NMR microscopy" (with a resolution of about 10 μm).

As an alternative to this theoretical approach it is possible to pursue a more empirical line[1] in which one takes published experimental details for the S/N, operating frequency, pixel volume, etc., and appropriately scales the quoted imaging times to apply to a common set of operating conditions. The attraction of this method is that it automatically includes the effects of all those technical problems which the mathematical model ignores. This is undoubtedly the approach to be recommended to prospective purchasers of commercial equipment when it becomes available. However, in the present context, many of the published images have been obtained under far from ideal conditions and are, in many cases, certainly not representative of the final quality to be obtained from a particular system. It is for this reason that we adopt the more rigorous mathematical approach in the hope that it will provide a better guide to those contemplating entry into the field.

We emphasize however that the results can only be considered approxi-

---

[1] P. A. Bottomley, *J. Magn. Reson.* **36**, 121 (1979).

mate and that there may well be technical difficulties, some of which are discussed here and in subsequent sections, which degrade the performance of certain imaging systems. We may also speculate that the interaction between the two sections of our argument (the analysis of the technique and the nature of the sample) plus experimental difficulties could lead to the situation where different techniques are preferable in different imaging situations.

In Section 5.2 we outline the mathematical basis of our analysis and make explicit the approximations inherent in it. We then proceed to a qualitative discussion of point, line, planar, and full three-dimensional methods in Section 5.3 and finally derive quantitative results in Section 5.4 for most of the imaging methods whose details have been described in the literature. This work is based mainly on the comprehensive review article of Brunner and Ernst.[2]

5.2. Mathematical Foundation

In what follows it is assumed that the noise is "white" (i.e., frequency independent) and that the noise per unit bandwidth is identical for all NMR imaging experiments. In practice this may not be rigidly true, particularly for the most efficient techniques such as echo–planar imaging which require a large overall bandwidth to simultaneously detect and discriminate between all image points. Nevertheless, it is a good first approximation and, with the further assumption that a matched filter[3] is employed, it is possible to write

$$S/N = \left(\frac{\text{total signal energy}}{\text{noise power/bandwidth}}\right)^{1/2} \tag{5.1a}$$

$$\propto (\text{total signal energy})^{1/2}. \tag{5.1b}$$

Rather than attempt to deal with a continuous distribution of spins, the object to be imaged is divided into a set of $n_x n_y n_z$ pixels with n_x, n_y, and n_z along the x, y, and z axes, respectively. It is further assumed that the full spin intensity within each pixel contributes at the central point, as illustrated in Fig. 5.1. If the spins have a constant spin–spin relaxation time T_2, then the application of a field gradient to the sample gives rise to a set of unbroadened Lorentzian lines of half-width $1/\pi T_2$.

The number of experiments N necessary to achieve complete resolution of all pixels will vary between 1 and $n_x n_y n_z$ depending on the efficiency of the technique. If the signal energy obtained in a single experiment is E_k, then the

[2] P. Brunner and R. R. Ernst, *J. Magn. Reson.* **33**, 83 (1979).
[3] R. R. Ernst, *Adv. Magn. Reson.* **2**, 1 (1966).

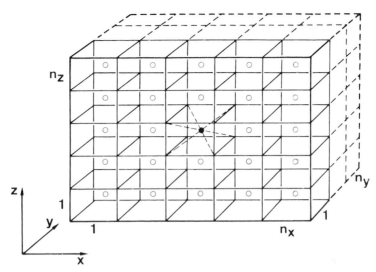

FIG. 5.1. Idealized orthorhombic object for the discussion of sensitivity and performance time. Each of the $n_x n_y n_z$ volume elements is represented by a single mass point. [From P. Brunner and R. R. Ernst, *J. Magn. Reson.* **33**, 83 (1979).]

total signal energy E achieved during the full imaging experiment is given by

$$E = \sum_{k=1}^{N} E_k. \tag{5.2}$$

Now, if the repetition interval for the individual experiments is T_r, the imaging time is

$$T_{tot} = N T_r, \tag{5.3}$$

so that the signal to noise per unit time $(S/N)_1$, which is the quantity we are interested in comparing, is given by

$$(S/N)_1 = (S/N)/(T_{tot})^{1/2} \tag{5.4a}$$

$$\propto (E/T_{tot})^{1/2}. \tag{5.4b}$$

The other quantity in which we shall be interested is the minimum time in which the image can be formed,

$$T_{min} = N T_{rmin}, \tag{5.5}$$

where T_{rmin} is the shortest repetition period consistent with the required resolution.

For imaging static objects we should favor those techniques with the highest $(S/N)_1$. However, in situations where there is motion over the time scale of the experiment (see Section 6.1), a short T_{min} may be of vital importance.

Returning to Eq. (5.2), the individual energy contributions E_k can be broken down into three factors. Thus,

$$E_k = M_v^2 G_k^2(T_r, T_1, \beta) f_k(T_2), \tag{5.6}$$

where M_v is the nuclear magnetization of a volume element and is given by

$$M_v = M_0/(n_x n_y n_z), \tag{5.7}$$

where M_0 is the total magnetization of the sample whose spin density is assumed uniform.

When a single volume element is repetitively observed, the initial height of the FID will depend on the repetition period T_r, the spin–lattice relaxation time T_1, and the flip angle of the rf pulse β. We discuss this further in Section 6.3.4 where it is shown that in favorable circumstances the factor $T_1 G_k^2/T_r$ may be as high as 0.5.

The factor f_k is a measure of the signal energy in a single FID of unit initial amplitude. It will depend on the spin–spin relaxation time T_2 and, if the signal amplitude is denoted by S, is given by

$$f_k(T_2) = \int_{T_{k_1}}^{T_{k_2}} S^2(T_2, t)\, dt, \tag{5.8}$$

where $S(T_2, T_{k_1}) = 1$.

If all experiments are performed under identical initial conditions then the G_k are all equal and

$$E = M_v^2 G_k^2(T_r, T_1, \beta) \sum_{k=1}^{N} f_k(T_2). \tag{5.9}$$

5.3. Sensitivity and Performance Time

5.3.1. QUALITATIVE INTRODUCTION

We refer back to Section 4.1 and Fig. 4.1 where we categorized the various imaging techniques into point, line, planar, and full three-dimensional (3D) methods. If we now consider the case in which we wish to image all $n_x n_y n_z$ volume elements (3D imaging), we note that a point method in which only a single volume element is observed at any one time will require $n_x n_y n_z$ separate experiments. Therefore,

$$T_{min}^{point} \propto n_x n_y n_z, \tag{5.10}$$

$$(S/N)_1^{point} \propto (n_x n_y n_z)^{-1/2}. \tag{5.11}$$

Similarly, for line-scanning methods (with the image line parallel to the x axis),

$$T_{\min}^{\text{line}} \propto n_y n_z, \tag{5.12}$$

$$(S/N)_1^{\text{line}} \propto (n_y n_z)^{-1/2}. \tag{5.13}$$

For planar methods, with the imaging plane parallel to xy, we have

$$T_{\min}^{\text{planar}} \propto n_z, \tag{5.14}$$

$$(S/N)_1^{\text{planar}} \propto n_z^{-1/2}. \tag{5.15}$$

Finally, for full 3D methods, the image is obtained in a single experiment so that in the appropriate units for which $T_{\min}^{\text{point}} = n_x n_y n_z$, $T_{\min}^{\text{3D}} = 1$ and $(S/N)_1^{\text{3D}} = 1$.

If we require the formation of an image of a single plane only, the corresponding results are

$$T_{\min}^{\text{point,2D}} \propto n_x n_y, \tag{5.16}$$

$$(S/N)_1^{\text{point,2D}} \propto (n_x n_y)^{-1/2}, \tag{5.17}$$

$$T_{\min}^{\text{line,2D}} \propto n_y, \tag{5.18}$$

$$(S/N)_1^{\text{line,2D}} \propto n_y^{-1/2}. \tag{5.19}$$

Planar methods will now obtain the image in a single experiment so that in the system of units for which $T_{\min}^{\text{point,2D}} = n_x n_y$, $T_{\min}^{\text{planar,2D}} = 1$ and $(S/N)_1^{\text{planar,2D}} = 1$.

We now proceed to derive quantitative estimates for the S/N per unit time and minimum performance time for the various imaging techniques.

5.3.2. PROJECTION RECONSTRUCTION

The basic projection reconstruction experiment has been discussed in Section 4.5.4 and is summarized in Fig. 5.2a. We assume that Fourier transform methods are used (see Section 4.5.4.7) to produce an image of a cubic object consisting of $n \times n \times n$ volume elements. (If conventional CW methods were applied then the sensitivity per unit time would, of course, be reduced by a factor of $n^{1/2}$.)

The signal following a simple excitation pulse is

$$S_k(T_2, t) = \exp(-t/T_2), \tag{5.20}$$

and, if the measuring interval T is made equal to the repetition period T_r, a condition which ensures maximum sensitivity,

$$f^{\text{PR}}(T_2) = (T_2/2)[1 - \exp(-2T/T_2)]. \tag{5.21}$$

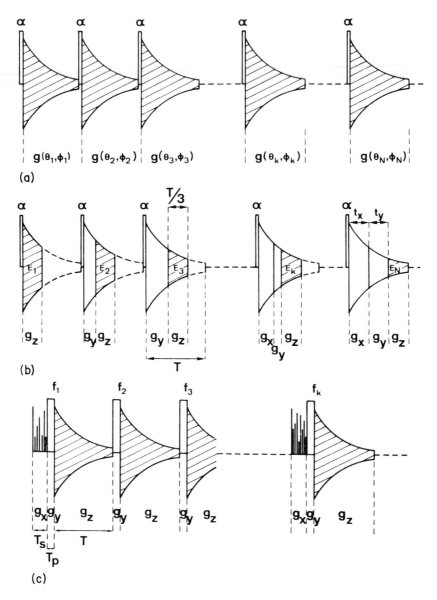

FIG. 5.2. Timing diagrams for (a) projection reconstruction, (b) Fourier imaging, and (c) line scanning. [From P. Brunner and R. R. Ernst, *J. Magn. Reson.* **33,** 83 (1979).]

Thus, for the full 3D imaging experiment

$$E^{3DPR} = (M_0^2/n^4)mG^2(T/T_1,\beta)f^{PR}(T_2), \qquad (5.22)$$

where m is the number of times the *full* experiment is repeated for signal averaging purposes and

$$T_{tot}^{3DPR} = mn^2 T. \qquad (5.23)$$

Planar studies may be performed if the initial nonselective pulse is replaced with a selective one. The corresponding expressions for the 2D or slice experiments are then

$$E^{2DPR} = (M_0^2/n^5)mG_1^{\ 2}(T,T_1,T_p,\beta)f^{PR}(T_2), \qquad (5.24)$$

$$T_{tot}^{2DPR} = mn(T + T_p), \qquad (5.25)$$

where T_p is the length of the selective pulse. We have written the equilibrium signal amplitude as $G_1(T, T_1, T_p, \beta)$ since it will depend in a rather subtle manner on the nature of the rf excitation pulse. Although this problem has not yet been fully discussed in the literature we take the view that it should be possible to maximize G_1 by suitable choice of T_p, T, and β in the same manner that one optimizes G through choice of T and β.

We also note that the argument of Brunner and Ernst[2] for the possible greater efficiency of 2D over 3D Fourier imaging also applies in this situation. Thus, for samples in which $T_1 \gg T_2$, one performs successive experiments on different planes. Then, since fresh spins are continually being excited, it is possible to use $90°$ pulses and faster repetition rates, provided the magnetization in the first plane has fully recovered before it is observed for the second time.

The above analysis ignores the effects of the reconstruction algorithm which can affect the noise spectrum and alter its distribution over the image. Further, it is not generally correct that one requires exactly n^2 experiments to construct a 3D image. Nevertheless, one might expect that Eqs. (5.22)–(5.25) would be reasonable approximations.

5.3.3. FOURIER ZEUGMATOGRAPHY

Fourier zeugmatography has been discussed in Section 4.5.1 and the basic experiment is summarized in Fig. 5.2b. It is necessary to ensure that the period T over which each experiment is performed remains constant in order to avoid introducing complicated differential relaxation effects. The FID is always sampled n_z times over a period $T/3$ during which a z gradient is applied. However, the x- and y-gradient periods t_x, t_y vary in n_x and n_y steps between 0 and $T/3$ where n_x, n_y, and n_z are the number of image points to be reconstructed along the x, y, and z directions, respectively. The f_k^{3DFI} terms are not therefore constant. For instance,

$$f_1^{3DFI} = T_2/2[1 - \exp(-2T/3T_2)], \tag{5.26}$$

$$f_2^{3DFI} = T_2/2 \exp(-2T/3T_2 n_x)[1 - \exp(-2T/3T_2)], \ldots \tag{5.27}$$

$$f_3^{3DFI} = \ldots, \text{ etc.}$$

Thus,

$$E^{3DFI} = (M_0^2/n_x^2 n_y^2 n_z^2)mG^2(T/T_1, \beta)f^{3DFI}(T_2), \tag{5.28}$$

where

$$f^{3DFI}(T_2) = \sum_{k=1}^{N} f_k^{3DFI} \tag{5.29}$$

$$= (T_2/2)\{1 - \exp(-2T/3T_2)\}$$

$$\times \{1 + \exp(-2T/3T_2 n_x)$$

$$+ \exp(-4T/3T_2 n_x) + \cdots + \exp(-2T/3T_2)\}$$

$$\times \{1 + \exp(-2T/3T_2 n_y) + \cdots + \exp(-2T/3T_2)\} \tag{5.30}$$

$$= (T_2/2)\{1 - \exp(-2T/3T_2)\}^3$$

$$\times \{1 - \exp(-2T/3n_x T_2)\}^{-1}$$

$$\times \{1 - \exp(-2T/3n_y T_2)\}^{-1}. \tag{5.31}$$

The total imaging time is given by

$$T_{tot}^{3DFI} = mn_x n_y T. \tag{5.32}$$

As is the case with projection reconstruction imaging, it is possible to perform a 2D variant of this experiment with the aid of a selective pulse (of length T_p). The corresponding results are

$$E^{2DFI} = (M_0^2/n_x^2 n_y^2 n_z^2)mG^2(T, T_1, T_p, \beta)f^{2DFI}(T_2), \tag{5.33}$$

with

$$f^{2DFI}(T_2) = (T_2/2)\{1 - \exp(-T/T_2)\}^2\{1 - \exp(-T/n_y T_2)\}^{-1}, \tag{5.34}$$

and

$$T_{tot}^{2DFI} = mn_y(T_p + T). \tag{5.35}$$

As previously noted this 2D technique may be more efficient than its 3D counterpart when used to image a 3D sample for which $T_1 \gg T_2$. The efficiency of both techniques can be improved by the use of larger g_x and g_y

gradients which permits the shortening of the periods t_x and t_y during which the spins are not observed. Modifications discussed in Section 4.5.1.4 in which t_x is held constant and g_x varied in amplitude, will also help to reduce imaging times.

Finally, we comment that we should expect the imaging times and sensitivity for Hoult's rotating-frame zeugmatography experiments (see Section 4.5.2) to be broadly similar.

5.3.4. SELECTIVE EXCITATION TECHNIQUES

5.3.4.1. *Line Scan*

In the simple line scan experiments discussed in Section 4.4.2 and illustrated in Fig. 5.2c, it is clear that each volume element contributes to a single FID only. Thus for a $90°$ selective pulse,

$$E^{LS} = \frac{1}{2} \frac{M_0{}^2 T_2}{n_x{}^2 n_y{}^2 n_z{}^2} m \left\{ 1 - \exp\left(-\frac{2T}{T_2}\right) \right\}, \qquad (5.36)$$

and the imaging time for a 3D experiment is

$$T_{tot}^{3DLS} = n_x \{ mn_y (T_p + T)[1 + \eta(T_s/T_1)] + 3T_1 \}, \qquad (5.37)$$

where $mn_y(T_p + T)$ is assumed $> T_1$; T_p is as usual the length of the selective pulse; T is the period during which the FID is observed. It is assumed that one waits $3T_1$ before preparing a new plane by selective saturation, that the process of saturation takes a time T_1 and is renewed η times every T_1 (in the evaluations which follow η has been set to unity).

When line scanning is used in its 2D mode the signal energy remains as given by Eq. (5.36), but the imaging time is reduced to

$$T_{tot}^{2DLS} = mn_y(T_p + T)[1 + \eta(T_s/T_1)] + T_s. \qquad (5.38)$$

5.3.4.2. *Planar and Multiplanar Imaging*

The planar imaging method discussed in Section 4.5.3 is equivalent to the line-scanning method analyzed above, except that a number of lines are simultaneously irradiated. One would naively expect therefore that the signal energies would remain the same but that the total imaging time would be reduced by a factor equal to the number of lines scanned per shot. Unfortunately, this is not the case since there is a loss in sensitivity due to a reduction in the size of the excited volume elements. This rather subtle point is perhaps best explained with reference to Fig. 5.3 which illustrates the n_y slices (lines) of relative width $1/q$ excited in a gradient g_y.

The spins are observed in an inclined gradient $\mathbf{g}$ (a combination of g_x and g_y) and it can be seen that the n_z volume elements along a particular slice are

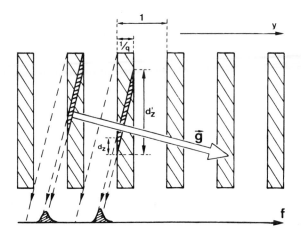

FIG. 5.3.   The principle of planar imaging. The selectively excited "lines" of relative width $1/q$ running along the z direction within the plane of choice lead after application of a suitable gradient **g** to a unique projection from which the planar spin density can be reconstructed. Note the elongation of the effective volume elements along the z axis by the applied inclined gradient. [From P. Brunner and R. R. Ernst, *J. Magn. Reson.* **33**, 83 (1979).]

parallelograms of width dz and length

$$dz' = dz[1 + n_z/(q - 1)]. \tag{5.39}$$

Thus, the z resolution depends on the relative width of the selected line $1/q$ in such a manner that it deteriorates rapidly as the lines are packed more closely together. If one makes a reasonable compromise by taking $q = n_z$, then $dz' \simeq 2dz$ and the signal energy is

$$E^{\text{PI}} = \frac{T_2}{2} \frac{M_0{}^2 m}{n_x{}^2 n_y{}^2 n_z{}^4} \left\{ 1 - \exp\left(-\frac{2T}{T_2} \right) \right\}, \tag{5.40}$$

where T is again the time during which the FID is observed.

In order to calculate the minimum performance time one assumes that successive experiments irradiate fresh spins in the same selected plane so that no waiting time between these experiments is required. Then, provided that $m(T_p + T) > T_1$,

$$T_{\text{tot}}^{\text{PI}} = n_x\{m(T_p + T)[1 + \eta(T_s/T_1)] + 3T_1\}, \tag{5.41}$$

where T_s and T_p are the saturation and selective pulse lengths, respectively. This estimate, however, includes a delay time $T_d = 3T_1$ which is necessary to allow the selectively saturated spins to recover before a new plane is selected. If one accepts an incomplete recovery of the spins, then a further reduction is

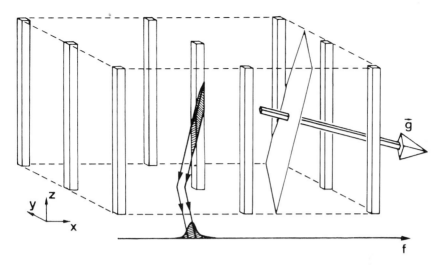

FIG. 5.4. The principle of multiplanar imaging. A two-dimensional array of selectively narrow "lines" is utilized to image the characteristic features of the three-dimensional object. By application of a suitably inclined gradient, a unique projection can be obtained without overlap of contributions from different lines. Resolution along the lines is reduced by the inclination angle of the gradient. [From P. Brunner and R. R. Ernst, *J. Magn. Reson.* **33**, 83 (1979).]

possible and the signal energy reduces to

$$E^{\mathrm{PI}} = \frac{T_2}{2} \frac{M_0^2 m}{n_x^2 n_y^2 n_z^4} \left\{ 1 - \exp\left(-\frac{2T}{T_2} \right) \right\} \left\{ 1 - \exp\left(-\frac{T_e}{T_1} \right) \right\}^2, \quad (5.42)$$

where T_e is the time elapsed since the last selective saturation. The corresponding imaging time is

$$T_{\mathrm{tot}}^{\mathrm{PI}} = n_x \{ m(T_p + T)[1 + \eta(T_s/T_1)] + T_d \}. \quad (5.43)$$

It is possible to optimize $T_{\mathrm{tot}}^{\mathrm{PI}}$ by adjusting m and T_d. This has been accomplished numerically in the plots which follow in Section 5.4.

If it is required to image a single plane only (2D imaging), then the signal energy remains as given by Eq. (5.40) but the imaging time reduces to

$$T_{\mathrm{tot}}^{\mathrm{2DPI}} = m(T_p + T)[1 + \eta(T_s/T_1)] + T_s, \quad (5.44)$$

again assuming $m(T_p + T) > T_1$.

Thus, planar imaging reduces the imaging time relative to line scanning by a factor of approximately n, but the sensitivity per unit time is reduced by a factor of the order of $n^{1/2}$, due to the diminished size of the volume elements.

The situation is even worse with multiplanar imaging as illustrated in Fig.

5.4. In this case the signal energy is

$$E^{\mathrm{MPI}} = \frac{T_2}{2} \frac{M_0{}^2 m}{n_x{}^4 n_y{}^2 n_z{}^6} \left\{ 1 - \exp\left(-\frac{2T}{T_2} \right) \right\}, \tag{5.45}$$

assuming we are prepared to tolerate a threefold elongation of the volume elements along the z axis. The corresponding performance time is

$$T_{\mathrm{tot}}^{\mathrm{MPI}} = m(T_{\mathrm{p}} + T)[1 + \eta(T_{\mathrm{s}}/T_1)] + T_{\mathrm{s}}, \tag{5.46}$$

provided that $m(T_{\mathrm{p}} + T) > T_1$ as usual. The performance time is thus improved by a factor around n^2 relative to line scanning, but the sensitivity per unit time is reduced by the same factor!

5.3.5 SENSITIVE-POINT AND SENSITIVE-LINE METHODS

The use of the steady-state free precession method allows one to establish an equilibrium magnetization

$$M_y = \tfrac{1}{2}(T_{\mathrm{a}}/T_2)(M_0/n_x n_y n_z), \tag{5.47}$$

where T_{a} is the time constant governing the approach to equilibrium

$$T_{\mathrm{a}}^{-1} = \tfrac{1}{2}(1/T_1 + 1/T_2). \tag{5.48}$$

Thus, for an observation time T, in the case of the sensitive-point method (see Section 4.3.1), the signal energy is

$$
\begin{aligned}
E^{\mathrm{SP}} = \frac{1}{8} &\frac{M_0{}^2}{n_x{}^2 n_y{}^2 n_z{}^2} T_2 \left(\frac{T_{\mathrm{a}}}{T_1} \right)^2 T \left\{ 1 - \exp\left(-\frac{2}{RT_2} \right) \right\} \\
&\times \left\{ R - \frac{1}{T} - \frac{2}{T} \frac{\exp[-(T/T_{\mathrm{a}})(1 - 1/RT)] - 1}{1 - \exp(1/RT_{\mathrm{a}})} \right. \\
&\left. + \frac{1}{T} \frac{\exp[-(2T/T_{\mathrm{a}})(1 - 1/RT)] - 1}{1 - \exp(2/RT_{\mathrm{a}})} \right\}
\end{aligned} \tag{5.49}
$$

where R is the repetition rate of the $90°$ phase alternated rf pulses.

For a full 3D experiment the performance time is

$$T_{\mathrm{tot}}^{\mathrm{3DSP}} = n_x n_y n_z T, \tag{5.50}$$

and for a 2D experiment, it is

$$T_{\mathrm{tot}}^{\mathrm{2DSP}} = n_x n_y T. \tag{5.51}$$

In the multiple sensitive-point method (Section 4.4.1) the signal energy remains the same, but the simultaneous observation of a line of spins leads to a reduction in imaging time of n_z. Thus for a 3D experiment,

$$T_{\mathrm{tot}}^{\mathrm{3DMSP}} = n_x n_y T, \tag{5.52}$$

and for a 2D experiment

$$T_{tot}^{2DMSP} = n_x T, \qquad (5.53)$$

5.3.6. Echo-Planar Imaging

The NMR signal in a switched magnetic field gradient is shown in Fig. 5.5 where the broken line indicates the response of a resonant volume of isochromatic spins. It is clear that the signal energy in the full 3D version of the echo-planar imaging experiments, as discussed in Section 4.5.5., is identical with that obtained by the projection reconstruction technique, i.e.,

$$E^{3DEPI} = (M_0^2/n^4)mG^2(T/T_1, \beta)f^{EPI}, \qquad (5.54)$$

with

$$f^{EPI} = (T_2/2)[1 - \exp(-2T/T_2)]. \qquad (5.55)$$

However, since it is possible to observe all spins in a single experiment of length $T \simeq T_2$, the performance time is reduced to

$$T^{3DEPI} = mT. \qquad (5.56)$$

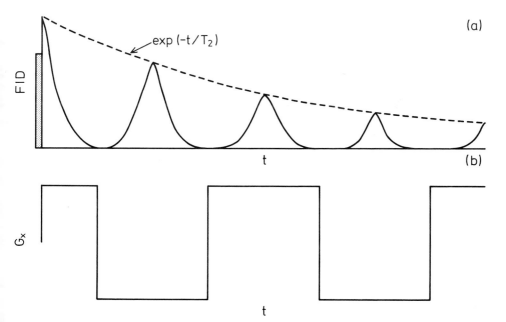

FIG. 5.5.   (a) The spin–echo train arising from the switched magnetic field gradient $G_x(t)$ shown in (b). The broken line indicates the response of resonant isochromatic spins.

For the 2D experiment the signal energy remains the same and the performance time becomes

$$T^{2\mathrm{DEPI}} = m(T + T_\mathrm{p}), \tag{5.57}$$

where T_p is the length of the selective pulse.

Although Eqs. (5.54) and (5.56) represent the best that is theoretically possible and echo–planar imaging is in this sense the optimum technique, we should recognize that there are a number of problems associated with its experimental implementation. Not least of these is the difficulty of achieving sufficient overall bandwidth without degrading the noise performance of the system. Nevertheless, we retain these expressions in our comparison as a standard by which other techniques can be judged.

5.4. Numerical Comparison and Discussion

In the comparisons which follow, the observation time T has been set to $3T_2$, $T_1 = 0.5$ sec, $T_\mathrm{s} = T_2$, $T_\mathrm{p} = 0$, and two relaxation time ratios, $T_1/T_2 = 3$ and $T_1/T_2 = 30$ have been considered for both 3D (Figs. 5.6 and 5.7) and 2D (Figs. 5.8 and 5.9) imaging. In all cases a cubic object with $n = 32$ volume elements in all directions has been assumed. The values have been arbitrarily scaled by setting the sensitivity and performance time of the projection reconstruction method to unity.

The labeling of the categories of techniques in the graphical results which follow is:

0	Sequential point methods	2RD	Techniques involving
1	Sequential line methods		a twofold reduction
2	Sequential plane methods		in dimensions
3	Full 3D methods	SCAN	Scanning methods,
1RD	Techniques involving		e.g., sensitive point
	a onefold reduction	FT	Fourier transform methods
	in dimensions		

The individual techniques are labeled as:

SP	Sensitive point	3DFI	Three-dimensional
LS	Line scan		Fourier imaging
MSP	Multiple sensitive point	2DPR	Two-dimensional projection
PI	Planar imaging		reconstruction
MPI	Multiplanar imaging	OT	Optimum technique
2DFI	Two-dimensional		(an idealized version
	Fourier imaging		of the echo–planar method)

5.4.1. SENSITIVITY

The first point to note in looking at Figs. 5.6–5.9 is the large spread in sensitivities—four orders of magnitude in the case of 3D methods and two in the case of 2D ones. This is related to the number of volume elements imaged, and is even more pronounced for matrix sizes with $n > 32$.

The slope of the line for a particular technique shows that, as expected, the sensitivity is proportional to the square of the time. There is some departure from linearity near the minimum performance time limit, particularly for the sensitive-point and planar imaging methods. It arises from a failure to recover completely from saturation and is more marked in the case of large T_1/T_2 ratios.

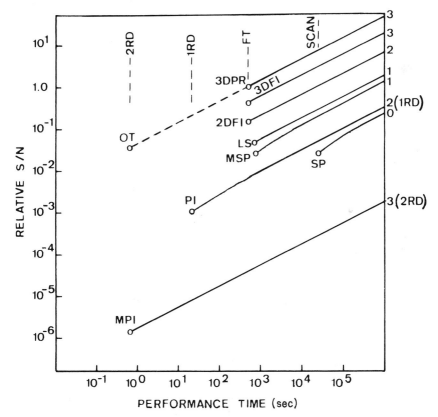

FIG. 5.6.   Relative sensitivity versus performance time for the three-dimensional imaging of a cubic object with $T_1/T_2 = 3$. [From P. Brunner and R. R. Ernst, *J. Magn. Reson.* **33**, 83 (1979).]

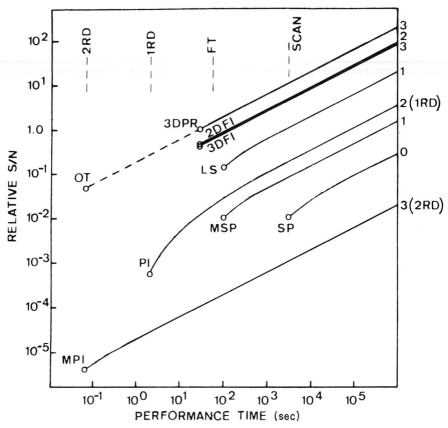

FIG. 5.7.   Relative sensitivity versus performance time for the three-dimensional imaging of a cubic object with $T_1/T_2 = 30$. [From P. Brunner and R. R. Ernst, *J. Magn. Reson.* **33**, 83 (1979).]

The most sensitive methods are echo-planar imaging and projection reconstruction, followed closely by Fourier zeugmatography; the slight loss in sensitivity of the latter method is due to the restricted observation period.

Note that 2D techniques such as 2D Fourier zeugmatography (or 2D echo-planar imaging) can approach or even surpass the efficiency of their corresponding 3D methods. This is particularly true when T_1/T_2 is large, since it is then possible to perform *n* planar experiments in the time it would take the spins to recover from a single 3D experiment. The same principle also applies to line scan and multiple sensitive-point methods vis-à-vis the 2D techniques.

One of the appeals of line and point methods is, of course, their comparative simplicity. There are no dynamic range or storage problems and the pro-

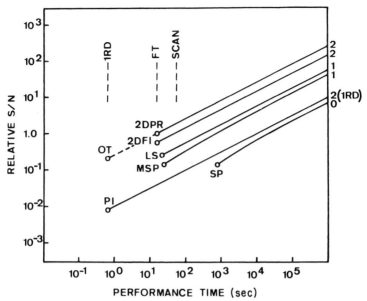

FIG. 5.8.   Relative sensitivity versus performance time for the two-dimensional imaging of a cubic object with $T_1/T_2 = 3$. [From P. Brunner and R. R. Ernst, *J. Magn. Reson.* **33,** 83 (1979).]

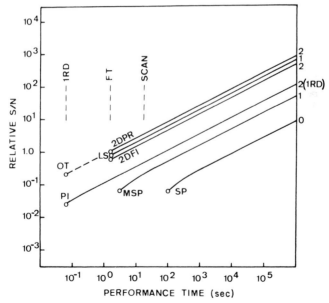

FIG. 5.9.   Relative sensitivity versus performance time for the two-dimensional imaging of a cubic object with $T_1/T_2 = 30$. [From P. Brunner and R. R. Ernst, *J. Magn. Reson.* **33,** 83 (1979).]

cessing necessary for image reconstruction is minimal or even nonexistant, as in the case of the sensitive point method.

The planar and multiplanar methods retain some of these virtues and are reasonably fast in situations where $T_1 \simeq T_2$. However, the sensitivity, as we have seen, is particularly low as a result of the reduction in size of the volume elements.

5.4.2. PERFORMANCE TIME

We may look alternatively at the various techniques from the viewpoint of performance time. As was the case with the sensitivities, there is a large spread in values—four orders of magnitude for 3D and two orders for 2D techniques. The slowest is the sensitive-point method. Multiple sensitive-point, line scan, and other Fourier transform techniques achieve an approximately n-fold reduction in the imaging time, although in the case of line scanning, additional periods are required for the selective pulse and to allow for recovery of the magnetization prior to selection of a new plane.

Planar and multiplanar imaging offer further reductions in imaging time via a one- and twofold reduction in dimensions, respectively. In the latter case all the information necessary to construct an image is obtained in a single FID. The same is also true of echo-planar imaging which, however, achieves this without loss of sensitivity.

5.5 Conclusion

In conclusion, the optimum S/N per unit time is achieved both by projection reconstruction and by echo–planar methods. The latter also has the advantage that it possesses the shortest theoretical imaging performance time. This may be of crucial importance if it is possible to obtain single-shot pictures of adequate resolution and S/N. Should extensive averaging ·be necessary, then it may be preferable to utilize the experimentally simpler projection reconstruction method. (See, however, Notes Added in Proof, Note 4.2.)

6. *Imaging Regimes*

6.1. General Principles

6.1.1. Introduction

In this chapter we direct our attention to the choice of operating conditions for a range of object sizes, divided for convenience into the following regimes:

Scale	Radius (m)
Large	~ 0.2
Intermediate	~ 0.05
Small	~ 0.01
Microscopic	$\lesssim 0.001$

Due to the presence of motion in living systems it is desirable to obtain an image of the required resolution and signal-to-noise ratio (S/N) in the minimum possible time. Taking the imaging of humans as an example, the important body motions fall into three categories:

(1) breathing motions with period ~ 5 sec;
(2) peristaltic motions with period ~ 2–5 sec; and
(3) cardiac motions with period ~ 1 sec.

Thus, whereas for heads and limbs imaging times of a few minutes may well be acceptable, artifact-free abdominal and thoracic imaging require times of <5 and <1 sec, respectively. It is possible to relax these conditions somewhat, for instance, by persuading the patient to hold his breath, by the administration of drugs to retard peristaltic motion, and, in the case of thoracic studies, by triggering the imaging cycle from the ECG. Nevertheless, these procedures are undesirable in what is a potentially noninvasive imaging system, and the dictum remains that one should obtain the images as fast as possible.

6.1.2. FREQUENCY LIMITATIONS

As is well known, the S/N of an NMR imaging experiment increases with frequency and it would therefore seem desirable, in view of the inherent insensitivity of the technique, to operate at the highest possible frequencies. However, a number of instrumental factors prevent the achievement of this aim in the case of large- and intermediate-scale regimes. Power dissipation in conventional electromagnets eventually renders their use impractical necessitating the use of expensive superconductive systems. It also becomes difficult to wind large rf coils of a sufficiently small inductance to tune.

One runs into more fundamental problems when the object being imaged is of comparable dimensions to the wavelength of the radiation. For instance, there will be phase variations over the sample and radiative losses in transmitter and receiver coils, reducing efficiency and imparing S/N.

The major limitation to the frequency, however, is imposed by the finite rf penetration depths of biological tissues. This is discussed in detail in Section 6.2 and, as we shall see, effectively limits the frequency for large-scale imaging to about 10 MHz.

Taking the maximum operating frequencies dictated by these results and assuming an optimized imaging system, the imaging time is calculated as a function of resolution and S/N ratio for each regime in Section 6.3. Representative images for these regimes are presented in Sections 6.4–6.7, where more specialized problems and future applications are also discussed.

6.1.3. RESOLUTION AND THE CHOICE OF FIELD GRADIENT

In order for an imaging system to achieve the required spatial resolution, it is necessary that the separation of each picture element, or pixel, in the frequency domain should dominate the line and spectral widths of the material contributing to that pixel. This, therefore, imposes on the field gradient G the condition that:

$$\gamma G \, \delta x \gtrsim 1/T_2^*, \tag{6.1}$$

where δx is the required degree of spatial resolution. The linewidth $1/T_2^*$ will contain contributions from the natural linewidth, the static field inhomogeneity, and variations in magnetic susceptibility, etc. Thus,

$$\frac{1}{T_2^*} = \frac{1}{T_2^{\text{natural}}} + \frac{1}{T_2^{\text{magnet}}} + \frac{1}{T_2^{\text{mag.sus.}}} + \cdots. \tag{6.2}$$

For biological samples, T_2^{natural} is typically about 50 msec so that a minimum requirement on the gradient strength is that $G > 20/\gamma \, \delta x$.

It is generally undesirable to increase the gradient strength more than is necessary to achieve the required resolution as this spreads the signal information over a wider bandwidth and can degrade the S/N in the final image. Thus, one should take steps to reduce those line-broadening effects which are experimentally variable. The dominating term in Eq. (6.2) can vary with the particular imaging regime and we therefore postpone discussion until Sections 6.4–6.7.

We should mention, however, that, when fine resolution is required, the diffusion term may be an important line-broadening mechanism. In a Hahn spin–echo experiment the echo height decays as

$$\exp\{-(2\tau/T_2) - \tfrac{2}{3}\gamma^2 G^2 \mathscr{D}\tau^3\},$$

where τ is the pulse separation and $\mathscr{D}$ the diffusion constant.[1] If the diffusion term dominates, then the time t for the echo amplitude to decay by $1/e$ is given by

$$\tfrac{2}{3}(t/2)^3\gamma^2 G^2 \mathscr{D} = 1, \tag{6.3a}$$

and the associated linewidth is

$$\Delta f = (\gamma^2 G^2 \mathscr{D}/12)^{1/3}. \tag{6.3b}$$

This dominance of the diffusion term occurs for $\delta x \lesssim (T_2\mathscr{D}/12)^{1/2}$ (typically about 2 μm), and we then require G to be chosen such that

$$\gamma G\, \delta x \gtrsim (\gamma^2 G^2 \mathscr{D}/12)^{1/3}, \tag{6.4a}$$

or

$$G \gtrsim \mathscr{D}/(12\gamma\, \delta x^3). \tag{6.4b}$$

6.1.4. Other Nuclei

The major application of NMR imaging has been, and is likely to remain for the foreseeable future, in proton mapping. Table 6.1 shows the NMR sensitivities of other potentially interesting nuclei calculated on the basis of the Hoult and Richards[1a] expression for the voltage S/N ratio. Thus, at constant field, the NMR sensitivity is proportional to $\gamma^{11/4}I(I+1)\times$ percentage abundance, whereas at constant frequency, it is proportional to $\gamma I(I+1)\times$ percentage abundance. If one seeks to optimize the sensitivity, then the maximum operating frequency consistent with complete rf penetration of the object (see below) should be chosen. It is then appropriate to compare the NMR sensitivities at constant frequency and these can be

[1] H. Y. Carr and E. M. Purcell, *Phys. Rev.* **94**, 630 (1954).
[1a] D. I. Hoult and R. E. Richards, *J. Magn. Reson.* **24**, 71 (1976).

TABLE 6.1

The Nuclear Sensitivities and Properties of Various Elements of Biological Interest

Nucleus	Spin	γ (kHz/G)	Abundance (%)	Relative NMR sensitivity (const. field)	Relative NMR sensitivity (const. frequency)	Typical human physiological conc. of the element	Relative imaging sensitivity (const. frequency)
$^{1}_{1}$H	1/2	4.2573	99.98	1	1	100 M^a	1
$^{2}_{1}$H	1	0.6537	0.02	2.4×10^{-6}	6.4×10^{-5}	100 M	6×10^{-5}
$^{13}_{6}$C	1/2	1.0705	1.11	2.5×10^{-4}	2.8×10^{-3}	—	—
$^{14}_{7}$N	1	0.3075	99.64	1.9×10^{-3}	1.9×10^{-1}	—	—
$^{17}_{8}$O	5/2	0.5771	0.04	1.9×10^{-5}	5.9×10^{-4}	50 M	3×10^{-4}
$^{19}_{9}$F	1/2	4.0052	100	8.5×10^{-1}	9.4×10^{-1}	4 μM	4×10^{-8}
$^{23}_{11}$Na	3/2	1.1263	100	1.3×10^{-1}	1.3	80 mM	1×10^{-3}
$^{31}_{15}$P	1/2	1.7237	100	8.3×10^{-2}	4.0×10^{-1}	75 mM	3×10^{-4}
$^{39}_{19}$K	3/2	0.1987	93.08	1×10^{-3}	2.2×10^{-1}	40 mM	9×10^{-5}

[a] M, molar.

multiplied by "typical" concentrations to yield estimates for the NMR imaging sensitivities as in the final column of Table 6.1. The results indicate that imaging at physiological concentrations will require substantial sacrifices in imaging time and/or spatial resolution.[2] Images of phantoms have, however, been obtained using phosphorus[3,4] and fluorine[5] compounds. The latter case is of interest since it has been demonstrated that certain fluorine-rich compounds such as perfluorotributylamine or perfluorotetrahydrofuran may be used as blood substitutes.[6] This would allow the vascular system to be selectively imaged with little background interference from the bulk tissues which contain virtually no natural fluorine.

More recently it has been possible to obtain high-resolution ^{31}P spectra from localized regions in experimental animals by the new technique of "topical" NMR spectroscopy.[6a] The resolution has to be rather coarse (~ 2 cm for humans) if reasonable signal averaging times are not to be exceeded. However, since the method yields information about metabolic pathways and energy conversion processes (see Chapter 7), it is likely to be of major future importance. ^{31}P images of excised animal hearts have also been produced recently,[6b] again with a necessarily low resolution. (See also Notes Added in Proof, Note 6.1.)

6.1.5. IMAGING OF SOLIDS

The condition that the field gradient should dominate the linewidth means that one is normally limited to observation of those regions of the sample which have fairly narrow linewidths: the mobile protons of fat, oil, and water for instance. Table 2.2 of Section 2.2 illustrates that the protons of the membrane lipids and macromolecules have T_2's of typically 1 msec and 100 μsec, respectively. Imaging of these components would thus require inordinately high gradients which, in turn, would impose severe requirements on the bandwidth of the system. In addition, one would be faced with the production of short high-power rf pulses over a sufficiently large volume and with overcoming the concomitant receiver recovery problems.

An elegant solution is to use the recently developed multipulse se-

[2] P. C. Lauterbur, in "NMR in Biology" (R. A. Dwek, I. D. Campbell, R. E. Richards, and R. J. P. Williams, eds.), pp. 323–335. Academic Press, New York, 1977.

[3] D. I. Hoult, J. Magn. Reson. 26, 165 (1977).

[4] D. I. Hoult, J. Magn. Reson. 33, 183 (1979).

[5] G. N. Holland, P. A. Bottomley, and W. S. Hinshaw, J. Magn. Reson. 28, 133 (1977).

[6] Symposium on Artificial Blood, Bethesda, Md., April 1974, Fed. Proc. Fed. Am. Soc. Exp. Biol. 34, 1428 (1975).

[6a] R. E. Gordon, P. E. Hanley, D. Shaw, D. G. Gadian, G. K. Radda, P. Styles, and L. Chan, Nature (London) 287, 736 (1980).

[6b] P. Bendel, C.-M. Lai, and P. C. Lauterbur, J. Magn. Reson. 38, 343 (1980).

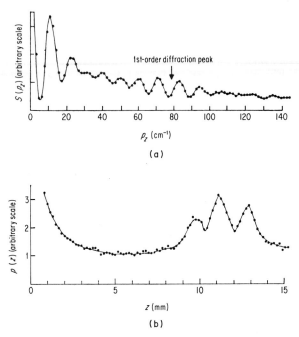

FIG. 6.1. (a) Transient nuclear signal from protons in a three-layer sample of synthetic camphor $C_{10}H_{16}O$ in response to the MREV8 multipulse sequence with $\tau = 6.4$ μsec and an applied gradient (G_z) of 0.77 G cm^{-1}. (b) Fourier transform of (a), the three camphor layers are clearly resolved. [From P. Mansfield and P. K. Grannell, *Phys. Rev.* **12**, 3618 (1975).]

quences[7-9] to reduce the natural linewidths of the solidlike components and therefore relax the gradient and bandwidth conditions (with corresponding beneficial effect on the S/N). However, one still requires short 90° pulses (typically, $\lesssim 1$ μsec), and this effectively limits the initial application to the smaller scale imaging regimes. Mansfield and Grannell[10,11] have demonstrated the use of the MREV8 cycle (132 : 1$\bar{3}$2 in their notation) in this context, producing one-dimensional (1D) projections of phantom samples. Figure 6.1a, taken from their work, shows the FID from a sample consisting of three regularly spaced camphor layers. The authors described this as NMR diffraction due to the similarity with the classical optical phenomenon (in this case a three-slit diffraction pattern). Zeroth and first-order diffraction

[7] J. S. Waugh, L. M. Huber, and U. Haeberlen, *Phys. Rev. Lett.* **20**, 180 (1968).
[8] A. N. Garroway, P. Mansfield, and D. C. Stalker, *Phys. Rev.* **11**, 121 (1975).
[9] U. Haeberlen, "High Resolution NMR in Solids: Selective Averaging" ["Advances in Magnetic Resonance, Suppl. 1" (J. S. Waugh, ed.)] Academic Press, New York, 1976.
[10] P. Mansfield and P. K. Grannell, *J. Phys. C* **6**, L422 (1973).
[11] P. Mansfield and P. K. Grannell, *Phys. Rev.* **12**, 3618 (1975).

peaks are discernible. Figure 6.1b shows the Fourier transform (FT) of the diffraction pattern and yields the 1D projection of the spin distribution. The camphor layers are clearly visible and their widths and separations correspond well to the actual dimensions when scaled by the known field gradient. The authors carried the optical analogy further[11] by deriving a pseudowave description of this NMR phenomenon using Green's functions. It is, of course, the wavelength of this pseudowave ($\propto 1/G$), rather than that of the rf, which determines the resolution of NMR imaging.

Wind and Yannoni[11a] have recently proposed an alternative method for imaging spin distributions in solids. It is based on the line-narrowing sequence of Yannoni and Vieth[11b] and relies on the strong offset dependence of the linewidth to define a sensitive region in the manner of Damadian's FONAR technique. One potential medical application for the NMR imaging of solids is the ^{31}P study of bone diseases.

6.1.6. FACTORS AFFECTING IMAGE DISTORTION

In order to produce distortion-free images, the static field, field gradients, rf field, and receiver coil response function have to obey quite stringent conditions. These depend on the particular imaging system and are generally more demanding for the faster, more efficient methods. Thus, projection reconstruction imaging requires that the static and field gradient uniformities be better than the frequency width of one pixel, that is, it requires

$$\Delta B < G\,\delta x, \tag{6.5}$$

$$\Delta G/G < \delta x/L, \tag{6.6}$$

where ΔB, ΔG are the respective field variations, L the extent of the object, and δx the width of a picture element in the direction of the field gradient G. The gradient condition is rather less strict, but may in practice be just as hard to meet. It is, however, possible to apply corrections for the known field inhomogeneities to the projections prior to reconstruction.[12] Line-scanning techniques are somewhat more forgiving of these effects, and it is possible to apply deconvolution procedures to remove distortions approaching 50%.[13]

[11a] R. A. Wind and C. S. Yannoni, *J. Magn. Reson.* **36**, 269 (1979).

[11b] C. S. Yannoni and H.-M. Vieth, *Phys. Rev. Lett.* **37**, 1230 (1976).

[12] P. C. Lauterbur, C. S. Dulcey, Jr., C.-M. Lai, M. A. Feiler, W. V. House, Jr., D. M. Kramer, C.-N. Chen, and R. Dias, *in* "Magnetic Resonance and Related Phenomena, Proceedings of the 18th Ampere Congress, Nottingham, 9–14th Sept. 1974" (P. S. Allen, E. R. Andrew, and C. A. Bates, eds.), Vol. 2, p. 431. North-Holland Publ., Amsterdam, 1975.

[13] J. M. S. Hutchison, R. J. Sutherland, and J. R. Mallard, *J. Phys. E Sci. Instrum.* **11**, 217 (1978).

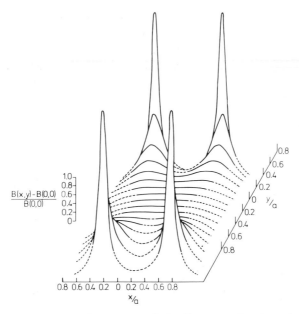

FIG. 6.2.   Field distribution for a saddle coil of radius a wound to the optimium Ginsburg and Melchner configuration.[14] As a result of the reciprocity theorem this is a direct measure of the receiver coil response function.

Point techniques are even less sensitive and, if time is not an important consideration, this represents a potential advantage of such methods.

It is possible to overcome the rf field inhomogeneity problem by increasing the transmitter coil dimensions. However, the same technique cannot generally be applied to the improvement of the receiver coil response due to the loss in S/N for filling factors less than one. One must therefore tolerate some distortion and Fig. 6.2 illustrates the receiver coil response function for a saddle coil wound to the optimum Ginsberg–Melchner[14] configuration. In practice, it is often necessary to depart considerably from this geometry so that Fig. 6.2 should be regarded as a lower limit to the distortion one is likely to encounter.

If the effects discussed above prove unacceptable they may be largely corrected by the use of a flood sample—a uniform volume of physiological saline, for instance. This may also compensate in part for effects due to rf penetration, although as we shall see, these are complex and are best avoided by operating at low frequencies. We return to the subject of coil design in Chapter 8.

[14] D. M. Ginsberg and M. J. Melchner, *Rev. Sci. Instrum.* **41,** 122 (1970).

Finally, we make the comment that computer core and dynamic range limitations may, in practice, restrict the achievable resolution, particularly if 3D imaging is envisaged.

6.2. Radio-Frequency Penetration

6.2.1. INTRODUCTION

It is possible in principle to determine exactly the electromagnetic field variations within an object by solution of Maxwell's equations if the distributions of permittivity and resistivity are known (we assume throughout the discussion below that the relative permeability is unity in biological tissues). These quantities have recently been measured for a variety of rat tissues by Bottomley and Andrew[15,16] using a vector impedance meter. Figures 6.3 and 6.4 illustrate some typical results taken from their work. Note that the resistivity is about 3 Ω m and has a fairly flat frequency dependence, whereas the relative permittivity is ≫ 1 and shows a strong dispersion due to dielectric relaxation in the region of interest. A more comprehensive collection of such data, covering a variety of animal and human tissues and extending over a wide frequency range, is available in Reference 48 of Chapter 9.

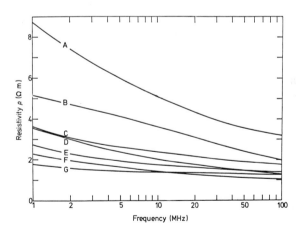

FIG. 6.3.   Resistivity of rat lung (A), brain (B), liver (C), kidney (D), heart (E), liver hepatoma D23 (F), and abdominal wall muscle (G) tissues at 37°C as a function of frequency. [From P. A. Bottomley and E. R. Andrew, *Phys. Med. Biol.* **23**, 630 (1978).]

[15] P. A. Bottomley, *J. Phys. E Sci. Instrum.* **11**, 413 (1978).
[16] P. A. Bottomley and E. R. Andrew, *Phys. Med. Biol.* **23**, 630 (1978).

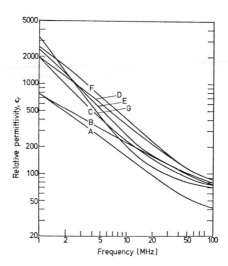

FIG. 6.4. Permittivity of rat lung (A), brain (B), liver (C), kidney (D), heart (E), liver hepatoma D23 (F), and abdominal wall muscle (G) tissues at 37°C as a function of frequency. [From P. A. Bottomley and E. R. Andrew, *Phys. Med. Biol.* **23**, 630 (1978).]

6.2.2. TISSUE MODELS

Rather than attempt to accurately model a real system, which is, of necessity, heterogeneous, one may hope to derive sufficient insight from the analytical solutions which may be obtained for homogeneous tissue samples. Bottomley and Andrew[16] have treated the cases of a semi-infinite plane and a cylinder, the latter being the more realistic model, particularly for the case of human imaging. Their choice of geometry was one in which the rf magnetic field B_1 was parallel to the cylinder axis (see Fig. 6.5a and b). This is technically quite feasible; one winds a solenoidal coil parallel to the cylinder axis which is itself perpendicular to the static field B_0. However, most whole-body imaging systems to date have employed an alternative geometry with B_1 (produced typically by a saddle coil assembly) perpendicular to the cylinder axis (Fig. 6.5c and d). The geometrical distinction is somewhat academic in the case of conventional electromagnets (Fig. 6.5b and d), although the solenoidal system will have S/N advantages.[17] (For a long solenoid enclosing the specimen, a lower rf absorption will also result, since the boundary conditions in this case require the continuity of the electric displacement D rather than the electric field E[18]. However, this is not the case for a finite solenoid around an extended specimen.) For the four-coil spherical systems,

[17] D. I. Hoult and R. E. Richards, *J. Magn. Reson.* **24**, 71 (1976).
[18] M. F. Iskander, Personal communication, Aug. 1979.

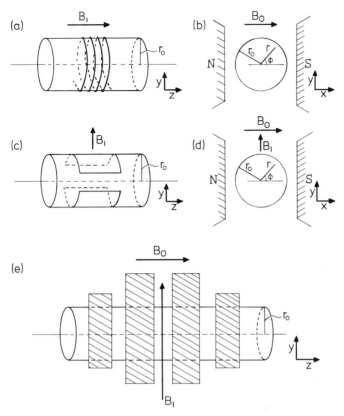

FIG. 6.5. Transmitter coil geometries for a cylindrical sample. (a) A solenoidal coil producing a B_1 field parallel to the cylinder axis. (b) The orientation of (a) in a conventional electromagnet. (c) A saddle coil producing a B_1 field perpendicular to the cylinder axis. (d) The orientation of (c) in a conventional electromagnet. (e) The preferred configuration for a four-coil
"spherical" system.

currently in vogue for human studies, the preferred patient configuration is
parallel to the magnet axis (Fig. 6.5e), since this allows maximum ease of
access. One is therefore obliged to have B_1 perpendicular to this axis in order
to maintain orthogonality with the static field.

6.2.3. CYLINDRICAL MODEL WITH B_1 AXIAL

We now present some analytical results for the magnetic field variations in
tissue samples. For the geometry of Fig. 6.5a and b, Bottomley and Andrew[16]
obtain a solution:

$$\mathbf{B}_1 = B_1[|I_0(Kr)|/|I_0(Kr_0)|] \exp\{i(\omega t - \xi(r))\}\mathbf{z}, \qquad (6.7)$$

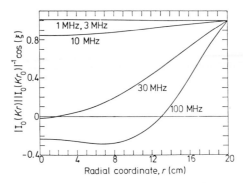

FIG. 6.6.   The amplitude phase variation $|I_0(Kr)||I_0(Kr_0)|^{-1} \cos \xi$ along the radius of a rat-muscle cylinder of diameter 0.4 m for a variety of frequencies. [From P. A. Bottomley and E. R. Andrew, *Phys. Med. Biol.* **23**, 630 (1978).]

where $I_0(Kr)$ is a modified Bessel function of the first kind of order zero,

$$\xi(r) = \arctan[\operatorname{Im} I_0(Kr_0)/\operatorname{Re} I_0(Kr_0)] - \arctan[\operatorname{Im} I_0(Kr)/\operatorname{Re} I_0(Kr)]. \quad (6.8)$$

and K, the complex wave number, is given by

$$K = \omega_0(\tfrac{1}{2}\epsilon\epsilon_0\mu_0\{[1 + 1/\rho^2\epsilon^2\epsilon_0^2\omega_0^2]^{1/2} - 1\})^{1/2}$$
$$+ i\omega_0(\tfrac{1}{2}\epsilon\epsilon_0\mu_0\{[1 + 1/\rho^2\epsilon^2\epsilon_0^2\omega_0^2]^{1/2} + 1\})^{1/2}, \quad (6.9)$$

where ρ is the resistivity of the medium, ω_0 the angular frequency, μ_0, ϵ_0 the permeability and permittivity of free space, respectively, and ϵ the relative permittivity or dielectric constant. These results are scaled relative to a value of $\mathbf{B}_1 = B_0 \exp(i\omega_0 t)\mathbf{z}$ at the cylinder surface and therefore make no allowance for phase and amplitude disturbances in the surrounding medium due to the presence of the tissue cylinder. Figure 6.6 shows a plot of the amplitude phase variation, $|I_0(Kr)||I_0(Kr_0)|^{-1} \cos \xi$, along the radial coordinate of a muscle cylinder of radius 0.2 m for a variety of frequencies.

6.2.4. CYLINDRICAL MODEL WITH B_1 PERPENDICULAR TO THE AXIS

We have considered the case in which B_1 is perpendicular to the cylinder axis (Fig. 6.5c, d, and e). The fields obey the following equations:

$$B_r = \frac{i}{\omega_0 r}\frac{\partial E_z}{\partial \phi}, \quad (6.10)$$

$$B_\phi = -\frac{i}{\omega_0}\frac{\partial E_z}{\partial r}, \quad (6.11)$$

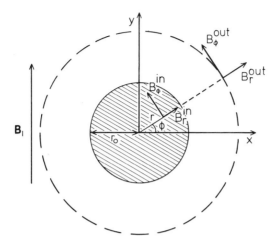

FIG. 6.7. The field coordinate system for a model in which B_1 is perpendicular to the cylinder axis. The tissue cylinder of radius r_0 is shown shaded and E_z is perpendicular to the plane of the paper.

$$\frac{\partial^2 E_z}{\partial r^2} + \frac{1}{r}\frac{\partial E_z}{\partial r} + \frac{1}{r^2}\frac{\partial^2 E_z}{\partial \phi^2} + K^2 E_z = 0, \qquad (6.12)$$

where B_r and B_ϕ are the radial and azimuthal components of the magnetic field and E_z is the z component of the electric field. The complex wave number K is given by Eq. (6.9) and the field coordinate system is illustrated in Fig. 6.7. Equation (6.12) can be separated into angular and radial functions:

$$E_z(r, \phi) = R(r)\Phi(\phi), \qquad (6.13)$$

where $\Phi(\phi)$ obeys the equation

$$\frac{1}{\Phi}\frac{\partial^2 \Phi}{\partial \phi^2} = -m^2, \qquad (6.14)$$

which has solutions of the form

$$\Phi(\phi) = A_m e^{im\phi} + B_m e^{-im\phi}, \qquad (6.15)$$

where A_m and B_m are constants and m is an integer. $R(r)$ satisfies Bessel's equation

$$\frac{\partial^2 R}{\partial r^2} + \frac{1}{r}\frac{\partial R}{\partial r} + \left(K^2 - \frac{m^2}{r^2}\right)R = 0, \qquad (6.16)$$

which has solutions

$$R(r) = I_m(Kr), \qquad (6.17)$$

where $I_m(Kr)$ are modified Bessel functions of the first kind of order m. Thus the general solution for the z component of the electric field inside the tissue cylinder is of the form

$$E_z^i = \sum_m I_m(Kr)\{A_m \cos m\phi + B_m \sin m\phi\}. \tag{6.18}$$

However, outside the tissue cylinder one can make the approximation that $K = 0$ since the conductivity is practically zero ($\rho \rightarrow \infty$) and $\mu_0\epsilon_0\omega_0^2 \ll 1$. Equation (6.16) then reduces to Laplace's equation

$$\frac{\partial^2 R}{\partial r^2} + \frac{1}{r}\frac{\partial R}{\partial r} - \frac{m^2 R}{r^2} = 0, \tag{6.19}$$

which has solutions

$$R(r) = C \log r + \sum_m [D_m r^m + (E_m/r^m)], \tag{6.20}$$

with C, D_m, and E_m as constants. The angular solution remains unchanged and therefore outside the cylinder,

$$E_z^e = C \log r + \sum_m [D_m r^m + (E_m/r^m)](F_m \cos m\phi + G_m \sin m\phi), \tag{6.21}$$

again with F_m and G_m as constants. The superscript "e" used here and "i" above refer respectively to external and internal fields.

The magnetic field components are readily derived from the general solutions [Eqs. (6.18) and (6.21)] via Eqs. (6.10) and (6.11). By applying the boundary conditions at the cylinder surface for E_z^e, E_z^i, B_ϕ^e, and B_ϕ^i, it is possible to evaluate all the constants, and we obtain as final solutions

$$E_z^e = i\omega_0 B_1\left\{r + \left[\frac{2I_1(Kr_0)/(Kr_0)}{I_0(Kr_0)} - 1\right]\right\}\cos \phi, \tag{6.22}$$

$$B_r^e = B_1\left\{1 + \left(\frac{r_0}{r}\right)^2\left[\frac{2I_1(Kr_0)/(Kr_0)}{I_0(Kr_0)} - 1\right]\right\}\sin \phi, \tag{6.23}$$

$$B_\phi^e = B_1\left\{1 - \left(\frac{r_0}{r}\right)^2\left[\frac{2I_1(Kr_0)/(Kr_0)}{I_0(Kr_0)} - 1\right]\right\}\cos \phi, \tag{6.24}$$

$$E_z^i = \frac{2iB_1}{KI_0(Kr_0)} I_1(Kr) \cos \phi, \tag{6.25}$$

$$B_r^i = \frac{2B_1}{I_0(Kr_0)} \frac{I_1(Kr)}{Kr} \sin \phi, \tag{6.26}$$

$$B_\phi^i = \frac{2B_1}{I_0(Kr_0)}\left[I_0(Kr) - \frac{I_1(Kr)}{Kr}\right]\cos\phi. \tag{6.27}$$

Note that as $r \to \infty$,

$$B_r^e \to B_1 \sin\phi, \tag{6.28a}$$

$$B_\phi^e \to B_1 \cos\phi, \tag{6.28b}$$

and **B** tends to a constant field B_1 in the y direction. As r is reduced, so the amplitude and phase of **B** start to vary. This is perhaps best illustrated by writing the fields in the manner of Eq. (6.7). Thus,

$$\frac{B_r}{B_1 \sin\phi} = A_r(r)\exp\{i(\omega_0 t - \xi_r(r))\}, \tag{6.29}$$

$$\frac{B_\phi}{B_1 \cos\phi} = A_\phi(r)\exp\{i(\omega_0 t - \xi_\phi(r))\}. \tag{6.30}$$

Figures 6.8a, b and 6.9a, b show plots of A_r, ξ_r, A_ϕ, and ξ_ϕ for a 0.2-m radius "liver cylinder" at a number of frequencies. (The values for ρ and ϵ were obtained from Figs. 6.3 and 6.4, respectively.) They indicate that even for quite modest amplitude variations, substantial phase differences can arise across the sample. Further, the phase at the cylinder surface is by no means negligible at frequencies of 10 or even 4 MHz. It may therefore not be possible to accurately tune a spectrometer using a phantom, and adjustments are likely to be necessary for patients of varying size and for images of different cross-sectional areas. Figures 6.10 and 6.11 show plots of $A_r(r)\cos\xi_r(r)$ and $A_\phi(r)\cos\xi_\phi(r)$, respectively, under the same conditions as for Figs. 6.8 and 6.9 and indicate the essential similarity with the results of Bottomley and Andrew (Fig. 6.6).

6.2.5. CONCLUSIONS

The conclusion with regard to large-scale imaging ($r_0 = 0.2$ m) is that one should generally restrict the operating frequency to about 10 MHz or less. If one insists on higher frequencies, then image correction will undoubtedly be required and such corrections may well need to encompass the reflection and diffraction effects of the full heterogeneous model. It is the authors' opinion that this is a problem best avoided.

Similar computations, again based on a "liver cylinder," indicate that the maximum operating frequencies for the various regimes are as shown in Table 6.2.

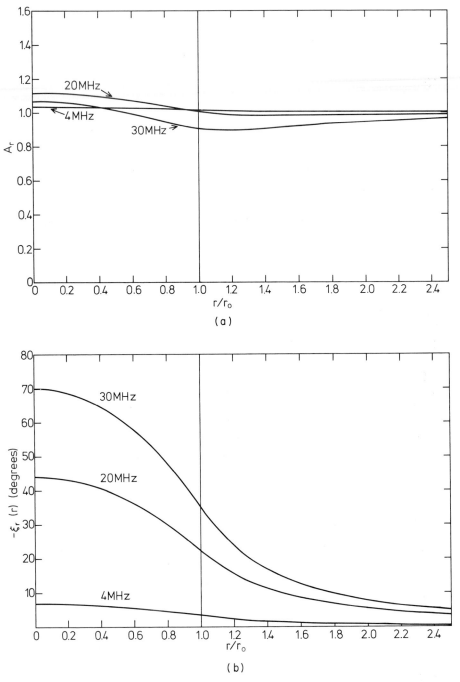

FIG. 6.8. The amplitude (a) and phase (b) of the radial magnetic field component at a variety of frequencies for a rat liver cylinder of radius 0.2 m.

188

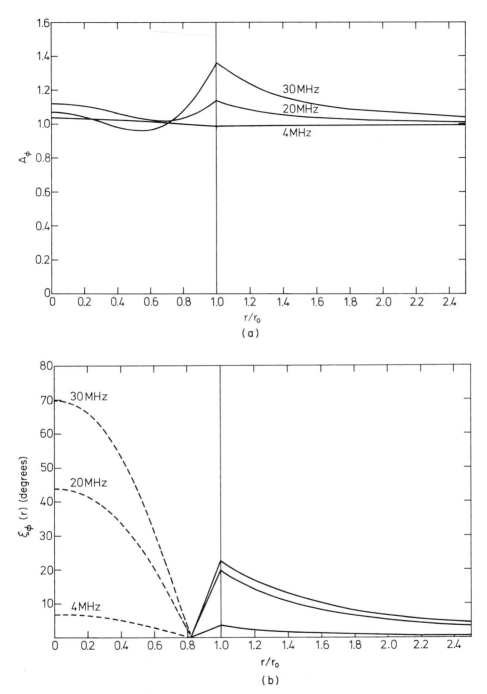

FIG. 6.9.   The amplitude (a) and phase (b) of the angular magnetic field component at a variety of frequencies for a rat liver cylinder of radius 0.2 m (dashed line denotes negative value).

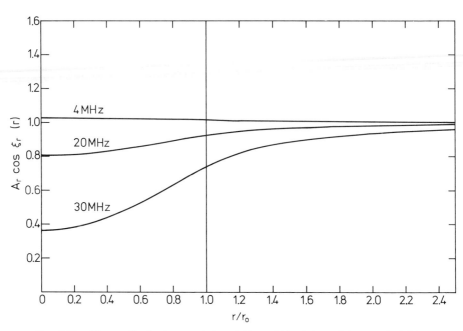

FIG. 6.10.   The amplitude phase variation $A_r(r) \cos \xi_r(r)$ of the radial magnetic field component at a variety of frequencies for a rat-liver cylinder of radius 0.2 m.

TABLE 6.2

IMAGING REGIMES IN TERMS OF SAMPLE RADIUS
AND PRACTICAL IMAGING OPERATING FREQUENCY

Regime	Radius (m)	Frequency (MHz)	$1/(a^4 v^{7/2})^a$
Large	0.2	4	4.88
		6	1.18
		8	0.43
		10	0.20
Intermediate	0.05	30	1.08
Small	0.01	150	2.42
Microscopic	0.001	600	189.00

^{a} This column is proportional to the voxel imaging time [Eq. (6.41)] and illustrates the advantage of large-scale imaging, especially above 6 MHz. N. B.: Above 10 MHz, rf penetration of the subject becomes problematical.

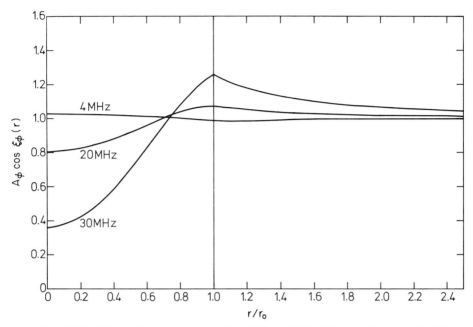

FIG. 6.11. The amplitude phase variation $A_\phi(r)\cos\xi_\phi(r)$ of the angular magnetic field component at a variety of frequencies for a rat-liver cylinder of radius 0.2 m.

We make the comment here, although defer the discussion until Chapter 9, which deals with biomagnetic effects, that the field expressions derived above provide a basis for estimating rf power absorption in an imaging experiment.

6.3. Signal-to-Noise Considerations

6.3.1. INTRODUCTION

In this section we calculate the imaging time for an optimized imaging system as a function of resolution and S/N. We use the maximum frequencies (as determined in Section 6.2) which allow distortion-free imaging for large, intermediate, and small-scale regimes. In addition, we have taken a frequency of 600 MHz for the microscopic regime as this represents the current technological limit for NMR spectrometers.[19]

[19] National Resource Center, Chemistry Dept., Carnegie-Mellon University, Pittsburgh, Pennsylvania 15213.

6.3.2. The Signal-to-Noise Ratio Following a 90° Pulse

An often-used expression for the voltage S/N ratio following a 90° pulse is:[20]

$$\frac{S}{N} = \frac{K\eta M_0}{F_n} \left(\frac{\mu_0 Q \omega_0 V_c}{4kT_c \Delta f}\right)^{1/2}, \tag{6.31}$$

where K is a numerical factor (approximately unity) depending on the receiver coil geometry, η is the filling factor, V_c is the coil volume, Q is the quality factor, k is Boltzmann's constant, T_c is the probe temperature, Δf is the bandwidth, and F_n the noise factor of the receiver [see Eq. (8.40)]. Thus, at first glance, one would expect the S/N to vary as the three-halves power of the frequency (since the equilibrium magnetization is proportional to the static magnetic field which is itself proportional to the frequency). However, as Hoult and Richards[17] have pointed out, many of the parameters in the above expression are interrelated and they derive an alternative expression:

$$\frac{S}{N} = \frac{K(B_1)_{xy} V_s N \gamma \hbar^2 I(I+1)}{7.12 k T_s F_n} \left(\frac{p}{kT_c l \xi \, \Delta f}\right)^{1/2} \frac{\omega_0^{7/4}}{[\mu \mu_0 \rho(T_c)]^{1/4}}, \tag{6.32}$$

where N is the number of resonant nuclei per unit volume of spin I, V_s and T_s are the sample volume and temperature, respectively, $(B_1)_{xy}$ is the effective field over the sample volume produced by unit current in the receiving coil, and p, l, and $\rho(T_c)$ are, respectively, the perimeter, length, and resistivity of the conductor. The unknowns η and Q of Eq. (6.31) have been replaced by the proximity factor ξ which, in practice, is fairly well known from experience (it is typically between 3 and 6) and is relatively invariant to coil geometry, provided the separation of the windings is much less than the coil dimensions. It therefore transpires that S/N is proportional to $\omega_0^{7/4}$. However, it should be noted that Eq. (6.32) can be reduced to Eq. (6.31) for the case of a solenoid, but is less than Eq. (6.32) by a factor of 3 to 4 for a coil with saddle geometry. This emphasizes the difficulty of employing Eq. (6.31) where the functional form of the variables in the expression is not explicit.

6.3.3. Optimizing the Signal-to-Noise Ratio

In order to optimize S/N, one needs to maximize $(B_1)_{xy}/(\text{coil resistance})^{1/2}$, and this can be achieved with the coil configuration illustrated in Fig. 6.12. Hoult and Lauterbur[21] have used such a coil (a solenoid of 6 turns, length

[20] A. Abragam, "The Principles of Nuclear Magnetism," Ch. 3. Oxford Univ. Press, London and New York, 1961.
[21] D. I. Hoult and P. C. Lauterbur, *J. Magn. Reson.* **34,** 425 (1979).

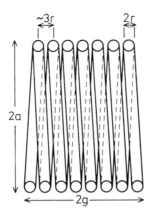

FIG. 6.12. The optimum winding geometry for a solenoidal receiving coil. [From D. I. Hoult and P. C. Lauterbur, *J. Magn. Reson.* **34**, 425 (1979).]

0.2 m, and diameter 0.25 m, constructed from copper tubing of diameter 0.02 m) to obtain an estimate for the S/N of the Lorentzian line following Fourier transformation of a single FID from a 1-ml water sample contained within a layer volume at 1 MHz. They find empirically,

$$S/N \simeq 213\sqrt{T_2{}^*}, \tag{6.33}$$

where $T_2{}^*$ is the experimental linewidth. This must be dominated by the field gradient. However, as discussed above, this dominance should not be excessive and we therefore assume that $T_2{}^*$ can be approximated by the known values of T_2 for biological tissues. Following Eq. (6.32), Eq. (6.33) can be extended to other frequencies and coil dimensions. Thus,

$$S/N \simeq (26.56\sqrt{T_2{}^* v^{7/4}})/a, \tag{6.34}$$

where a is the radius of the coil [assumed solenoidal since this yields the best $(B_1)_{xy}/(\text{coil resistance})^{1/2}$ ratio] and v is the radio frequency in MHz. Taking $T_2 \sim 100$ msec, Hoult and Lauterbur finally obtain

$$S/N \simeq (8.4v^{7/4})/a. \tag{6.35}$$

6.3.4. SIGNAL AVERAGING

Now, if one wishes to perform signal averaging to improve the S/N, one should optimize the pulse nutation or flip angle β and the interpulse delay T. In fact, the maximum sensitivity is achieved for β, $T \rightarrow 0.$[22] However, interference between successive FIDs occurs for short times $T \lesssim T_2$ (an effect which is, of course, exploited in the SSFP technique), and the frequency resolution improves as T is lengthened (see Section 3.4.7). It is therefore necessary to come to some compromise. The height of successive FIDs, once equilibrium has been established, is[22,23]

$$H\left(\frac{T}{T_1}, \beta\right) = \frac{\{1 - \exp(-T/T_1)\}}{\{1 - \cos\beta \exp(-T/T_1)\}} \sin\beta. \tag{6.36}$$

In the optimum case when β, $T \to 0$, the function $C = T_1 H^2/T$ approaches its limiting value of 0.5. Fortunately, the maximum of this function is fairly broad allowing reasonable flexibility in the choice of β and T.

Thus the S/N obtained in the frequency domain per FID is

$$\text{S/N} \simeq [(2.656 \times 10^7 v^{7/4})/a]\sqrt{T_2^*} H \, \Delta x \, \Delta y \, \Delta z, \tag{6.37}$$

where we have introduced the volume element dimensions Δx, Δy, and Δz expressed in meters. For N FIDs the signals add coherently, the noise adds incoherently, and there is therefore a $\sqrt{N}$ improvement in S/N. Thus,

$$\text{S/N} \simeq [(2.656 \times 10^7)/a]\sqrt{T_2^*} v^{7/4} H \, \Delta x \, \Delta y \, \Delta z (t/T)^{1/2}, \tag{6.38}$$

where $t = NT$.

6.3.5. IMAGING TIME

For a cubic volume element of side Δx, the time t_{vol} taken to achieve a particular S/N with given resolution is

$$t_{\text{vol}} \simeq \left(\frac{S}{N}\right)^2 a^2 \left(\frac{T_1}{T_2^*}\right) \frac{1.418 \times 10^{-15}}{v^{7/2} C} \left(\frac{1}{\Delta x}\right)^6. \tag{6.39}$$

Note that this is proportional to a^2, $v^{-7/2}$, and $(\Delta x)^{-6}$. In particular, the inverse sixth-order dependence of imaging time on linear dimension emphasizes the extreme price which must be paid for increased resolution.

Suppose now that we reduce the scale of the object but maintain the same fractional resolution. Then,

$$\Delta x/a = \text{constant} \ (= 128, \text{ typically}), \tag{6.40}$$

$$t_{\text{vol}} \propto 1/(a^4 v^{7/2}). \tag{6.41}$$

In order to keep the imaging time constant one must compensate for the decrease in a by a corresponding increase in operating frequency. As the final column of Table 6.2 demonstrates, however, it is generally a disadvantage to reduce the object size.

[22] J. S. Waugh, J. Mol. Spectrosc. **35**, 298 (1970).
[23] P. Brunner and R. R. Ernst, J. Magn. Reson. **33**, 83 (1979).

If instead of having a cubic volume element we have a slice of constant thickness and vary only the "in-plane" resolution, then the corresponding expression for the imaging time becomes

$$t_{\text{slice}} \simeq \left(\frac{S}{N}\right)^2 a^2 \left(\frac{T_1}{T_2*}\right) \frac{1.418 \times 10^{-15}}{v^{7/2} C} \left(\frac{1}{\Delta z}\right)^2 \left(\frac{1}{\Delta x}\right)^4 . \qquad (6.42)$$

The dependence on resolution is thus relaxed to inverse fourth order. Such an experiment is feasible in situations where the structure has a high degree of correlation in one dimension, e.g., plant stems or the long bones of limbs.

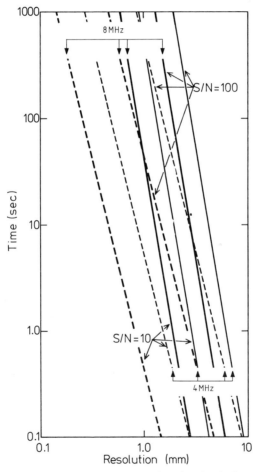

FIG. 6.13.   Imaging time (t_{vol}, t_{slice}) versus resolution plots for the large-scale regime ($r_0 = 0.2$ m) for the two operating frequencies 4.0 and 8 MHz (solid line, cubic volume element; dashed line, 10-mm slice).

Figures 6.13–6.16 are plots of expressions (6.39) and (6.42) for the four imaging regimes and frequencies of Table 6.2. (The values $C = 0.5$, $T_2^* = 100$ msec, and $T_1 = 600$ msec have been used in these calculations). We assume that in an optimal imaging system, simultaneous reception and discrimination between signals from all volume elements is achieved. The times t_{vol} and t_{slice} of Eqs. (6.39) and (6.42) are therefore the times to produce a complete NMR image of specified S/N and resolution.

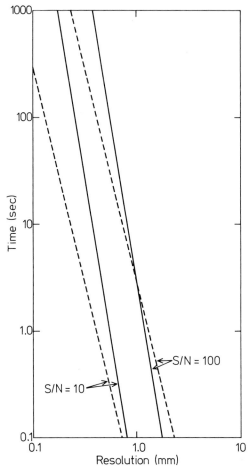

FIG. 6.14.   Imaging time (t_{vol}, t_{slice}) versus resolution plots for the intermediate-scale regime ($r_0 = 0.05$ m; frequency, 30 MHz; solid line, cubic volume element; dashed line, 1-mm slice).

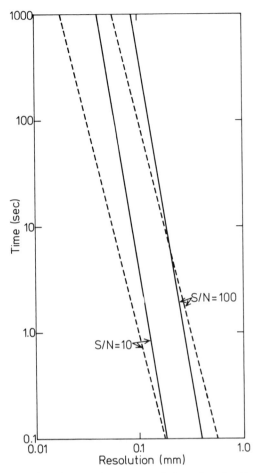

FIG. 6.15. Imaging time (t_{vol}, t_{slice}) versus resolution plots for the small-scale regime ($r_0 = 0.01$ m; frequency, 150 MHz; solid line, cubic volume element; dashed line, 0.2-mm slice).

6.3.6. REAL IMAGING SYSTEMS

We have discussed in Chapter 5 how closely the imaging schemes, which have so far been proposed, approach the optimum. For instance, in the case of projection reconstruction, one simultaneously receives information from all volume elements, but there is a minimum performance time corresponding to the acquisition of a sufficient number n of projections to enable reconstruction of the image. Thus,

$$t_{min} \simeq nT \simeq (a/\Delta x)T. \qquad (6.43)$$

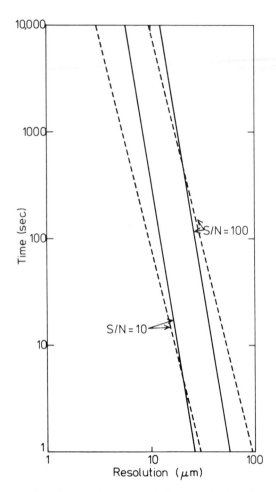

FIG. 6.16.   Imaging time (t_{vol}, t_{slice}) versus resolution plots for the microscopic scale regime ($r_0 = 0.001$ m; frequency, 600 MHz; solid line, cubic volume element; dashed line, 20-μm slice).

For projection reconstruction then, the slope of the graphs in Figs. 6.13–6.16 would be given by Eq. (6.43) for t_{vol} or $t_{slice} < t_{min}$.

6.3.7. SSFP Techniques and Tissue Contrast

The above calculations have assumed an optimized FID approach to signal acquisition. However, it is possible to take advantage of the allegedly more efficient methods such as the Carr–Purcell, DEFT and SSFP techniques and, as we have seen, a number of imaging schemes do just this.

The maximum signal which can be maintained in the x, y plane for an SSFP sequence with optimized flip angle approaches the limit[24] $(M_0/2)(T_2/T_1)^{1/2}$. The analogous expressions to Eqs. (6.39) and (6.42) for this technique are thus,

$$t_{vol}^{SSFP} \simeq \left(\frac{S}{N}\right)^2 a^2 \left(\frac{T_1}{T_2^*}\right) \frac{2.836 \times 10^{-15}}{v^{7/2}} \left(\frac{1}{\Delta x}\right)^6, \tag{6.44}$$

$$t_{slice}^{SSFP} \simeq \left(\frac{S}{N}\right)^2 a^2 \left(\frac{T_1}{T_2^*}\right) \frac{2.836 \times 10^{-15}}{v^{7/2}} \left(\frac{1}{\Delta z}\right)^2 \left(\frac{1}{\Delta x}\right)^4, \tag{6.45}$$

and the ratio of imaging times t^{SSFP}/t is therefore,

$$(2T_1 H^2)/T = 2C. \tag{6.46}$$

Thus, provided we are able to select T and β such that C approaches its maximum value of 0.5, there is essentially no difference between the two methods. (Note, however, that one is not free to choose any arbitrary value for T. If sufficient spatial resolution is to be maintained, then $T \geq T_2^*$ and an acceptable value might be $T \sim 3T_2^*$.) It is worth noting that the imaging time is proportional to T_1/T_2, a factor which, as we have noted in Section 2.3, is heavily frequency dependent through the strong dispersion of T_1 and relative invariance of T_2. Typically, it may vary by a factor of ~ 5 on going from 1 to 100 MHz and will somewhat offset the S/N advantages of the higher frequencies.

6.3.8. OTHER FACTORS AFFECTING THE IMAGING TIME

We emphasize that the above imaging-time calculations are intended only as rough guidelines. There are, for instance, a number of ways in which one might improve or degrade the performance:

(1) We have assumed a sample and receiver coil temperature of 293 K. If it is possible to cool the coil to liquid nitrogen (77 K) or liquid helium (4.2 K) temperatures, then one can achieve improvements in S/N of 3 and at least 12, respectively, for lossless samples.

(2) A further $\sqrt{2}$ improvement can be obtained by using quadrature detection of the signal.

(3) We have assumed an optimum coil geometry which maximizes $B_1/(\text{coil resistance})^{1/2}$. It may well be necessary, for ease of access or other reasons, to use a saddle coil and this would result in an S/N degradation of ~ 3. Additional factors are discussed in Section 8.6.3.

(4) The preamplifier system will have a noise factor > 1.

[24] P. A. Bottomley, Ph.D. Thesis "NMR Imaging by the Multiple Sensitive Point Technique," p. 37, Nottingham, 1978. See also Eq. (3.89) of this book.

(5)   If the signal is to be simultaneously received from a number of volume elements, the receiver coil Q may need to be reduced in order to achieve a sufficient bandwidth to accommodate the required spatial region. In this case it may well be a relatively simple matter to cool the additional damping resistor in liquid nitrogen or helium, even if the cooling of the whole probe as suggested in (1) is not feasible.

(6)   There may be problems arising from the loss of signal in the receiver dead time.

(7)   As noted in Section 6.1.6, the receiver coil radius may need to be substantially greater than that of the object to be imaged in order to achieve sufficient homogeneity of response (see also Section 8.4.3).

(8)   As well as giving rise to rf penetration problems, sample losses can be an important source of noise which we have hitherto assumed to arise entirely from the receiver coil resistance. This aspect is discussed more fully in the following section and also in Section 8.6.

When it is remembered that the imaging time is proportional to $(S/N)^2$ it is clear that many of the above factors could alter our estimates by at least an order of magnitude. We thus reemphasize the approximate nature of these results.

6.4. Large-Scale Imaging

6.4.1. INTRODUCTION

As we have seen in Section 6.2, the optimum frequency for large-scale (0.2-m radius) imaging is in the 4–10 MHz range. For slightly smaller objects, such as human heads, however, it may be possible to work at up to 20 or 25 MHz. Figure 6.13 shows the imaging-time estimates for an optimum imaging system at 4 and 8 MHz.

6.4.2. SAMPLE LOSSES

In Section 6.3.8 we commented that for human imaging, sample losses may well prove important in terms of their effect on the S/N. Hoult and Lauterbur[21] have derived expressions for the effective series resistance arising from these effects. For the inductive loss, they obtain

$$R_m = \frac{\pi \omega_0^2 \mu_0^2 n^2 b^5}{30\rho[a^2 + g^2]}, \tag{6.47}$$

for a solenoid, where a is the coil radius, g its half-height, n the number of turns, and b is the radius of a spherical sample of conductivity ρ. The analogous result for a saddle coil is

$$R_m = \frac{1.37\pi\omega_0{}^2\mu_0{}^2n^2b^5}{30\rho a^2}. \qquad (6.48)$$

For the dielectric loss they obtain

$$R_e = \tau\omega_0{}^3L^2C_d, \qquad (6.49)$$

where τ is the loss factor, L the coil inductance, and C_d its distributed capacitance. Recently, however, Gadian and Robinson[25] have raised objections to this latter result and have proposed their own alternative expression. The dielectric loss, though less than the inductive loss is certainly not negligible, but it can be eliminated by interposing a Faraday shield between the coil and the sample.

The magnetic (or inductive) losses are intrinsic and have to be lived with. Hoult and Lauterbur[21] estimate, for example, that for head scans at 4 MHz, the S/N will be degraded by a factor of about 1.22, whereas for body scans the degradation is about 3.33.

We comment that, whereas Eqs. (6.47)–(6.49) are valid only when the magnetic field is uniform over the sample volume, our results derived in Section 6.2.4 can be used to obtain estimates even when the penetration depth is comparable with the object dimensions. However, we reiterate the fact that real samples are strongly heterogeneous and so the only safe estimates must inevitably be based on practical experience. Although we make no further mention of the fact, the above considerations will also apply to the remaining imaging regimes discussed below.

6.4.3. Imaging Times

Even allowing for a possible degradation of S/N due to inductive loss, it should still prove possible to achieve 1-cm slice resolutions of about 3 mm in a time of the order of 1 sec. This would therefore allow one to obtain images of the human abdomen which were relatively free from motional artifacts (see the comments of Section 6.1.1). In order to image the thoracic region and, in particular, the heart, one requires imaging times of rather less than 1 sec. These can only be obtained with some loss in resolution, say to 4 mm, or with a slice thickness of about 4 cm or with a lower S/N.

[25] D. G. Gadian and F. N. H. Robinson, *J. Magn. Reson.* **34**, 449 (1979).

6.4.4. Whole-Body Imaging

The medical application of NMR imaging is the subject of Chapter 7. In the discussion which follows, we shall therefore be predominantly concerned with technical aspects when considering human studies. In some cases we have found it necessary to refer ahead to images which it was felt were better included in Chapter 7.

A number of whole-body images have been produced recently, although none with anything approaching an optimal imaging system. Figure 7.5 shows a cross-sectional image through the abdomen of one of the authors (P.M.) at the level of L2–L3[26] It was obtained with a selective irradiative line-scan procedure at 4 MHz.[27] The in-plane resolution is about 4 mm and the slice thickness is estimated to be 4 cm. The total imaging time was 40 min, although this was largely a result of computer speed limitations and could have been reduced to about 5 min. The image represents an average over breathing and peristaltic motions which, coupled with a progressive phase shift, give rise to major distortions in the region of the stomach. Distortions also arise from the static magnetic field inhomogeneity (due to the presence of structural ironwork in the laboratory), field gradient nonlinearities, and a nonuniform receiver coil response (see Section 6.1.6). These latter effects are more clearly apparent in the phantom image of Fig. 6.17. This consists of a supporting matrix containing 3/4-in. water-filled tubes placed at 1-in. intervals to form the letters "IP." Such phantoms, consisting of grids of high-proton-density material, can be used to correct for intensity and geometric distortions in other images. The subject of image enhancement is beyond the scope of the present work, however.

Figure 6.18 shows a cross section through the human chest. This, again, was obtained with a variant of the selective irradiative line-scan procedure, but at a frequency of 1.7 MHz.[28] The resolution is 1 cm, the slice thickness 1.85 cm, and the total scan time 2 min. The severe motional artifact is due to the beating heart and is a consequence of the imaging procedure as discussed in Section 4.4.2.4.

Figure 6.19 also shows a human thoracic section, obtained in this case by the FONAR method.[29,30] It was, in fact, the first human cross section to be

[26] P. Mansfield, I. L. Pykett, P. G. Morris, and R. E. Coupland, *Br. J. Radiol.* **51**, 921 (1978).

[27] P. G. Morris, P. Mansfield, I. L. Pykett, R. J. Ordidge, and R. E. Coupland, *IEEE Trans.* **26**, 2817 (1979).

[28] J. Mallard, J. M. S. Hutchison, W. A. Edelstein, C. R. Ling, M. A. Foster, and G. Johnson, *Phil. Trans. R. Soc. London Ser. B* **289**, 519 (1980).

[29] R. Damadian, M. Goldsmith, and L. Minkoff, *Physiol. Chem. Phys.* **9**, 97 (1977).

[30] R. Damadian, L. Minkoff, M. Goldsmith, and J. A. Koutcher, *Naturwissenschaft* **65**, 250 (1978).

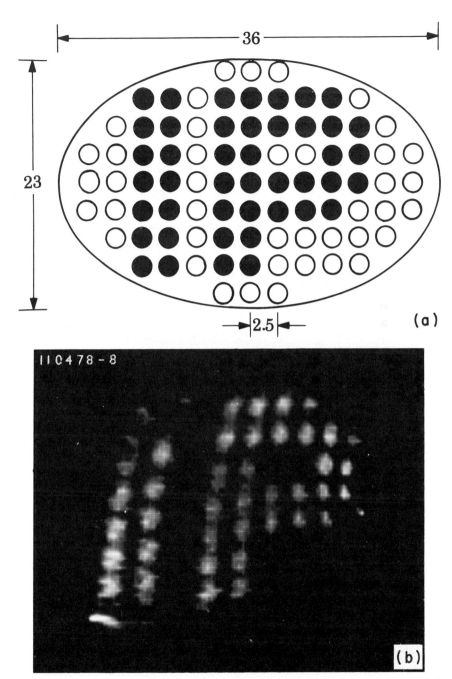

FIG. 6.17. (a) The "IP" test-tube matrix phantom. Shaded holes contained water-filled test tubes (dimensions in centimeters). (b) NMR image of the phantom illustrated in (a). [From P. G. Morris, P. Mansfield, I. L. Pykett, R. J. Ordidge, and R. E. Coupland, *IEEE Trans.* **26,** 2817 (1979).]

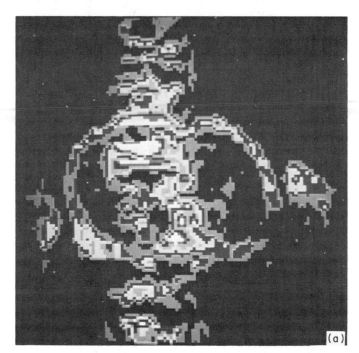

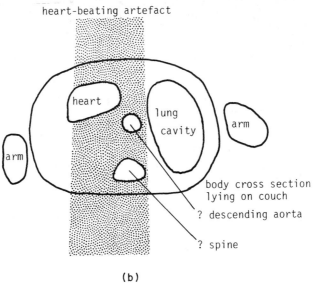

heart-beating artefact

heart

lung
cavity

arm

arm

body cross section
lying on couch

? descending aorta

? spine

(b)

Fig. 6.18. A human whole-body cross-sectional image taken with the Aberdeen NMR imaging machine. It shows a cross section through the lungs of one of the NMR team (J. M. S. Hutchison): The lungs are clearly seen, together with the heart, descending aorta, and the spine. The lines which run from top to bottom of the picture arise from the beating movement of the heart during image formation. (a) The actual NMR scan. (b) A diagram identifying the salient features of the scan. [From J. R. Mallard, private communication (1980). See also Mallard *et al.*[28]]

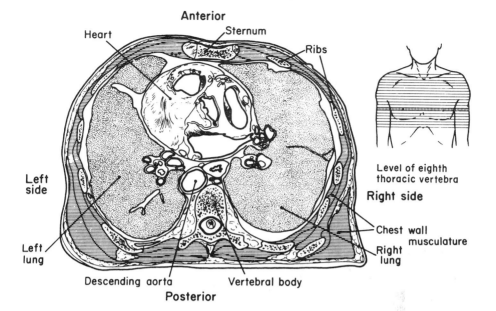

Anterior

Heart   Sternum

Ribs

Left side

Right side

Level of eighth thoracic vertebra

Chest wall musculature

Left lung

Right lung

Descending aorta   Vertebral body

Posterior

(a)

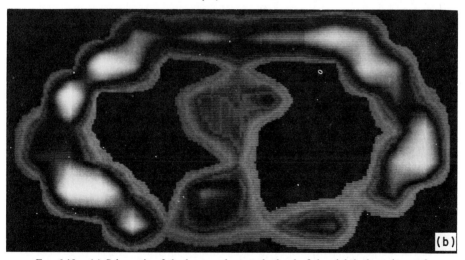

(b)

FIG. 6.19.   (a) Schematic of the human chest at the level of the eighth thoracic vertebra. (b) FONAR cross section of the live human chest at this level. Proton-signal intensity is coded with black assigned to zero-signal amplitude, white assigned to signals of strongest intensities, and intermediate gray scales assigned to intermediate intensities. The top of the image is the anterior boundary of the chest wall. The left area is the left side of the chest. Proceeding from anterior to posterior along the midline, the principal structure is the heart seen encroaching on the left lung field (black cavity) which is diminished in size relative to the right lung as it should be. More posteriorly and slightly left of the midline is a gray elliptical structure corresponding to the descending aorta. In the body wall, beginning at the sternum (anterior midline) and proceeding round the ellipse, the alternation of high intensity (white) with intermediate intensity (gray) could correspond to alternation of intercostal muscles (high intensity) with rib (low intensity) as shown in (a). The image is a black and white photograph of the original fourteen-color video display. [From R. Damadian, L. Minkoff, M. Goldsmith, and J. A. Koutcher, *Naturwissenschaft* **65**, 250 (1978).]

imaged by NMR. The resolution was estimated to be 6.3 mm in a slice of thickness 6.3 mm and the total imaging time was about 4.5 hr.

6.4.5 CONCLUSION

The imaging times quoted above, when compared with those for an optimal imaging method (Fig. 6.13), emphasize the inefficiencies of the line and point techniques. Thus, following the discussion of Section 5.3.1, a line-scan image will take longer than an optimally produced image of comparable resolution by a factor roughly equal to the number of lines scanned. Similarly, a point method will take longer by a factor equal to the total number of pixels. One could use the ratio of actual imaging times to times estimated on this basis as a measure of the efficiency with which an experimenter has implemented his particular technique.

Finally, we refer the reader to Chapter 7 for an NMR scan of the human head obtained by projection reconstruction. Detailed anatomical description of this and some discussion of Fig. 6.18 can also be found there.

6.5. Intermediate-Scale Imaging

6.5.1. IMAGING TIMES AND MOTION

The final column of Table 6.2 and the graphs of Fig. 6.14, which are of imaging times for an object size of 0.05 m at 30 MHz, indicate that it should take about five times as long to achieve the same fractional resolution as in the large-scale imaging case at 10 MHz. At 4 MHz, however, large-scale imaging is about five times slower. In both regimes it may therefore prove difficult to avoid motional artifacts, especially with experimental animals which have characteristic cardiac, peristaltic, and breathing times correspondingly shorter than those of humans. Nevertheless, the use of drugs and other invasive techniques will obviously be more acceptable in this regime and, coupled with the relative experimental simplicity, means that small animal studies will continue to be of medical value.

6.5.2. FRUITS AND VEGETABLES

The imaging of nonmobile objects can be achieved with extremely good resolution in quite modest imaging times: the lemon cross section of Fig. 6.20 being an excellent example of this. It was obtained with the multiple

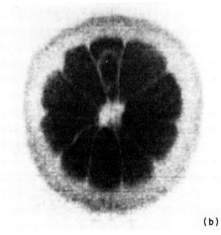

Fig. 6.20. Thin-section proton NMR image of an intact lemon: (a) cross section corresponding to the image shown in (b). [From E. R. Andrew, P. A. Bottomley, W. S. Hinshaw, G. N. Holland, W. S. Moore, C. Simaroj, and B. S. Worthington, *Proc. Congr. Ampere, 20th, Tallinn, U.S.S.R.* p. 53 (*1978*).]

sideband method at 30 MHz by Hinshaw and co-workers.[31,31a] The resolution was about 0.4 mm, the slice thickness 3 mm, and the scanning time 9 min. Note particularly that the seed case and the very fine septa have been clearly resolved. (In this image, black corresponds to high and white to low mobile-proton density.)

[31] W. S. Hinshaw, P. A. Bottomley, and G. N. Holland, *Nature (London)* **270**, 722 (1977).
[31a] E. R. Andrew, P. A. Bottomley, W. S. Hinshaw, G. N. Holland, W. S. Moore, C. Simaroj, and B. S. Worthington, *Proc. Congr. Ampere, 20th* p. 53 (1978).

Figure 6.21 is an NMR "shadowgraph" of one-half of a green pepper.[2]
It was obtained by reconstruction from 65 projections, each one the Fourier
transform of the average of 100 FID signals.

The imaging of fruits and vegetables has more value than its pure aesthetic
appeal; the ripening and decay processes are associated with changes in both
the amount and the state of the water, the latter being reflected through
variations in relaxation times. It may therefore be possible to use NMR
imaging as an aid to the selection of picking and marketing times and to the
optimization of storage conditions.

Food products may also be studied, for example, one can investigate the
seepage of water through pie cases. Such diffusion experiments may also be
of interest in the study of building materials such as bricks, concrete, and
wood, and the improvement of damp-proofing methods.

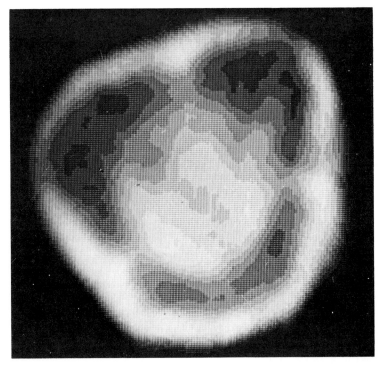

FIG. 6.21.   Proton NMR shadow zeugmatogram of one-half of a green pepper. [From P. C.
Lauterbur, *in* "NMR in Biology" (R. A. Dwek, I. D. Campbell, R. E. Richards, and R. J. P.
Williams, eds.), pp. 323–335. Academic Press, New York, 1977.]

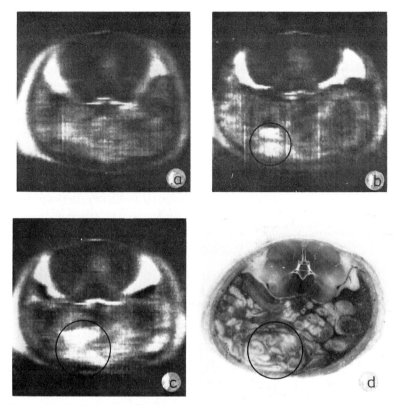

FIG. 6.22. NMR images of a thin transverse section through the abdomen of a Wistar rat taken *in vivo*: (a) at commencement of study; (b, c) after development of hepatoma D23 (circled) on days 6 and 7, respectively; (d) postmortem anatomical section corresponding to the NMR images. [From P. A. Bottomley, *Cancer Res.* **39**, 468 (1979).]

6.5.3. HUMAN LIMBS AND ANIMALS

Human limbs, if they can be held sufficiently steady, or clamped, can likewise be imaged with good accuracy. As an example, Fig. 7.3 shows an *in vivo* wrist cross section[31] obtained in similar time and with similar resolution to Fig. 6.20. Note the transition from maximum to minimum mobile-proton density over a single sensitive line at the left-hand edge of the central wrist bone (the lunate) emphasizing the excellent modulation transfer function of the system. For a full anatomical description, we again refer the reader to Chapter 7 (see also Hinshaw *et al.*[32]).

[32] W. S. Hinshaw, E. R. Andrew, P. A. Bottomley, G. N. Holland, W. S. Moore, and B. S. Worthington, *Br. J. Radiol.* **52**, 36 (1979).

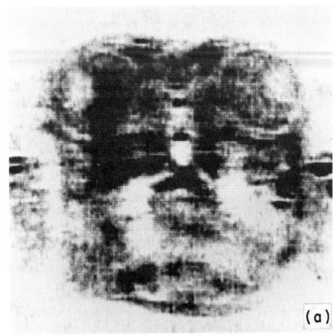

(a)

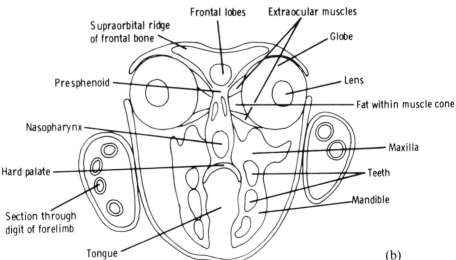

(b)

FIG. 6.23.   (a) Thin midorbital cross-sectional proton NMR image of the head of a rabbit. (Black corresponds to high proton density.) (b) Annotated diagram showing recognizable anatomical features. [From E. R. Andrew, P. A. Bottomley, W. S. Hinshaw, G. N. Holland, W. S. Moore, C. Simaroj, and B. S. Worthington, *Proc. Congr. Ampere, 20th, Tallinn, U.S.S.R.* p. 53 (*1978*).]

As we indicated above, motional artifacts may well present problems in the imaging of live animals. These are evident in the images of Fig. 6.22 which show the development of a D23 hepatoma in the abdomen of a live rat.[33] The gut region in particular has received strong motional averaging due to breathing and peristalsis. The occasional vertical striations arise as a result of information loss from individual sensitive lines due to sudden movement during the scanning of those particular lines. (The central horizontal artifact, however, is the consequence of an imbalance between the two channels of the quadrature detector).

One way in which to avoid such problems is to sacrifice the animal prior to imaging, as was the case for the rabbit in Fig. 6.23.[31a,34] One then, of course, faces the uncertainty with regard to possible changes in the NMR relaxation times and other parameters due to the rigor process. However, these changes may themselves become the subject of study in forensic medicine, as well as within the meat-producing industries where improvements of quality reap large financial benefits. For instance, approximately one in seven pigs slaughtered produce a very poor quality, pale, soft exudative meat which might be detectable by NMR. Imaging work with animals could also be of value to the pharmaceutical industry in the study of drug efficacy.

6.6. Small-Scale Imaging

6.6.1. IMAGING TIME

Table 6.2 indicates that even at 150 MHz, one loses a further factor of two in imaging speed, and therefore the introductory comments of Section 6.5 apply here *a fortiori*. Figure 6.15 illustrates the time resolution dependence at 150 MHz for a coil of radius 0.01 m. Unlike the situation in the previous two regimes, however, small-scale imagers have not generally taken full advantage of the gain in sensitivity which can be achieved by working at or near the maximum frequency of Table 6.2. The reasons for this are largely historical in that the first imagers demonstrated their technique with the equipment they happened to have available before moving "up-scale" toward human studies and building their equipment for that purpose.

[33] P. A. Bottomley, *Cancer Res.* **39**, 468 (1979).
[34] W. S. Hinshaw, E. R. Andrew, P. A. Bottomley, G. N. Holland, W. S. Moore, and B. S. Worthington, *Br. J. Radiol.* **51**, 273 (1978).

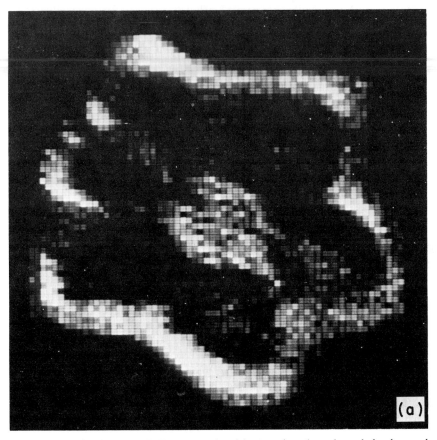

FIG. 6.24.   (a) Line-scan NMR cross-sectional image taken through a whole okra seed pod.

6.6.2. FINGERS

Figure 7.1 shows a cross-sectional image through the midphalanx of a human finger.[35] This was the first demonstration of clearly resolved human anatomy *in vivo* and was obtained by a selective irradiative line-scanning technique at 15 MHz. The in-plane resolution was 0.25 mm and the slice thickness about 8 mm. Each line was averaged 48 times with a delay of 0.5 sec giving a total imaging time of 23 min.

6.6.3. PLANTS

Figure 6.24 shows a cross section through an okra seed pod obtained under similar experimental conditions[36] to Fig. 7.1. In this case, 100 averages were performed with a delay of 0.73 sec giving a total imaging time of 78 min.

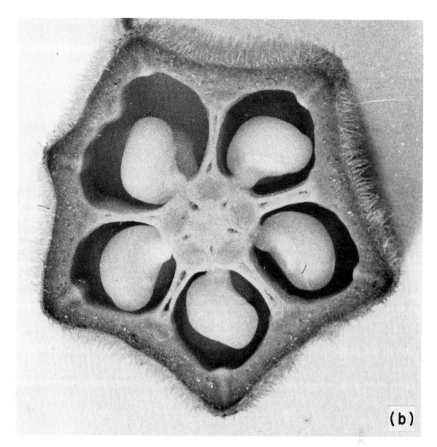

FIG. 6.24. (b) Cross-sectional photograph taken through the same region of the pod after the NMR scan. [From P. Mansfield and I. L. Pykett, *J. Magn. Reson.* **29**, 355 (1978).]

Some intensity distortions are apparent in these images. They are believed to be a result of the phase shifts caused by gradient-switching transients induced in the adjacent pole faces of a conventional electromagnet. The long imaging times for these relatively thick slices indicate the penalties paid for working at too low a frequency.

Figure 6.25 shows an early cross-sectional image through a spring onion obtained with the sensitive-point method at 60 MHz.[37] The concentric rings are clearly visible and the resolution is 0.3 mm in a plane of thickness ~0.3 mm. Even though the frequency approaches the optimum, the inefficiency of the single-point technique required an imaging time of about

[35] P. Mansfield and A. A. Maudsley, *Br. J. Radiol.* **50**, 188 (1977).

[36] P. Mansfield and I. L. Pykett, *J. Magn. Reson.* **29**, 355 (1978).

[37] W. S. Hinshaw, *J. Appl. Phys.* **47**, 3709 (1976).

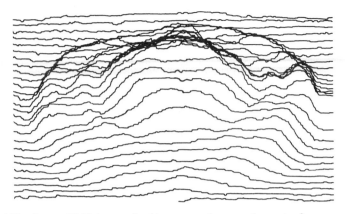

FIG. 6.25. Proton NMR image of a thin cross section near the roots of a reasonably fresh spring onion. The apparent height of the image is proportional to the density of free water at the corresponding point in the sample. The outer rings and wet central region are clearly visible. [From W. S. Hinshaw, *J. Appl. Phys.* **47**, 3709 (1976).]

2 hr. Nevertheless, this image serves to demonstrate the possible study of water in plant tissues; others are available in the literature.[38,39] There are, for instance, a number of problems associated with water transport in plants which might be usefully investigated.

6.6.4. SEEDS

The study of seeds such as wheat is of major importance, since milling and other processes are critically dependent on the stage of development and, hence, on the water content. NMR imaging might therefore be useful in this context. The effect of water content on the storage of grain is also of crucial importance. It is, for instance, responsible for the cracking of rice grains and leads to a huge annual wastage of this commodity.

The study of oil-bearing seeds is another possible application of NMR imaging. Conventional NMR has already had great success in selecting those specimens which have high oil-bearing characteristics.[40] As an illustration, Fig. 6.26 is a cross-sectional image showing the oil distribution in a pecan nut obtained with a continuous-wave projection reconstruction technique at 9.13 MHz.[41] The reconstruction was from 12 projections giving a resolution

[38] P. Mansfield, A. A. Maudsley, and T. Baines, *J. Phys. E Sci. Instrum.* **9**, 271 (1976).

[39] P. Mansfield and A. A. Maudsley, *Phys. Med. Biol.* **21**, 847 (1976).

[40] H. Weisser, *Bruker Rep.* **1**, 9 (1978).

[41] P. C. Lauterbur, *Pure Appl. Chem.* **40**, 149 (1974).

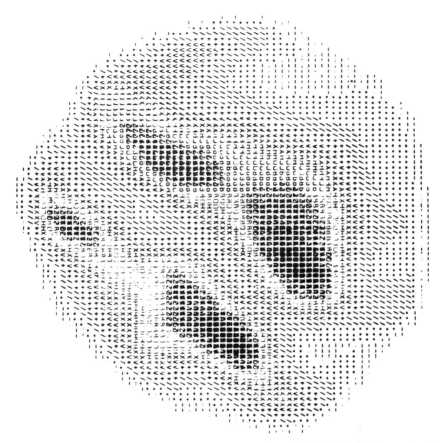

FIG. 6.26. Proton NMR shadow zeugmatogram of the oil distribution within an intact pecan nut (*Carya illinoensis*). [From P. C. Lauterbur, *Pure Appl. Chem.* **40**, 149 (1974).]

of about 0.2 mm. Again, the nonideal conditions lead to a long scan time of 20 min.

6.6.5. OTHER APPLICATIONS

Water plays an important role in the food industry. It would seem, therefore, most likely that NMR in general and NMR imaging in particular will make further major impacts in food science and technology. Finally, we mention that one could image perfused organs to obtain useful metabolic information[42] if it proved impossible to achieve sufficient sensitivity with the organ *in situ*.

[42] D. G. Gadian, *Contemp. Phys.* **18**, 351 (1977).

6.7. Microscopic-Scale Imaging

Figure 6.16 illustrates that even with the highest frequencies currently available, it will be difficult to achieve resolutions of better than a few microns in reasonable times. Nevertheless, it may be possible to perform useful histological studies and, as recent experience with the acoustic microscope[43] has indicated, such images may differ profoundly from their optical counterparts, adding a new dimension to the study of such systems.

Undoubtedly one of the major problems to be encountered with such experiments will arise from the distribution of anisotropic magnetic susceptibilities which, at frequencies of 600 MHz, may give rise to linewidths of the order of 1 kHz. Nevertheless, we feel that the pursuit of such a microscope is a worthwhile aim and will provide a useful complement to optical microscopy.

[43] R. A. Lemons and C. F. Quate, *Appl. Phys. Lett.* **24,** 163 (1974).

7. Potential Use in Medicine

7.1. Introduction

In this chapter we wish to review some of the results already achieved by NMR imaging on human subjects and discuss their relevance and potential as clinical diagnostic aids. First, however, the results of NMR imaging should be seen in context with the range of other imaging modalities. It is doubtful, for example, that NMR imaging will ever be envisioned as a supplementary, let alone replacement, for normal X-ray shadowgraph techniques.

Since their discovery by Röntgen in 1890, X-ray shadowgraphs have found an increasingly important role in clinical diagnosis for a whole range of disorders. More recently, with the introduction of computerized tomographic (CT) X-ray techniques[1,2] by EMI in 1972, the range of usefulness and therefore value of X-ray methods has been immensely increased. Of course, X rays are one form of ionizing radiation, and in using them the clinician has always to weigh the risk of damage against the potential benefit to the patient in a particular situation. The hazard inherent in the use of X rays[3] has led people to consider other forms of imaging to which we refer later. However, before leaving the subject it should be pointed out that apart from morphological considerations, X-ray methods, CT or otherwise, find increasing use in the study of physiology. Since X rays reflect the variations of electron density in biological tissue, physiological function is studied by the invasive introduction of radio-opaque dyes or contrast media, which, it should be remembered, carry their own problems of allergic side effects in some patients. In spite of these shortcomings, it must be emphasized that

---

[1] G. N. Hounsfield, *Br. J. Radiol.* **46,** 1016 (1973).
[2] J. Ambrose, *Br. J. Radiol.* **46,** 1023 (1973).
[3] B. J. Culliton, *Science* **193,** 555 (1976).

the quality of current CT X-ray images has improved beyond all expectation. Both the speed and resolution of these images make it somewhat unlikely that NMR imaging will ever be regarded as a true competitor,[3a] although NMR may well find a role eventually as an initial screening method. However, our view is that NMR imaging should not be regarded as a competitive imaging technique, but rather a supplementary technique, for, as we shall discuss in some detail later, NMR measures quite different parameters.

Radioisotope scanning,[4] though inherently invasive, is now widely used for the study of physiology. Recent important advances in this area now make it possible to study cardiac motions in real time. Although most of these latter uses rely on image projections of the heart, valuable information can nevertheless be obtained on ejection fraction and edge-enhanced pictures are able to reveal abnormalities in the ventricular motion symptomatic of myocardial ischemia. However, as with X-ray techniques, there is the double hazard of the ionizing γ-ray emitter, although used in very small doses, and the possible allergic effect of the radioisotopic dye. This latter effect is of course very small. It is fair to say, that despite the relative crudity of all radioisotope images, the method has considerable clinical value and is used widely as a diagnostic aid.

A third imaging method which has emerged as an extremely important diagnostic technique, is of course, ultrasonic imaging.[5,6] This technique, which at low sonic powers is considered hazardless, quickly found initial application in live-fetal scanning.

The applications have broadened considerably and now include tumor detection in the liver, kidney, and breast, and more recently, echocardiography. Ultrasonic images can be produced quickly and modern imaging equipment produces quite good quality pictures at relatively low cost. The cardiographic work is produced in real time and in many cases does not produce a full image, but a cross-sectional slice through the heart, which may be used together with blood pressure (invasive) or the electrocardiograph (ECG) r wave, to study cardiac ejection fraction and ventricular-wall motion abnormalities. Real-time phased-array techniques have also improved fetal scanning to the point where the cranial midline of the fetus can be clearly seen and is a valuable indicator of possible abnormality due to hydrocephalus. The fetal heart can also be seen beating and in some cases the ventricular septum may be discerned.

[3a] See Notes Added in Proof, Note 7.1.
[4] T. F. Budinger, in "Recent Advances in Nuclear Medicine" (J. H. Laurence, ed.), Ch. 2. Grune & Stratton, New York, 1974
[5] P. N. T. Wells, ed., "Ultrasonics in Clinical Imaging." Churchill, London, 1972.
[6] C. R. Hill, *J. Phys. E Sci. Instrum.* **E9**, 153 (1976).

Another potentially hazardous imaging method not so widely used at the moment is based on positron emission. This involves invasive inhalation of specially prepared short-lived radioactive isotopes, usually of oxygen or nitrogen. Because of the short half-life, these isotopes must be produced on site and this requires a cyclotron. Quite apart from the hazard of the γ rays which accompany the positron emission and which, of course, are detected and used to form the image, there is the additional risk to the operator and possibly the patient of neutron emission during irradiation in the cyclotron if the apparatus is not properly screened.

All the methods discussed have associated with them an intrinsic hazard and/or limitation. In the case of ultrasonic imaging it is very difficult to obtain images in parts of the anatomy where there are large tissue-density differences. For example, bony regions like the adult head. Equally, it is difficult to obtain images through tissue/air or gas interfaces, for example, the thorax and the abdomen. Thus for cardiac imaging it is necessary to launch the ultrasonic wave through a natural "window" above the sternum, roughly along the line of the aorta, or alternatively from below the rib cage.

Set then within the context of this existing formidable range of valuable diagnostic imaging modalities,[6a] we hope in this chapter to demonstrate the potential value of NMR imaging from the standpoints of the morphology, physiology, and pathology of human subjects.

7.2. Morphology

A variety of imaging techniques have been used to produce the pictures discussed here. The merits or otherwise of the various techniques used will not be emphasized here. Full discussion and comparison of all the techniques referred to is given in Chapters 4 and 5, respectively.

The first NMR images to reveal live human anatomy were produced by the line-scanning technique.[7,8] The region scanned was a cross section through a finger. Figure 7.1a and b shows the images obtained by a slow

---

[6a] A valuable up-to-date collection of articles on the various imaging modalities briefly touched upon here is contained in the proceedings of an international symposium "Towards Safer Cardiac Surgery" (D. B. Longmore, ed.). M. T. P. Press Ltd., Lancaster, U. K., 1980.

[7] P. Mansfield and A. A. Maudsley, *Proc. Congr. Ampere, 19th* (H. Brunner, K. H. Hauser, and D. Schweitzer, eds.). Groupment Ampere, Heidelberg, 1976.

[8] P. Mansfield and A. A. Maudsley, *Br. J. Radiol.* **50,** 188 (1977).

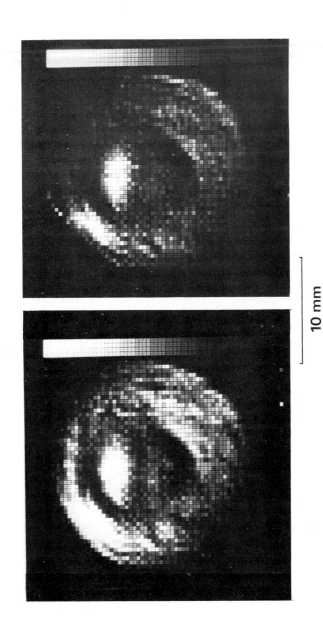

(a)   (b)

10 mm

Fig. 7.1. Cross-sectional NMR images of a live finger obtained *in vivo* (see text for full details): delay time (a) τ = 0.5 sec, and (b) τ = 0.3 sec. [From P. Mansfield and A. A. Maudsley, *Proc. Congr. Ampere, 19th, Heidelberg* p. 247 (1976); and *Br. J. Radiol.* **50**, 188 (1977).]

scan and a faster scan, respectively, through the midphalanx of the third digit (see Section 4.4.2 for experimental details). Bright zones correspond to high mobile-proton density. Figure 7.2 is an annotated tracing, taken from Fig. 7.1a, indicating the recognizable anatomical features. Apart from the general outline of the finger cross section, a most striking feature is the upper dark ring of the bone surrounding the brighter bone-marrow region. Immediately below this is a second dark circular structure corresponding to the flexor tendon sheath which surrounds the flexor profundus and flexor sublimis tendons. Also visible above the bone, toward the dorsal surface, is a dark region corresponding to the thinner extensor tendon. Around the perimeter of the finger can be seen, here and there, the fascial region between skin tissue and the subcutaneous areolar supportive tissue matrix. Within this supportive tissue and to the left of the bone and tendon sheath is one branch of the digital artery and digital nerve. There is a hint of the other branches to the right, as indicated in Fig. 7.2.

The scan delay time for line averaging was 0.5 sec for Fig. 7.1a, so that most of the spins were close to the fully relaxed equilibrium magnetization in this case. In Fig. 7.1b the same section was scanned with the delay time shortened to 0.3 sec. A general reduction of picture intensity is seen since much of the specimen is clearly characterized by a $T_1(x, y, z)$ greater than 0.3 sec. Significantly, the bone-marrow region remains bright indicating a zone of short relaxation time. In an adult, the phalanges contain essentially yellow bone

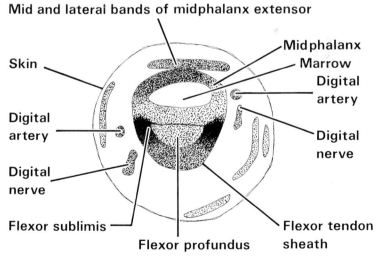

Mid and lateral bands of midphalanx extensor

Skin

Midphalanx

Marrow

Digital artery

Digital artery

Digital nerve

Digital nerve

Flexor sublimis

Flexor profundus

Flexor tendon sheath

FIG. 7.2. A labeled tracing taken from Fig. 7.1a showing the recognizable anatomical features. [From P. Mansfield and A. A. Maudsley, *Proc. Congr. Ampere, 19th, Heidelberg* p. 247 (1976); and *Br. J. Radiol.* **50,** 188 (1977).]

marrow, which is largely fatty tissue. Adipose or fatty tissue is known from other work[9] to possess short relaxation times of around 100–200 msec, (see Section 2.3, for example), and this is consistent with the contrast observed in Fig. 7.1b. The imaging times for Figs. 7.1a and b were 23 and 15 min, respectively at a frequency of 15.0 MHz.

We shall return to T_1 variations again, since it is illustrative of the way in which tissues of similar mobile-proton content may be usefully differentiated. We hasten to add, however, that unique tissue typing based on the two parameters $T_1(x, y, z)$ and proton density, does not appear to be feasible at the present time. Additional parameters, such as $T_2(x, y, z)$ diffusion constant and chemical shift, may of course change the situation.

Typical anatomical detail obtained from another line-scanning technique, the multiple sensitive-point method is shown in Fig. 7.3. Figure 7.3a is a thin section scan through a live human wrist.[10] The plane thickness was approximately 0.3 cm and the scan time 9 min at a frequency of 30.0 MHz. Figure 7.3b is an annotated cross-sectional diagram of a typical wrist section through the carpal bones. Comparison of image and sketch shows the wealth of morphological detail that can easily be distinguished, even by the untrained eye.

With the palmar aspect uppermost in the image, the five carpal bones, pisiform, triquetrium, capitate, lunate, and scaphoid, from left to right, all show strong signals from the bone-marrow regions, the bone tissue itself producing virtually no signal. (N.B.: In this image, black corresponds to high signal which is a function of both mobile-proton density and the factor $T_2/(T_1 + T_2)$, see Section 4.4.1.4.) Again, as with the finger section, the tendons and sheaths appear as low-signal regions outlined by stronger signals from the muscle and other tissue rich in mobile protons. Interestingly, in this image the radial artery to the right of the scaphoid bone is clearly resolved as a region of low-image intensity. The ulnar artery and also the nerve are visible just above and to the right of the pisiform bone. A number of other scans through various sections of the hand, wrist, and forearm have also been reported.[11,12]

As with Fig. 7.1, the wrist image is not a pure mobile-proton-density map, but a composite of density $T_1(x, y, z)$ and $T_2(x, y, z)$ effects. No attempt was made to distinguish the various contributions in this case. We take the view

[9] D. P. Hollis, L. A. Saryan, J. C. Eggleston, and H. P. Morris, *J. Natl. Cancer Inst.* **54,** 1469 (1975).

[10] W. S. Hinshaw, P. A. Bottomley, and G. N. Holland, *Nature (London)* **270,** 722 (1977).

[11] W. S. Hinshaw, E. R. Andrew, P. A. Bottomley, G. N. Holland, W. S. Moore, and B. S. Worthington, *Neuroradiology* **16,** 607 (1978).

[12] W. S. Hinshaw, E. R. Andrew, P. A. Bottomley, G. N. Holland, W. S. Moore, and B. S. Worthington, *Br. J. Radiol.* **52,** 36 (1979).

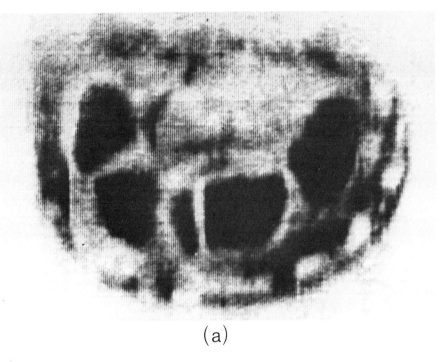

(a)

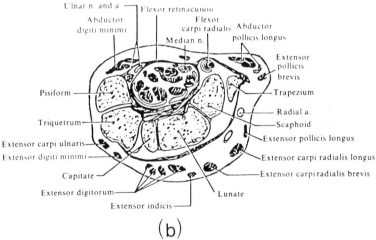

(b)

Fig. 7.3.   (a) An NMR image of the distribution of mobile protons in a thin transverse section through the left wrist of a live subject. The view is into the arm with palmar surface uppermost and was taken through the distal tip of the anterior horn of the lunate bone. Dark zones correspond to high concentrations of mobile protons as in the bone marrow of the carpals and also subcutaneous fat. Light zones correspond to regions of low mobile-proton density. (b) An annotated diagram of a typical anatomical specimen cross section through a cadaver wrist. (N.B.: All labeled features are not observable in the NMR image.) [From W. S. Hinshaw, P. A. Bottomley, and G. N. Holland, *Nature* (*London*) **270**, 722 (1977).]

that such parameters should be extricated, and indeed, some efforts have already been made which we discuss later. Of course, such admixtures of NMR parameters enhance picture contrast in some cases and may thus produce useful images which are of real clinical value. Unfortunately, different techniques will depend on the NMR parameters in different ways, and this could make standardization and interpretation of NMR images rather difficult for future radiologists. It is our view, therefore, that eventually it will be a prerequisite for all NMR imaging machines to measure unambiguously, at least, density $\rho(x, y, z)$, $T_1(x, y, z)$, and possibly $T_2(x, y, z)$.

As already discussed in Chapter 6, the first whole-body thoracic image was obtained by Damadian et al.[13] and is reproduced in Fig. 6.19. Pictures of improved quality have since been obtained and compared with CT images.[14] An NMR image of the thorax has also been obtained by Mallard et al.[15] and is shown in Fig. 6.18a; Fig. 6.18b is their annotated interpretation. The image was obtained at a frequency of 1.7 MHz by a modified line-scan selective irradiative procedure using switched field gradients, (see Section 4.4.2.4). Motional artifacts attributed by the authors to the heart beat are present, but fortunately fall outside the thorax thus making the interpretation easier in this case. It should be remembered, however, that intensity artifacts outside a picture correspond to loss of signal and, hence, information within the image.

The general outline of the chest is clearly seen with an arm visible on each side. The anterior aspect is uppermost. Within the image, the pleural cavities show as the two dark regions to each side of the bright central zone. Anterior and to the left of the zone is the heart mass. Posterior and slightly to the right is a triangular region which could be the vertebral column. Anterior to the vertebra is a small oval region which could be the aorta (left) and possibly oesophagus (right). The slice thickness is believed to be about 1.0 cm.

A more recent thoracic image by the same group[16] is shown in Fig. 7.4. This image was produced using a minor modification of the Fourier zeugmatography technique of Kumar et al. fully described in Section 4.5.1. Figure 7.4a is the thoracic section proton-density image. Figure 7.4b is a T_1 map of the same region, and Fig. 7.4c is an annotated diagram indicating the clearly assignable features of the image. Note that, in general, there is more detail revealed in the T_1 map than in the proton-density image. In-

[13] R. Damadian, M. Goldsmith, and L. Minkoff, *Physiol. Chem. Phys.* **9,** 97 (1977).
[14] C. L. Partain, M. E. James, J. T. Watson, R. R. Price, C. M. Coulam, and F. D. Rollo, *Radiology* **136,** 767 (1980).
[15] J. Mallard, J. M. S. Hutchison, W. A. Edelstein, C. R. Ling, M. A. Foster, and G. Johnson, *Phil. Trans. R. Soc. London Ser. B* **289,** 519 (1980); also private communication (1980).
[16] W. A. Edelstein, J. M. S. Hutchison, G. Johnson, and T. Redpath, *Phys. Med. Biol.* **25,** 751 (1980).

terestingly, despite the 2-min total imaging time for both pictures, and the consequent cardiac motions, it would seem that clear delineation of the ventricular cavities and septum is possible. The slice thickness, which is approximately 2.0 cm, is achieved by a focused selective irradiative technique.

Other parts of the human anatomy have also been scanned by NMR. Figure 7.5 shows the first line-scan image through the abdomen at the L2–3 level.[17–19] This scan took 40 min and corresponds to a slice thickness of 4.0 cm. The image was produced at 4.0 MHz. The original number of data points was 90^2, which has been interpolated to 360^2. The picture data depth is 8 bits and this is displayed on both a single- and double-cycle gray scale using an Optronix photowriter. In Fig. 7.5a, the full data range corresponds to about 150 levels and this is displayed on a 0–255 gray-scale range. In Fig. 7.5b the data are displayed using two cycles of the same 128-level gray scale. Data levels 0–127 correspond to gray levels black through to white. Data level 128 is reset to black and higher levels to corresponding grays in the second cycle. This procedure is useful, since it artificially enhances the picture contrast, especially in the high-density range, where it is especially difficult to differentiate adjacent levels as photo prints, even though the film negative may contain differentiable levels. Figure 7.5c is an annotated picture corresponding to the uninterpolated version of Fig. 7.5a, and indicates many of the morphological details that can be seen in the image. A difficulty of interpretation and clear assignment arises through motional artifacts caused mainly through breathing in this case. The subject was scanned while standing. The considerable abdominal wall motion which occurs in this position is thought to be the cause of signal loss in the anterior region (top) of the image, though some intestine is visible. The arrow indicates the midline posteriorly. The liver mass is seen as the bright zone to the right of both pictures (Figs. 7.5a and b. The dark zone within the liver mass (Fig. 7.5b) is indicative of the exceptionally high signals arising from this region. The contrast enhancement of Fig. 7.5b allows the gall bladder, spleen, pancreas, ascending colon, and right kidney to be more clearly discerned. The unusual shape of the right kidney may also be ascribed to motion, since during the imaging process, it will have undergone a complex periodic gyration causing enhancement of the displacement extrema, where the kidney is moving more slowly.

[17] P. Mansfield, I. L. Pykett, P. G. Morris, and R. E. Coupland, *Br. J. Radiol.* **51**, 921 (1978).
[18] P. G. Morris, P. Mansfield, I. L. Pykett, R. J. Ordidge, and R. E. Coupland, *IEEE Trans. Nucl. Sci.* **26**, 2817 (1979).
[19] I. L. Pykett, P. Mansfield, P. G. Morris, R. J. Ordidge, and V. Bangert, "Imaging Processes and Coherence in Physics," p. 453. Workshop Proc., Centre de Physique, Les Houches, France, 1979 (M. Schlender, M. Fink, J. P. Goedgebuer, C. Mulgrange, J. Ch. Vienot, and R. H. Wade, eds.). Springer-Verlag, Berlin and New York, 1980.

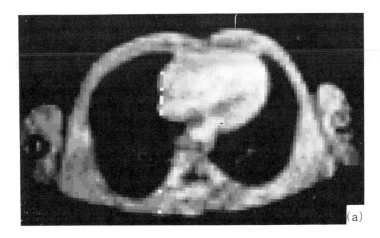

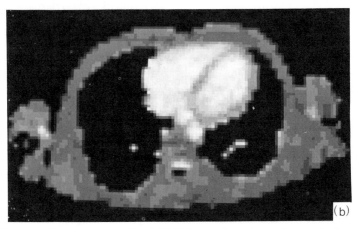

FIG. 7.4.   Live human cross-sectional NMR image through the thorax, obtained at 1.7 MHz. The image comprises 64^2 pixels of 1-cm^3 volume. The slice thickness was 18.5 mm. Imaging time to obtain both proton density and relaxation time was 2.0 min: (a) proton-density image; (b) T_1 map of same cross section as in (a); (c) annotated diagram giving morphological assignments. [From W. A. Edelstein, J. M. S. Hutchison, G. Johnson, and T. Redpath, *Phys. Med. Biol.* **25,** 751 (1980).]

The vertebral column in this scan does not show the contrast that one might expect from bone surrounded by muscle. This is thought to be due to the slice thickness of 4.0 cm, which means that probably one intervertebral joint is included in this scan. Cartilage and nucleus pulposus in the joint will effectively alter the mean contrast. Nevertheless, the general outline of the vertebra can be discerned. More interesting is the clear dark ring of the

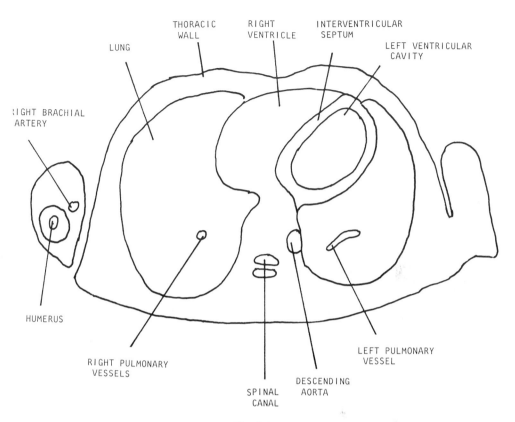

THORACIC   RIGHT   INTERVENTRICULAR
WALL   VENTRICLE   SEPTUM

LUNG   LEFT VENTRICULAR
                                                  CAVITY

RIGHT BRACHIAL
ARTERY

HUMERUS

                                                 LEFT PULMONARY
                                                 VESSEL
RIGHT PULMONARY
VESSELS
                                     DESCENDING
                      SPINAL   AORTA
                      CANAL

FIG. 7.4c

aorta anterior and to the left of the vertebra. A region of lower intensity adjacent and to the right of the aorta seems likely to be the vena cava.

One of our more recent whole-body NMR scans[20] is shown in Fig. 7.6 and corresponds to a 2.0-cm cross-sectional slice through the abdomen at the level L1–2. This image was obtained by a filtered back-projection method and took 9.5 min to produce. Figure 7.6a is the NMR scan in which bright zones correspond in general to high mobile-proton densities. Figure 7.6b is an annotated line tracing from Fig. 7.6a indicating the morphological detail visible. The cross-sectional view is from below (the accepted radiological convention for the body). Clear delineations of the liver (L), gall bladder (GB), kidneys (K), vertebra (V), ribs (R), and some gas pockets (G) in the stomach and intestines are indicated. More significant is the clear

[20] V. Bangert, P. Mansfield, and R. E. Coupland, *Br. J. Radiol.* **54,** 152 (1981); also V. Bangert, P. Mansfield, and R. E. Coupland, *Proc. ISMAR-Ampere Conf., 22nd, Delft* (1980).

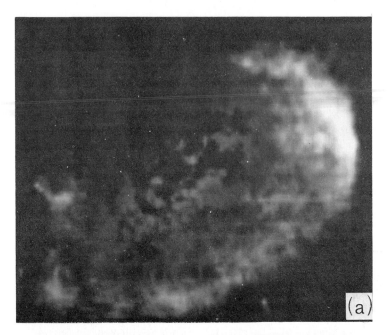

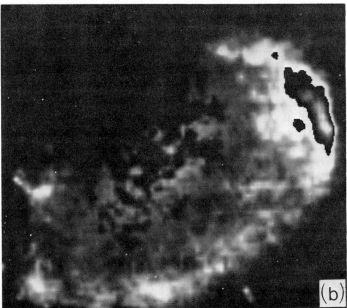

FIG. 7.5. An early live, cross-sectional line-scan NMR image through the abdomen at L2–3. Arrow indicates the midline posterior. Left side of subject lies to left of the illustration. Bright zones correspond in general to high mobile-proton content: (a) single-cycle photowriter image comprising 360^2 points; (b) double-cycle display as in (a); (c) annotated image from (a) and (b).

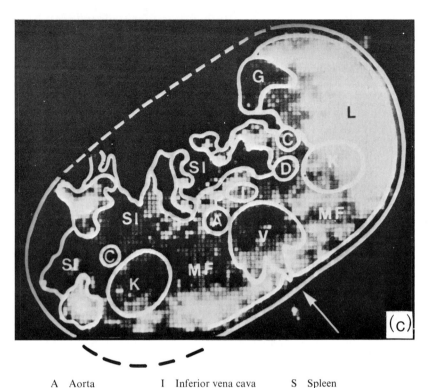

A	Aorta	I	Inferior vena cava	S	Spleen
C	Colon	K	Kidneys	SI	Stomach
D	Duodenum	L	Liver		and intestines
G	Gall bladder	P	Pancreas	V	Vertebra

Abdominal muscles and retroperitoneal fat (MF) are seen adjacent to the vertebra. [From P. Mansfield, I. L. Pykett, P. G. Morris, and R. E. Coupland, *Br. J. Radiol.* **51**, 921 (1978).]

indication of the inferior vena cava (ic), aorta (a), superior mesenteric artery (sa), and vein (sv). For other details, see the figure caption.

Figure 7.7 is an NMR projection shadowgraph of a live hand,[21] and is in striking contrast to that expected from a corresponding X-ray picture. This serves to emphasize the complementary character of soft- and hard-tissue imaging in the cases of NMR and X-ray, respectively. Only slight density changes can be discerned corresponding to the metacarpals within the palmar region. Changes in intensity corresponding to the digital phalangeal joints can just be seen, although the signals from the fingers are rather weak. The soft-tissue mass of the palmar region shows particularly well and suggests

[21] P. Mansfield, P. G. Morris, R. J. Ordidge, I. L. Pykett, V. Bangert, and R. E. Coupland, (to be published).

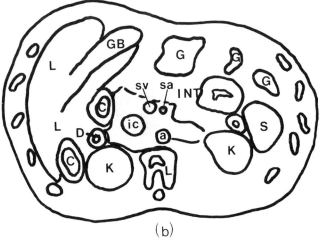

FIG. 7.6. A more recent abdominal NMR scan produced by the method of filtered back projection. Slice thickness is approximately 2.0 cm and corresponds to level L1–2. (a) Interpolated image (360^2 array); bright zones correspond in general to high mobile-proton content (e.g., water and fat). (b) Annotated tracing from (a) above:

C	Colon	R	Rib
D	Duodenum	S	Spleen
G	Gas in stomach	V	Vertebra
GB	Gall bladder	a	Aorta
INT	Intestines	ic	Inferior vena cava
RK, LK	Right and left kidneys	sa	Superior mesenteric
L	Liver		artery
P	Pancreas	sv	Superior mesenteric
			vein

[From V. Bangert, P. Mansfield, and R. E. Coupland, *Proc. ISMAR-Ampere Conf., 22nd, Delft* (1980); also *Br. J. Radiol.* **54**, 152 (1981).]

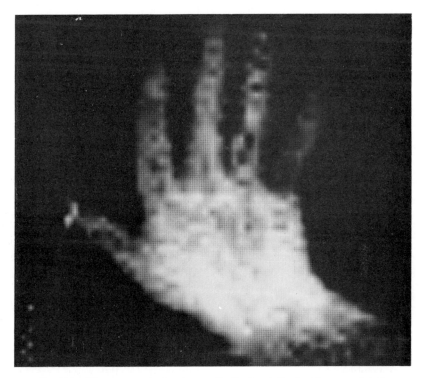

Fig. 7.7.   Line-scan projection image of a live hand. See text for more details.

that this type of image might be of value in the study of disorders in the vascular system, *vide infra*. This image took 15 min to scan at 4.0 MHz.

Finally, we turn to live images of the head. Figure 7.8 shows the first NMR image of the head[22] obtained in 1978 in an experimental machine developed at Thorn-EMI. Although not stated, it seems obvious from the radial streak artifacts, that some type of projection reconstruction method was used to produce this image (see Section 4.5.4). The imaging time was 6.0 min at 4.0 MHz with a slice thickness of 1.0 cm. In spite of some spatial distortion, this head scan through the orbits shows considerable recognizable ana-tomical detail. The eyes and eye sockets are clearly seen at the top of the picture. Part of the brain midline is clearly visible, together with the anterior horns of the lateral ventricle. The cerebrospinal fluid within the ventricle shows dark, indicating a longer relaxation time than that of the white

[22] H. Clow, and I. Young, EMI Central Research Laboratories. Images presented at the British Institute of Radiology Meeting, Nov. 1978. *Br. J. Radiol.* **52,** 680 (1979); see also *New Sci.* **80,** 588 (1978).

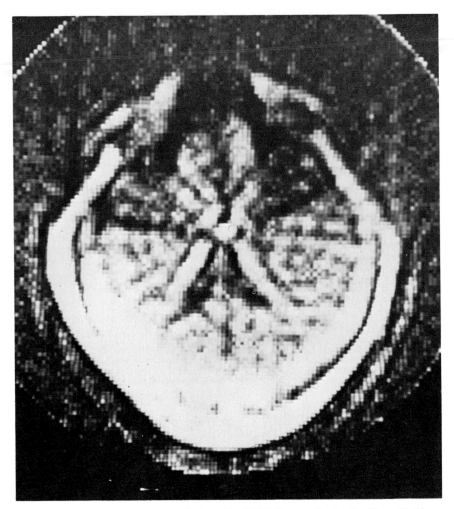

FIG. 7.8.   NMR head scan through the orbits (1978). See text for details. [From H. Clow and I. Young; see also *New Sci.* **80**, 588 (1978).]

and gray nervous tissue. Some indicaton of the convolutions is also apparent. The cranial bone appears by virtue of its low water content. The outer skin layer serves to define the periphery of the head. Other studies of the head by both line scanning and projection reconstruction have recently been reported.[23-26] The scan times are around 2.0 min and the slice thick-

[23] G. N. Holland, W. S. Moore, and R. C. Hawkes. *J. Comput. Assist. Tomogr.* **4**, 1 (1980).
[24] G. N. Holland, W. S. Moore, and R. C. Hawkes, *Br. J. Radiol.* **53**, 253 (1980).

ness 1.0 cm. Examples of these and other very high-quality scans[22] have not been made available for reproduction in this book. (See Notes Added in Proof, Note 7.2.)

7.3. Physiology

7.3.1. INTRODUCTION

Although the initial impact of NMR imaging has so far been made mainly through its ability to resolve anatomical detail, we consider that a major role will be realized in physiological applications. In this section we shall discuss potential areas of application including the vascular and digestive systems, cardiophysiology, flow, and metabolism.

7.3.2. VASCULAR STUDIES

Cross-sectional images, or as we have seen, projection shadowgraphs (Fig. 7.7) could well turn out to be of value in the study of circulatory defects, which, if allowed to progress could lead to ischemia of localized tissue regions in various organs of the body.

In muscle tissue, ischemia   is usually associated with increased water content[27-29] and increased T_1, although the latter may not generally be the case.

In addition to relying on the naturally occurring water content and T_1 differences, it would seem logical and prudent to consider the use of contrast agents, administered in much the same way as those used in X-ray imaging. In X-ray studies such commercial agents as Conray comprise a water-soluble solution of a heavily iodated organic molecule designed to produce strong X-ray scattering. Some types are tissue or organ selective. These so-called radio-opaque dyes are not paramagnetic and therefore have little influence on T_1 in NMR experiments. Since NMR images are to some extent complementary to X-ray images, the equivalent contrast agent for proton density

---

[25] W. S. Moore, G. N. Holland, and L. Kreel, *J. Comput. Tomogr.* **4,** 1 (1980).

[26] G. N. Holland, R. C. Hawkes, and W. S. Moore, *J. Comput. Assist. Tomogr.* **4,** 429 (1980); also R. C. Hawkes, G. N. Holland, W. S. Moore, and B. S. Worthington, *J. Comput. Assist. Tomogr.* **4,** 577 (1980).

[27] P. C. Lauterbur, J. A. Frank, M. J. Jacobson, *Dig. Int. Conf., 4th, Med. Phys.* Abstract 33.9 (1976).

[28] P. C. Lauterbur, *Conf. Comp. Product. Techniques Non-Invasive Med. Diag.,* New Engl. Coll., Henniker, New York, *Aug.* (1976).

[29] J. A. Frank, M. A. Feiler, W. V. House, P. C. Lauterbur, and M. J. Jacobson, *Clin. Res.* **24,** 217A (1976).

would be something which substitutes for ^1H in water or for ^1H in a water-soluble organic molecule. Deuterium would be ideal in the first case, were it not harmful. Of course, a few cubic centimeters of heavy water administered locally may not present a particular hazard. An example of the second kind would be ^{19}F used as a tracer in suitable salts or as a fluorinated blood substitute.[30] With fast imaging methods, time course studies of ^{19}F or the lack of ^1H (if D_2O is used) associated with the injected bolus could be monitored.

Other, less hazardous time course studies of an injected bolus of material could be undertaken using saline or sugar solutions. In these cases, changes of the T_1, though slight, might be the parameter studied.

The dramatic effect on T_1 of paramagnetic impurities of the order of a few parts per million, is of course, well known to NMR practitioners[31] (see also Section 2.3.4). The search for suitable nontoxic paramagnetic compounds would thus be very worthwhile, especially organ-specific agents. Some preliminary work on the effect on T_1 of introducing paramagnetic agents to animals has already been started.[32] The ions of copper, iron, and manganese are perhaps the best known relaxation centers, but it remains to be seen if ionic salts of these elements can be tolerated by the body at the requisite concentration levels.

7.3.3. DIGESTIVE SYSTEM

Yet another possibility for studying the stomach and intestines is the oral administration of *ferrum redactum*, fine particles of pure iron. Swallowed in the form of an iron meal, its passage could be tracked by the strong localized inhomogeneity produced. Colloidal iron is, of course, an established medicament used in the treatment of anemia and is tolerated by the digestive system in small doses. Really large doses could be suitably encapsulated in small millimeter-size capsules designed to pass straight through the system with no interaction with the digestive juices. (See Notes Added in Proof, Note 7.3.)

7.3.4. CARDIAC STUDIES

A full discussion of high-speed echo-planar imaging has already been given in Chapters 4, 5, and 6, and it is clear from Fig. 6.13 that for 10-mm

[30] G. N. Holland, P. A. Bottomley, and W. S. Hinshaw, *J. Magn. Reson.* **28**, 133 (1977).

[31] A. Abragam, "The Principles of Nuclear Magnetism." Oxford Univ. Press (Clarendon) London and New York, 1961.

[32] P. C. Lauterbur, M. H. M. Dias, and A. M. Rudin, in "Electrons to Times, Frontiers of Biological Energies" (P. L. Dutton, J. S. Leigh, and A. Scarpa, eds.), Vol. 1, p. 752. Academic Press, New York, 1978.

slices, for example, single-shot images with an S/N of 10 are possible in times less than 100 msec, provided the picture element is 3 mm or more. That corresponds approximately to a 128^2 whole-body image with an object field of 36^2 cm^2. Of course, if one is prepared to accept single-shot images with a S/N less than 10, together with the possibility of real-time display, or even a loop display, then the eye will average the signal, bringing the viewed or subjective S/N back to, or even greater than 10, while at the same time allowing the motions of the object, which occur from frame to frame, to be studied. It is with this type of imaging in mind that we discuss the potential of NMR for cardiac studies. (See Notes Added in Proof, Note 7.4.)

For a practical operating system, the imaging time should not exceed 64 msec. Of course, the shorter the time the better from the motional point of view. But possibly a more important constraint on single-shot imaging will be the tissue T_2. However, we point out that cross-sectional images from live unanesthetized rats have already been obtained in times as long as 64 msec. Blood is not likely to be troublesome on this score, although some muscle tissue might well be discriminated against, showing up as a dark rather than bright zone. Fatty regions and cholesterol deposits rich in mobile protons are generally characterized by short T_1's of the order of 100–200 msec with similar T_2's. Thus these regions should be clearly visible as bright zones on fairly rapid scans, as already demonstrated in the case of yellow bone marrow in a live finger. Blood T_1's are usually much longer and could be distinguished both from the known morphology of arteries and veins and T_1 maps. In addition, the blood is moving and it is expected on theoretical grounds that blood flow will be measurable by NMR imaging, *vide infra*.

Valuable parameters necessary for a meaningful prognosis and therapy planning in heart disease include the ejection fraction defined as the ratio of stroke volume at systole to left ventricular end-diastolic volume, as well as cardiac output. Multislice, high-speed imaging should allow these parameters to be directly measurable, as well as allowing studies of the ventricular wall movements, valuable in locating and assessing infarcted regions of the myocardium. In addition to the purely hemodynamic approach outlined, water content and T_1 of the myocardial tissue should supplement the diagnosis.

We have stressed the moving-picture possibilities, but of course, it is entirely feasible to produce images triggered from the ECG r wave. This approach, combined with an advancing delay facility, gives intriguing possibilities for stroboscopic studies. (See Notes Added in Proof, Note 6.1.)

7.3.5. FLOW

The idea of studying fluid flow by NMR dates back to the early pioneers who quickly saw the potential for studying random microscopic flow or diffusion by NMR[33-35] The study of nonrandom macroscopic flow has also been extensively discussed over the years,[36] and attempts made to apply NMR to the study of blood flow. Experiments on fingers and limbs where relatively exposed arteries can be easily reached using surface search coils have been of limited value,[37] the prime reason being, of course, the lack of spatial localization. Various methods for measuring nonlocalized flow have been proposed[38-41] and reviewed[42] elsewhere. Among these, perhaps the simplest is a time-of-flight method involving two separated coils. The first coil, upstream, is the transmitter, and downstream from this is the receiver coil. The delay time between delivering an rf pulse to the moving spins, and receipt of the signal is inversely proportional to the fluid flow rate, assumed to be constant.

With fast echo-planar imaging methods, it is easy to see that such a prepulse applied downstream from an image "read" experiment would directly yield flow information, while at the same time retaining the spatial distribution information. This type of flow measurement is clearly best done with the imaging plane normal to the velocity vector, a situation not uncommon with cross-sectional views of the body. Of course, for a generalized velocity vector the situation becomes more complicated.[43]

In general, flow and motion affect all NMR images and thus we can expect the number of NMR flow methods to be roughly equal to the number of different imaging methods. The artifact in Fig. 6.18a attributed by the authors[15] to heart motion, exemplifies the point. Naturally, the problem will be extricating the motional artifacts from the static regions. With regard to this, the fact that blood flow is pulsatile may turn out to be a helpful factor.

Although there is much discussion of combining flow methods with imaging methods, the only experimental work so far reported which actually attempts this is that of Lauterbur et al.[44] In this work a model pulsatile system was constructed with a flow bifurcation surrounded by static water as indicated in Fig. 7.9a. Using the sensitive-point technique (see Chapter

[33] N. Bloembergen, E. M. Purcell, and R. V. Pound, *Phys. Rev.* **73**, 679 (1948).

[34] E. L. Hahn, *Phys. Rev.* **80**, 580 (1950).

[35] H. C. Torrey, *Phys. Rev.* **76**, 1059 (1949).

[36] A. I. Zhernovoi and G. D. Latyshev, "NMR in a Flowing Liquid" Plenum, New York, 1965.

[37] T. Grover and J. R. Singer, *J. Appl. Phys.* **42**, 938 (1971).

[38] K. J. Packer, *Mol. Phys.* **17**, 355 (1969).

[39] R. J. Haywood, K. J. Packer, and D. J. Tomlinson, *Mol. Phys.* **23**, 1083 (1972).

[40] A. N. Garroway, *J. Phys. D Appl. Phys.* **7**, L159 (1974).

[41] P. A. de Jager, M. A. Hemmisga, and A. Sonneveld, *Rev. Sci. Instrum.* **49**, 1217 (1978).

[42] J. R. Singer, *J. Phys. E Sci. Instrum.* **11**, 281 (1978).

[43] R. Rzedzian and P. Mansfield, (to be published).

[44] P. C. Lauterbur and C.-M. Lai, *Proc. NHLBI Div. Heart Vasc. Dis. Devices Technol. Annu. Contract. Meet.* p. 158 (1977).

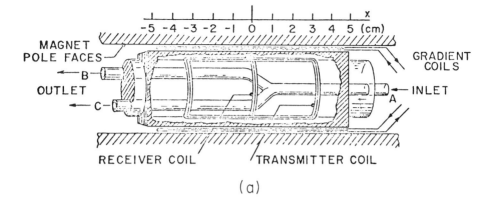

(a)

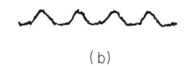

(b)

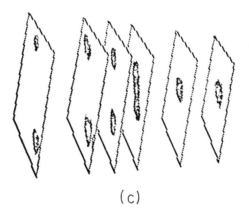

(c)

FIG. 7.9.   (a) Probe and phantom arrangement used to detect pulsatile flow through a bifurcated glass tube surrounded by static aqueous solution. Measurements were performed in a field of 2 kG. (b) Time-modulated sensitive-point signal. (c) Sensitive-point image scans performed through a number of transverse planes as indicated. [From P. C. Lauterbur and C.-M. Lai, *Proc. NHLBI Div. Heart Vasc. Dis. Devices Technol. Annu. Contract. Meet.* p. 158 (1977).]

4), the pulsatile region is mapped out for several planes and gives good agreement with the tube positions. The flow magnitude is *not* measured in this experiment. It is essentially the differentiation of moving and static spins.

Clearly, much has to be done in this new and potentially valuable area. The seeming lack of progress in flow mapping, we believe, is related to the rapid and continuous development and refinement of the basic imaging techniques themselves. However, we are confident that when NMR images are produced routinely by high-speed techniques, flow studies will follow automatically. (See Notes Added in Proof, Note 7.5.)

Immediate clinical potential would be in the study of arteriosclerosis. Since its effects are generally widespread, studies of the femoral or even the carotid arteries and their respective branches could well be a useful starting point for the evaluation of the efficacy of drug treatment.

7.3.6. METABOLISM

One of the most exciting applications of NMR, which has only recently emerged, is the study of ^{31}P nuclei in biological systems.[45-47] Phosphorus occurs naturally, for example, in membrane structures in the form of phospholipids and in many other tissues including nervous tissue, muscle, and liver where it appears in several compounds including adenosine triphosphate (ATP), which provides the driving energy of the cell.

In spite of its 100% isotopic abundance, phosphorus has a somewhat low relative sensitivity (see Table 6.1). Superficially, it might therefore seem unthinkable to consider phosphorus as a candidate for imaging at all. However, as we shall see, ^{31}P resonances contain extremely interesting and potentially useful medical diagnostic information. To understand its importance, we must consider the details of cell metabolism.

ATP contains three phosphate groups, α, β, and γ. The molecule is broken down by enzymes in muscle, for example, to form adenosine diphosphate (ADP) and inorganic phosphate P_i. ADP contains only the α and β phosphate groups. ATP is regenerated, under the action of enzymes, from phosphocreatine (PCr) plus ADP, yielding ATP and creatine (Cr). The net effect is to break down phosphocreatine to form creatine plus inorganic phosphate. This is summarized in the expression

$$PCr \rightarrow Cr + P_i + E$$

[45] D. I. Hoult, S. J. W. Busby, D. G. Gadian, G. K. Radda, R. E. Richards, and P. J. Seeley, *Nature (London)* **252**, 285 (1974).
[46] C. T. Burt, T. Glonek, and M. Barany, *Biochemistry* **15**, 4850 (1976).
[47] D. G. Gadian, *Contemp. Phys.* **18**, 351 (1977).

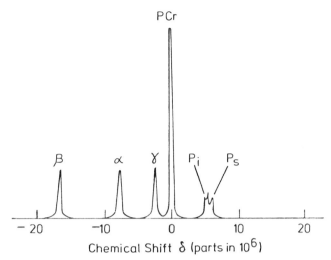

where E is the energy required by the cell. Some ATP is synthesized from the breakdown of glycogen (glycolysis) which produces sugar phosphate P_s as an intermediate step. However, most ATP is produced by oxidative phosphorylation.

The large numbers of hydrogen atoms in these bio-organic compounds and surrounding proteins make the high-resolution proton resonances extremely complicated and virtually impossible to interpret. Conversely, the small numbers of phosphorus nuclei yield relatively simple and interpretable ^{31}P spectra. The reaction metabolites are resolvable by virtue of their different chemical shifts. (See also Notes Added in Proof, Note 7.6.)

A typical ^{31}P spectrum in normal muscle is shown in Fig. 7.10 and indicates the positions of the α, β, and γ ATP phosphate groups, the strong phosphocreatine line and the weaker inorganic and sugar phosphate resonances. An important fact is that the chemical shift of the inorganic phosphate resonance is dependent on the local pH value, thus providing a new and direct way of measuring this parameter. In myocardial ischemia, for example, lack of nutrient and oxygen, resulting from an infarct, prevents synthesis of ATP by oxidative phosphorylation so that the metabolic pathway stops at glycolysis, causing lactic acid to accumulate. This decreases the local pH, causing a measurable shift of the P_i peak and an increase in the sugar phosphate peak.[48,49] As well as these effects, there is an even stronger effect on the PCr peak in ischemia observed in simulated infarcts by ligation of the

[48] D. P. Hollis, R. L. Nunally, W. E. Jacobus, and G. J. Taylor, *Biochem. Biophys. Res. Commun.* **75**, 1086 (1978).
[49] P. B. Garlick, G. K. Radda, and P. J. Seeley, *Biochem. J.* **184**, 947 (1979).

coronary arteries in both perfused beating hearts and in rat heart studies of anoxia performed *in vivo*.[48,50,51] These particular studies were performed invasively by conventional high-resolution experiments on exposed or removed hearts enclosed in the NMR coil. The results indicate clearly that following ligation and the onset of ischemia, the PCr peak drops while the P_i and P_s peaks increase. During this phase the ATP levels stay fairly constant. Finally, as glycogen reserves diminish, the α, β, and γ peaks drop. Similar results occur following respiratory arrest.[51]

Figure 7.11 shows two ^{31}P spectra from the same excised specimen of frog gastrocnemius muscle.[51a] Figure 7.11a is the spectrum obtained immediately following excision and shows the ATP peaks, strong PCr, and weaker P_s and P_i resonances. Figure 7.11b shows the spectrum 2 hr after excision ($\times 2$ scale) and clearly illustrates the diminished PCr peak and much increased P_i and P_s resonances as the tissue dies.

More recently, methods of achieving noninvasive localized ^{31}P chemical analysis have been described. In one method,[52] a surface coil is used to locally excite and receive signals. For large objects, this clearly has resolution and depth limitations. These problems are largely overcome in a new technique recently described.[53] In this experiment the ^{31}P resonance is observed at a single sensitive point by profiling the static magnetic field to give a small volume of high homogeneity (see also Damadian *et al.*[54] and Fig. 4.11c). High-resolution spectra from live whole rats and localized liver regions within the rat demonstrate the virtual absence of the PCr peak in liver, where large reserves of glycogen are stored. This result confirms the low liver phosphocreatine levels measured by chemical means. The spectra were obtained at 32.5 MHz corresponding to a field of 1.89 T produced by a wide-bore superconductive magnet.

The implications of these exciting developments are clear, for we now have a nondestructive and noninvasive method of following the time course of the phosphorus metabolites in living organs. In other words, chemical analysis relevant to the disease states of various organs can be performed *in vivo*.

[50] D. P. Hollis and R. L. Nunnally, *Phil. Trans. R. Soc. London Ser. B* **289,** 437 (1980).

[51] J. J. H. Ackerman, P. J. Bore, D. G. Gadran, T. H. Grove, and G. K. Radda, *Phil. Trans. R. Soc. London Ser. B* **289,** 425 (1980).

[51a] These results were obtained at NIMR. We wish to thank Professor D. R. Wilkie and Dr. J. Dawson for supplying the specimen.

[52] J. J. H. Ackerman, T. H. Grove, G. G. Wong, D. G. Gadian, and G. K. Radda, *Nature (London)* **283,** 167 (1980).

[53] R. E. Gordon, P. E. Hanley, D. Shaw, D. G. Gadian, G. K. Radda, P. Styles, and L. Chan, *Nature (London)* **287,** 736 (1980).

[54] R. Damadian, *Phil. Trans. R. Soc. London Ser. B* **289,** 489 (1980).

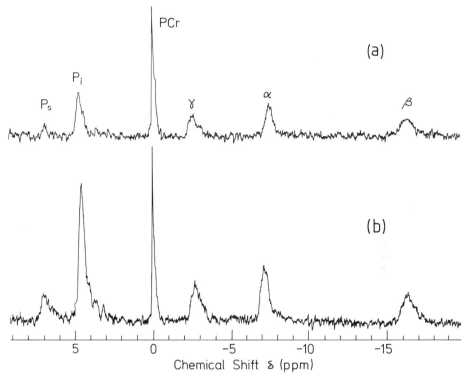

FIG. 7.11.   High-resolution ^{31}P spectra of an excised specimen of frog gastrocnemius muscle obtained at 80.94 MHz; (a) spectrum immediately following excision; (b) result obtained 2 hr following excision (vertical scale ×2). In both spectra, the α, β, and γ ATP peaks are clearly resolved, together with the phosphocreatine (PCr), inorganic phosphates (P_i), and sugar phosphate (P_s) resonances. Note the increased P_i and P_s peaks in (b).

The above work relates to a sensitive point or volume, and from the NMR imaging viewpoint, as we have seen in Chapter 5, point scanning is the least efficient method. In order to accurately locate the region of interest for chemical analysis, we believe it will be necessary to image the subject for specific "landmarks." We therefore envisage a rapid ^{1}H scan of medium resolution followed by a ^{31}P scan at the same frequency and of suitably coarsened resolution, enabling the picture-point S/N to be maintained at the same value as that of the proton scan. Of course, this latter requirement would necessitate drastically reducing the ^{31}P picture matrix size. For example, if we take the proton picture matrix as 128^2, the corresponding ^{31}P matrix would have to be reduced to $(128/10\sqrt{10})^2 = 4^2$ in order to maintain the same S/N for the same slice thickness. If we allow the slice thickness to vary in addition to the array size, the resulting picture is not so coarse as in the above case. Suppose, for example, that the proton scan corresponds to a 1.0-cm-thick slice of 128^2 square pixels of length 3.0 mm, that is to say, a square object field of side 38.4 cm. Each volume element is therefore 90 mm^3. The corresponding volume element for a ^{31}P image, treated as a cube, is therefore $(10^3 \sqrt{90})^3$ mm^3 yielding a picture voxel of 4.48^3 cm^3. The array size is therefore reduced to approximately 9^2. The matrix reduction factor arises because we have assumed that there are roughly 10^3 protons for every phosphorus nucleus present. If the factor is greater than 10^3, then the ^{31}P picture matrix becomes still coarser.

Implicit in this discussion is the assumption that chemical shift information can be obtained from, for example, echo-planar images without recourse to the field-profiling technique referred to above. We believe this to be possible. We also mention that by different means, spatially distributed chemical shift information may be obtained from suitably modified imaging experiments[55] (see Section 4.4.2.2). The effects of ^{19}F chemical shifts on line-scan images have also been reported.[30]

Finally, we point out that the high-resolution ^{31}P spectra discussed above have been obtained at high frequencies on relatively small specimens where rf penetration problems are not so important. However, it is clear from our discussion in Chapter 6 that the prospect of whole-body imaging at frequencies much in excess of about 5.0 MHz looks somewhat bleak. The real question, yet unanswered, is whether such ^{31}P chemical shift information can be obtained at these lower, but more practical frequencies. It seems clear that much more research needs to be done in this most exciting new area.

7.4. Pathology

7.4.1. INTRODUCTION

Many maladies are, of course, pathogenic in origin, and are caused by bacterial or viral infection. NMR imaging can in no way be regarded as some sort of diagnostic panacea. Indeed, it seems highly improbable that NMR, *in vitro* or *in vivo* will ever compete with established methods of, for instance, blood and urine analysis for general screening purposes. Nevertheless, some workers[56,57] have used surface coils to study *in vivo* T_1's of accumulated

---

[55] P. C. Lauterbur, D. M. Kramer, W. V. House, and C.-N. Chen, *J. Am. Chem. Soc.* **97,** 6866 (1975).

[56] G. J. Bené, B. Borcard, E. Hiltbrand, P. Maguin, and R. Sechehaye, *C.R. Acad. Sci. Paris* **284,** B 141 (1977).

[57] G. J. Bené, B. Borcard, E. Hiltbrand, and P. Maguin, *Phil. Trans. R. Soc. London Ser. B* **289,** 501 (1980).

urine in the bladder. These are not imaging experiments as such, but the implications for imaging are clear if large T_1 changes can be unequivocally related to disease states. Related work, for example, on samples of amniotic fluid, show clear differences in T_1, at low field strengths, due to the presence of excessive meconium, which can be an indicator of fetal abnormality. For example, at 2 kHz pure water possesses a $T_1 \simeq 3.5$ sec, for amniotic fluid $T_1 \simeq 2.2$ sec, and for meconial tainted amniotic fluid $T_1 \simeq 1.8$ sec. However, for frequencies in excess of 100 kHz these T_1 differences vanish. All three fluids are then characterized by the appropriate T_1 for pure water.[57a]

Whether in a general sense, a systemic effect on T_1 of the bulk body fluids, and the water balance could be general pointers to the state of health remains to be seen. Certainly, specific water distribution and/or T_1 changes due to localized infection, hemorrhage, such as subdural hematoma, edema, and hydrocephalus, particularly in young children or even in the fetus, should be detectable, thereby enabling earlier therapy and treatment. Sclerotic changes in organs, especially the liver, might well be detectable at an early stage by NMR. Scar tissue, especially in the pleural cavities may also show.

Much of this discussion is at present speculative. Clinical evaluation of NMR for a whole range of abnormal conditions is urgently required.

7.4.2. ONCOLOGICAL APPLICATIONS

As we have discussed at some length in Chapter 2, statistically, there seems to be a longer T_1 value for the water associated with tumor tissue as opposed to that of the normal host tissue. However, malignant tissues do not have a uniform composition and show marked variations in cell types and content of fibrous tissue. It is therefore not surprising that some malignant tumors do not show such an increase, so that hopes of uniquely detecting cancerous tissue by T_1 alone do not seem to be substantiated by experiments. Nevertheless, many tumors, both malignant and benign, *do* affect T_1. Even where no T_1 change is apparent, as in the case of human breast tumors,[58] NMR images do show differences due to apparent T_1 changes.[59]

Figure 7.12 shows two line-scan projection images through a whole breast performed $1\frac{1}{2}$ hr following mastectomy. The line-scan repetition period was 0.3 sec in Fig. 7.12b. Since much of the breast material is adipose tissue with a T_1 around 100 msec, one sees essentially the mobile-proton distribution in the specimen. The breast spread to a roughly elliptic shape being

[57a] G. J. Bené, Data presented at the 4th EENC Conf. at Grenoble, France, June 1979 (unpublished).

[58] W. M. Bovee, K. W. Getrener, J. Smidt, and J. Lindeman, *J. Natl. Cancer Inst.* **61**, 53 (1978).

[59] P. Mansfield, P. G. Morris, R. Ordidge, R. E. Coupland, H. M. Bishop, and R. W. Blamey, *Br. J. Radiol.* **52**, 242 (1979).

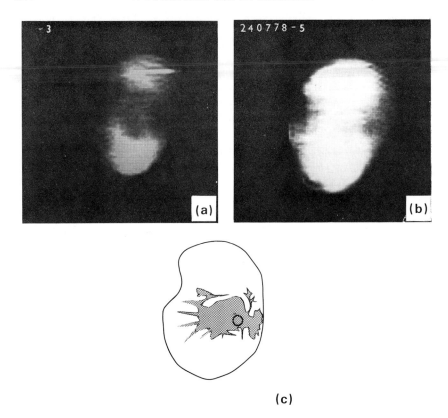

FIG. 7.12.   NMR line-scan images of an excised breast. (a) Delay time, 0.15 sec. Dark region to right corresponds with the nipple. Dark central region corresponds to the tumor site. (b) Delay time increased to 0.30 sec, thus increasing the image brightness and lowering contrast. This image corresponds more to the proton content. The dark region to the left of both images corresponds to a missing notch in the tissue mass. (c) Reconstruction of the position of the scirrhous carcinoma which was thicker centrally and tapered toward the left and right margins. The black circle denotes the nipple position. The axillary tail is uppermost in the diagram. [From P. Mansfield, P. G. Morris, R. Ordidge, R. E. Coupland, H. M. Bishop, and R. W. Blamey, *Br. J. Radiol.* **52**, 242 (1979).]

approximately 15 cm along the major axis, the axillary tail being uppermost in the scans. The breast was around 4.0-cm thick in the central region, tapering down to about 1.0 cm at the edges. The maximum signal is therefore expected at the center of the image and in Fig. 7.12b this is more or less the case. However, when the repetition period is halved to 0.15 sec, the same object produced the scan shown in Fig. 7.12a. The central region, now darker, clearly indicates a longer T_1 than that of the remaining bright zones at the top and bottom of Fig. 7.12a. Later histological examination of sections through the breast indicated the extent of a scirrhous carcinoma (Fig. 7.11c), the palpable core of which was about 1.5-cm long and slightly to the left of the nipple.

According to T_1 results on biopsy specimens of breast tumors,[58] there is no significant difference between normal and abnormal mammary glandular tissue. Yet there is clearly a large repetition delay time dependence of the NMR image with rather good correspondence to the observed extent of the tumor. Furthermore, approximate T_1 values extracted from the image data of Fig. 7.12a and b correspond to long T_1's in the tumor-core region of up to 800 msec ranging down to around 100 msec for the "host" tissue.[60]

It would appear that the explanation of these interesting results is a progressive interpenetration of the tumor tissue into the supportive adipose connective tissue matrix. The apparent range of T_1 values could therefore be a simple weighting effect of two T_1's producing a nonexponential spin–lattice relaxation recovery. The longest "effective" T_1 of 800 msec in the image is close to the value expected from glandular tissue alone. The predominance of the long effective T_1 in the tumor-core region is attributed to the high proportion of glandular tumor relative to that of fat. In a normal breast the ratio of glandular tissue to fat is reported to be about 2:1, but this obviously must depend on the position within the breast. Interestingly, the region around the nipple and underlying the areolar is distinguishable as a dark circular region to the left in Fig. 7.12b, and reflects its somewhat lower mobile-proton content.

It should, perhaps, be emphasized that unpublished images of ours of a breast containing a benign fibroadenoma also indicated the tumor, but it was by no means as clearly shown as the carcinoma. A T_1 analysis was not performed. Other unpublished images on cadaver brains look promising for tumor detection.

In summary, our view is not as pessimistic as some authors[9,61] who believe that NMR has virtually no role to play in the detection of tumors. On the other hand, we do not see NMR imaging primarily as a general diagnostic aid for cancer, although it may well find a place in the detection of certain tumors.

[60] P. Mansfield, P. G. Morris, R. J. Ordidge, I. L. Pykett, V. Bangert, and R. E. Coupland, *Phil. Trans. R. Soc. London Ser. B* **289**, 503 (1980).

[61] H. M. Schwartz, *J. Magn. Reson.* **29**, 393 (1978).

8. Some Hardware Considerations

8.1. Static Magnetic Field Requirements

8.1.1. INTRODUCTION

Considerable expertise has gone into the design and manufacture of magnet systems for NMR over the last 20–30 years. It is true that other scientific disciplines use magnets but, in general, the stability and homogeneity requirements of magnets for cyclotrons and beam bending are nowhere near as demanding as those in high-resolution NMR, for example.

In the past, the NMR requirements have been for very high homogeneity over quite small volumes of the order of a few cubic centimeters, and although difficult and expensive to achieve, the cost has been mitigated by the relatively small size of the installation. Thus, for example, a typical medium resolution magnet with 9-in. pole faces can produce a homogeneity of 2 parts in 10^6 over a pill-box volume 1 in. diam. $\times$ 1/2 in. thick, with a working magnetic field strength of up to 10 kG.

A typical high-resolution electromagnet will operate with magnetic field strengths of up to 25 kG with a field homogeneity of 1 part in 10^7 over a volume of 1.0 cm^3, whereas superconductive magnets, designed for very high-resolution NMR experiments will operate at fields up to 100 kG (10 T), producing static field homogeneities of the order of 1 part in 10^8 over a volume of 1.0 cm^3 or so.

The achievement of high and very high resolution is made possible by the use of shimming or correction fields added to the main magnetic field. The design of such shim coils has become quite a sophisticated business, and the actual process of shimming the field, or adjusting the currents to minimize the inhomogeneities, is quite an art. Nevertheless, with experience and a proper shimming procedure, the independent current controls can be fairly quickly aligned to compensate for imperfections in the main magnet.

We therefore see that many of the problems associated with the design of large magnets for whole-body imaging are already well understood and, in some cases, solved. For such imaging the general requirements of a magnet are that it have a large working volume, a medium resolution, and a relatively low magnetic field strength. The design problems of such a system then reduce considerably to those of simple mechanical scaling of the magnet dimensions. However, even though we are speaking of fairly modest magnetic fields of perhaps a few kilogauss, the intercoil stresses imposed on the structure in an air-cored electromagnet can be quite high depending on the strength of the field required, the intercoil spacing, etc.

We shall return to a discussion of air-cored magnets in later sections. There are, of course, other possibilities for the design of low field magnets and in the next section we turn to a discussion of iron-cored electromagnets.

8.1.2. IRON-CORED ELECTROMAGNETS

It is fair to say that the majority of standard NMR spectrometer installations utilize a traditional iron-cored electromagnet. There are also some permanent magnets in use in some laboratories and the general design considerations are much the same as those for electromagnets with regard to the pole gap and pole face design. Both types of magnet may be represented by a magnetic circuit as shown in Fig. 8.1. For an electromagnet the magnetomotive force M is given by

$$M = NI = \phi \left[\frac{L}{\mu\mu_0 A} + \frac{g}{\mu_0 A_g} \right], \tag{8.1}$$

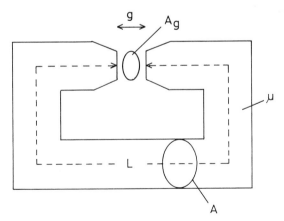

FIG. 8.1.   Sketch of magnetic circuit for either a permanent magnet or an electromagnet. A_g and A are the cross-sectional areas of the air gap and the iron core, respectively.

where μ is the relative permeability of the iron, μ_0 the permeability of free space, the other symbols being indicated in Fig. 8.1. Of course, the total magnetic flux ϕ passing through any cross section of the magnetic circuit should really be an integrated sum throughout that section; that is to say

$$\phi = \int \mathbf{B} \cdot d\mathbf{S}, \tag{8.2}$$

but we shall assume that the iron core is homogeneous and is not saturated and also that any stray magnetic fields, especially in the region of the gap, can be ignored. In other words, we shall assume that the magnetic fields both within the iron core and in the gap are uniform throughout the magnetic circuit cross section and zero everywhere outside. For a permanent magnet, the magnetomotive force applied to the circuit is zero.

These simple considerations show that when using soft iron yokes with a maximum saturation field B_s of 20 kG, it is necessary to have substantial amounts of iron of large cross section to close the magnetic flux return path.

The homogeneity of such a magnet system will depend on the characteristics of the pole shoes. Important factors are their shape, the type and texture of the magnetic material, and in particular, the ratio of pole-face diameter to gap. For a ratio of three, one can expect medium resolution with homogeneities of up to a few parts in 10^6 over spherical volumes of diameter up to one-sixth the gap width. For better homogeneity over the same volume, or for the same homogeneity over a larger fraction of the gap width, one must increase the ratio of pole diameter to gap width.

So far we have discussed the conventional magnet geometry, which in a symmetric form would require excessive amounts of iron to form a return path for the magnetic flux. A modification of the geometry, which is particularly suited for the generation of low magnetic fields over large volumes, leads to a magnetic design proposed by Watson.[1] A symmetric form of this type of magnet is sketched in Fig. 8.2 and comprises two field-energizing coils wrapped around two soft iron cores. Two soft iron plates cap each end of the system and form extended pole pieces. The magnetic circuit, as sketched, shows the flux return path which passes between the plates. If the plates are parallel and fabricated from homogeneous magnetic material, the field over most of the enclosed volume is uniform. Of course, the same conditions of continuity apply to this magnetic circuit as with that discussed previously in connection with Fig. 8.1, and that means that the iron in the relatively thin plates must not saturate. To ensure this requires typically that the fields produced in the working space are limited in the case of soft iron

[1] I. J. Richmond and D. W. Parker, *Proc. Int. Conf. Magnet Technol., 2nd, Oxford (Rutherford Lab.)* p. 129 (1967).

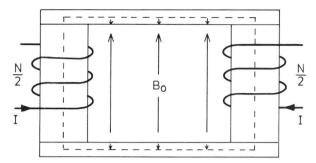

FIG. 8.2. Schematic of a low-field electromagnet based on the Watson principle.

to about 500 G. Of course, if special magnetic materials are employed, for example iron–cobalt alloys the working magnetic field could be increased somewhat. This type of magnet has considerable promise. However, at the present time, bearing in mind the high cost of special magnetic materials, it does not appear to be a practical method of generating fields large enough for the first few generations of whole-body imaging machines. The sheer weight of the magnet may also cause structural problems.

Another point which weighs against the use of iron-cored magnets is the effect of the iron on the generation of magnetic field gradients. All imaging methods require magnetic field gradients generated by additional coils within or external to the main uniform magnetic field. If the gradients are to be modulated or switched, the presence of large volumes of iron will increase the inductance of the gradient coils and, in general, impair their performance when being switched. This is because of induced eddy currents in the pole faces or iron structure. However, if really large working volumes could be produced using the Watson design, then it should be possible to switch gradients within the working volume, provided that the gradient coils are well away from any iron.

A considerable advantage of an enclosed iron magnet is that it is self-screening against the effect of other magnetic material in the vicinity of the apparatus; for example, iron girders and reinforcing mesh used in concrete building structures, together with electrical conduits and fittings, drainage and service pipes, and ventilation ducting, inherent in most modern hospital buildings where such imaging instruments might be used. The housing racks and consoles for most ancillary electronic equipment will contain ferrous material unless otherwise specified, and it would clearly be cheaper to use standard materials such as iron if it were not a problem.

With open-coil air-cored structures, the magnetic fields for whole-body imaging systems can extend several meters in all directions so that the effect must be considered of large stray fields on the ancillary electronic apparatus like TV displays, magnetic cores for computer memories, computer storage disks and cassettes, and not least, the personnel operating the system. However, in the latter case, static magnetic fields, especially fields as low as 1 kG, are not thought to have any harmful biological effects. But we shall discuss this subject more fully in Chapter 9.

8.1.3. AIR-CORED ELECTROMAGNETS

For reasons outlined in Section 8.1, the only large-volume magnets made so far for whole-body NMR imaging have been air-cored coil systems. The first electromagnet system was a four-coil ellipsoidal resistive magnet made by Walker Scientific. More recently, Damadian et al.[2] have reported the use of a home-built superconductive magnet. There are naturally considerable advantages in the use of both resistive and superconductive magnets and, in the long run, it may reduce to a matter of personal choice or more likely, economics, which decides ultimately which system is adopted. Technically, both types are feasible but we shall consider superconductive magnets first.

By their nature, supercons as they are often referred to, require special low-temperature cryogenic apparatus for their operation. The coils may be wound from special copper-clad wire made from a niobium–tin alloy whose electrical resistance vanishes at a critical temperature in the range 4–10 K. The coils are therefore enclosed in special nonmagnetic stainless-steel Dewar vessels containing liquid helium, which boils at 4.2 K at atmospheric pressure. The Dewar is a vacuum-walled vessel, and to reduce the helium evaporation loss, it in turn is placed in a second Dewar vessel filled with liquid nitrogen which boils at 78 K.

Each coil of the final magnet system is thus separately or collectively surrounded by a double Dewar vessel system designed to give room temperature access to the working volume. The vacuum vessels can be pumped down to high vacuum and sealed, thereby obviating the use of permanently sited, specialized high-vacuum equipment. However, constant supplies of liquefied nitrogen and helium are required, and although in an experimental situation one could rely quite satisfactorily on deliveries of these liquefied gases, it could be an expensive business, especially if the helium gas is blown off to waste. Liquid nitrogen costs are also becoming significant.

For hospital use, one must think in terms of a closed-cycle refrigerator for both liquid nitrogen and liquid helium supplies and in this circumstance the initial cost of the magnet system could be prohibitive.

[2] R. Damadian, M. Goldsmith, and L. Minkoff, *Physiol. Chem. Phys.* **9,** 97 (1977).

The use of superconductors requires a number of safeguards and precautions. For example, flux jumps in the superconducting wire can cause the magnet to revert suddenly to a normal resistive conductor state. If this happens, the entire magnetic energy of the magnet is suddenly dissipated in the helium bath with a consequent rapid evaporation of the liquid helium. Safety precautions must therefore be taken to ensure that the gas from the evaporating liquid does not develop dangerously high pressures which could cause an explosion.

Reversion to the normal state will also result in a sudden quenching of the magnetic field, and if this happens too rapidly, harmful induced currents might be produced in the body of a patient being imaged. It is true that a similar situation obtains in resistive magnets during power failure, or worse still, if a power connection to the magnet should come loose. Precautions will have to be taken against such an eventuality, and this is discussed in Section 8.3.6.

Although no power supply is required for a superconductive magnet running in the persistent mode, there can be a slow decay of the current due to resistance of the solder joints on the input and output electrical connection leads. This current decay results in a slow magnetic field drift which is compensated by a topping up power supply of much lower power rating than the main initializing current supply. This initializing power supply is required to charge the magnet in the first place and it must be capable of supplying the full working current. In the persistent mode, the power supply is removed and replaced by a superconductive link. Thus, savings in power running costs are to be expected from a cryogenic magnet and, in addition, the charging power supply stability requirements are less stringent than those required of a stabilized power supply for a resistive magnet system.

In spite of the advantages of supercons, however, we feel that the additional cryogenic expertise required to operate the system, added to the risks in usage make them somewhat less attractive as magnet systems for hospital use then resistive systems. Low-impedance, high-current semiconductor power supplies have made resistive magnets an attractive proposition, particularly since a whole-body magnet can be powered with as little as 20 kW to produce a 1-kG magnetic field. The power is dissipated mainly in the coil windings and one must ensure that this is conducted away via water cooling, otherwise the coils will heat up and quite apart from the possibilities of the winding insulation melting or deteriorating, the magnet structure will assume an unacceptable temperature resulting in the heating of the patient. This aspect will be taken up again in Section 8.3.7.

We now turn to the important question of selection of the conductor geometry in choosing the coil design.

8.2. Coil Design

In a very general sense, the coil design must be one which produces a sufficiently uniform magnetic field over a large volume. The actual volume over which substantial uniformity is required will depend in part on the imaging technique to be used and, of course, on the size of the object. For a whole-body cross section, one is typically interested in an axial cylindrical region of uniformity about 36 cm in diameter and for imaging in the magnet symmetry plane, the cylinder length could be 3 or 4 cm only, corresponding to a relatively thin cross-sectional slice through the body. If, however, the imaging plane is transverse to the magnet axis, then the desired homogenous region necessarily extends over a sphere 36 cm in diameter.

8.2.1. SOLENOIDS

The simplest magnet system is the solenoid. For an infinitely long solenoid, the magnetic field is everywhere uniform. For long but finite coils, however, the axial magnetic field drops from its highest value at the coil center approaching a half of this maximum value at each end of the coil. This falloff in the field can be compensated to some extent by increasing the current density toward each end of the solenoid. This can be done by either winding a separate sleeve coil at each end of the solenoid either in series with the main coil or powered separately, or simply by decreasing the pitch of the windings toward each end of the solenoid. Both methods preserve the coil diameter $2a$ and therefore do not restrict the axial access. Generally speaking the coil volume and, hence, cost of material becomes prohibitively large if one tries simply to increase the coil length l to achieve the desired performance.

8.2.2. HELMHOLTZ COILS

Two flat coils of diameter $2a$ and separation a, will produce a region of uniform magnetic field at the coil center, which can be calculated analytically allowing for the finite size of the coil cross section. In general the diameter of such a coil system which is required to produce a given homogeneity is much larger than for an equivalent compensated system. However, in considering compensated systems, one should bear in mind the additional adjustments required with each additional set of compensation coils necessary to affect improvements in homogeneity.

8.2.3. ELLIPSOIDAL COILS

It is well known that an electromagnet made by winding a coil on the surface of a sphere in such a way that the surface current density follows a

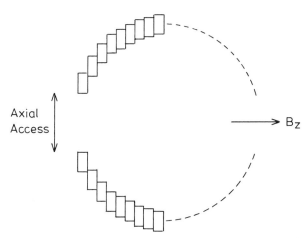

FIG. 8.3.   Section through a spherical magnet with removed end caps allowing axial access.
The current density per rectangular winding is constant. Only one hemisphere is shown.

cosine distribution (Fig. 8.3) will produce a uniform magnetic field throughout the entire spherical volume.[3] If the spherical pole caps are removed to
allow axial access, then the field inside will be degraded by exactly the sum
of the magnetic field produced by the missing current distributions of the
two pole caps. This is the principle of superposition of magnetic fields.
Similar arguments apply to coils wound in ellipsoidal form of which the
sphere is a special case. We see that a Helmholtz coil may be regarded as a
very crude approximation to a spherical magnet. A better approximation is
to use two pairs of coils with diameters and spacing chosen to lie on the
surface of a sphere. For this combination of coils, the spacing between the
large pair is less than a. A typical arrangement is shown in Fig. 8.4. This type
of magnet has the advantage of being a relatively open structure with a large
enclosed volume for ancillary equipment like the rf probe and various
gradient coil arrangements. Calculations show, however, that the design is
relatively sensitive to the coil spacings and therefore could present adjustment
problems unless particular attention is paid to the support structure and coil
adjustment mechanism.

   Franzen[4] has shown that for this type of magnet, nominally an eight-order
system,[5] the finite coil cross sections effectively reduce the magnet to a sixth-
order corrected system, and in this case particular ratios of coil cross-
sectional thickness to width apply for each coil pair. Both Walker Scientific
and Oxford Instruments make magnets similar to this design. Figure 8.5

[3] J. E. Everett and J. E. Osemeikhian, *J. Sci. Instrum.* **43,** 470 (1963).
[4] W. Franzen, *Rev. Sci. Instrum.* **33,** 933 (1962).
[5] M. W. Garrett, *J. Appl. Phys.* **22,** 1091 (1951).

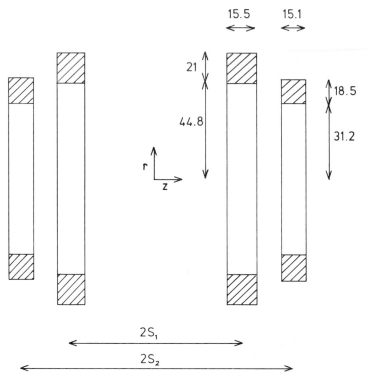

FIG. 8.4.   Section through a typical four-coil electromagnet used for human whole-body imaging (dimensions are in centimeters). The shaded regions correspond to uniform current-density windings.

shows the expected axial homogeneity for such a system as a function of distance along the axis. The coil parameters in this example are those indicated in Fig. 8.4, the coil spacings corresponding closely to a spherical configuration for the optimum spacings S_1 and S_2. Other values of S_2 show how critical this design is to small misalignments.[5a]

8.2.4. SPLIT SOLENOID

Yet another design of magnet is shown in Fig. 8.6. It is essentially one of the compensated forms of the solenoid previously discussed, but with the main solenoid split into two separate coils.[5a] This allows both axial and transverse access to the working volume and is in fact a common magnet design used in split-coil supercons. Figure 8.7 shows a graph of the radial

[5a] We are grateful to Bruker A. G. for supplying details of this coil design.

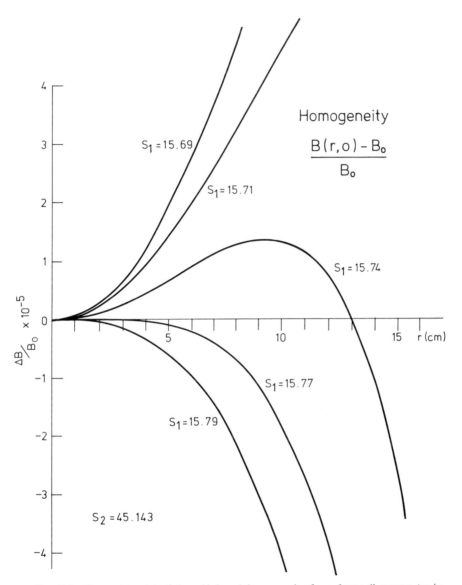

FIG. 8.5. Parametric plot of the midplane inhomogeneity for a four-coil magnet (as in Fig. 8.4) as a function of radius r from the center. Separation S_2 is fixed and S_1 varied. The optimum S_1 appears to be around $S_1 = 15.74$ cm giving ~ 1 in 10^5 homogeneity out to $r = 13$ cm in this particular design.

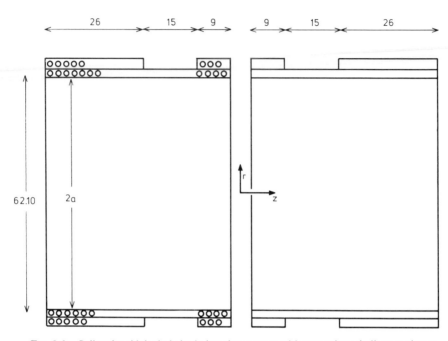

FIG. 8.6. Split-solenoidal whole-body imaging magnet with correction windings as shown (all dimensions are in centimeters).

homogeneity obtainable as a function of distance r off the axis. The curve is plotted for the given coil dimensions and optimized shim currents. It would appear that this type of magnet can easily achieve high homogeneity over large volumes, but may be just as sensitive to errors in construction or adjustment as the four-coil system. However, as we have repeatedly stressed, the choice revolves ultimately around costs. This type of magnet will produce a magnetic field and hence create magnetic energy over a much larger volume than that of a spherical or approximately spherical magnet. If it is accepted that the diameter $2a$ must correspond to the largest diameter of a four-coil system, then the cost of materials plus running costs will be substantially higher for a split solenoid magnet.

The above statement on equal diameters, however, might well be questioned. Why is it necessary to have the largest diameter of a spherical magnet equated to the diameter of the split solenoid? Could not the solenoid diameter equal that of the small coils in a spherical system? After all, one only requires axial access large enough for a patient and coil probe. The answer to this question depends ultimately on the precise details of the imaging method to be used. If room for gradient coils is allowed for totally within the main magnet volume, then the solenoid diameter must be large. This

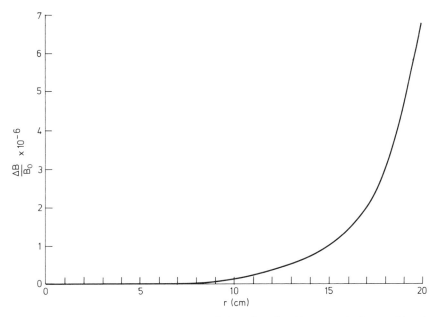

FIG. 8.7 Radial inhomogeneity ($z = 0$) of the split solenoid magnet (coil gap, 5.0 cm) shown in Fig. 8.6. According to this design, 1 in 10^6 homogeneity is obtainable out to $r = 15$ cm.

would be the case, for example, if the gradients needed to be modulated or switched rapidly. If, on the other hand, static gradients are to be employed, then the magnet diameter could be reduced substantially. However, this choice cannot be taken in isolation, since the type of gradient used depends very much on the imaging speed which, in turn, depends on the choice of imaging method. A diagram showing a typical whole-body magnet assembly based on a spherical magnet design is shown in Section 8.5 (Fig. 8.19).

8.3. Main Magnet

8.3.1. VOLUME

It may be desirable in general to have the volume of the magnet greater than the proposed working volume of high homogeneity. That is to say, it could well be desirable to have regions within the magnet where the patient can be partially or fully magnetically polarized, but where the homogeneity is not so important. Certain types of fluid-flow experiment require this kind of spin prepolarization and it could turn out to be useful for blood-flow measurements as well. Other reasons for having a magnet with a large enclosed volume include provision of space for the necessary rf probe assembly and also the magnetic gradient coils. All this can be housed in low-grade field regions. The prime space at the magnet center would be reserved for the region of the subject to be imaged.

8.3.2. STATIC HOMOGENEITY

We consider a thin disk coaxial with the magnet axis and centrally placed. Let the radius of the disk be r_0. For a properly aligned magnet, the field and hence the Larmor angular frequency of the spins in that field will be a function of radial position from the center of the disk, that is $\omega(r)$. The maximum deviation in angular frequency across the disk is therefore

$$\Delta\omega(r_0) = \omega(r_0) - \omega. \tag{8.3}$$

If all the spins in a uniformly distributed disk of material contribute to the observed absorption line, then the effect of inhomogeneity is to broaden the absorption linewidth. Whatever the line shape, the linewidth $\delta\omega$ will be less than $\Delta\omega$, the precise relationship depending on the distribution function $\omega(r)$ or its inverse $r(\omega)$. The absorption line shape will therefore be given by

$$g(\omega) = 2\pi r(\omega)(dr/d\omega). \tag{8.4}$$

The important point is the degree of broadening $\delta\omega$ that can be tolerated in $g(\omega)$ for an imaging experiment.

All NMR imaging methods employ field gradients in one form or another to achieve spatial resolution of the spin distribution. If a linear gradient G_x is applied across the disk, then in the absence of static-field broadening, isochromatic planes of spins are created. Let us suppose that we are interested in resolving a row of points with separation Δx in a line of spins selected from the plane. The angular frequency separation of the points is $\Delta\omega_x = \gamma G_x \Delta x$. It is clear that if mixing or overlap of signals from adjacent points is to be prevented, we require ideally that $\Delta\omega(r_0) < \Delta\omega_x$. A less stringent but useful guide is when $\delta\omega < \Delta\omega_x$. We see that the worst case corresponds to a selected line through the disk diameter. Here, the maximum deviation in the static field $\Delta\omega(r_0)$ occurs.

The maximum tolerable inhomogeneity can also be expressed in terms of the final picture element size. That is to say, for a row of l picture elements spaced Δx apart along the x axis, we have that

$$l\,\Delta x = r_0, \tag{8.5}$$

and thus from the above considerations we obtain

$$\delta\omega < (\gamma G_x r_0)/l. \tag{8.6}$$

As a numerical illustration we consider the case when $r_0 = 15$ cm and $l = 64$. Let us take a typical gradient value of $G_x = 0.1$ G cm^{-1} and consider the example with protons where $\gamma = 8.4\pi \times 10^3$ rad sec^{-1} G^{-1}. In this case the incremental frequency between successive picture points is 210 Hz. The percentage inhomogeneity of the magnet allowable should thus be less than 5×10^{-5} at 4.0 MHz. For resolution of 128 picture elements using the same gradient, the inhomogeneity should be less than 2.5×10^{-5}.

In the case of echo-planar imaging, the most stringent conditions result if we require $\delta\omega$ to be equal to the resolution of a single picture point, thus resulting in a distortion-free picture. In this case we require for a square array of l^2 pixels

$$y_{max} G_y = l^2 \, \Delta\omega_x / \gamma. \tag{8.7}$$

If we take the maximum field change $y_{max} G_y = 10$ G across a specimen of diameter 30 cm, a modest gradient, $G_y = 0.3$ G cm^{-1} is required. From Eq. (8.7) with $l = 100$ and with $\Delta\omega_x = \delta\omega$, the required static field homogeneity is $1.0 \, 10^{-6}$. We note from the equations that use of a larger field gradient reduces the static-field homogeneity requirement. However, generation of a larger gradient may itself be difficult to achieve, especially when modulated or switched. In addition, there may be a loss of S/N (see Section 6.1.3).

Our discussion so far has been concerned with maintaining parallel isochromatic planes within the specimen. If this can be achieved, then picture distortions are minimized and a direct correspondence between the resulting picture and the geometry of the specimen is maintained. An altogether less stringent but more practical approach is to allow some picture distortion. In such a circumstance the question is whether a unique recovery of the undistorted picture is possible. This reduces to finding a condition of frequency uniqueness for all picture elements. For, clearly, if this exists, production of an undistorted image reduces to stretching or compressing the picture points in real space, thus compensating for the known magnetic field distortions. Whether such a scheme can be made to work or not depends to a large extent on the precise details of the imaging method to be employed. A discussion of some aspects of compensation for picture distortions has been given by Hutchison et al.[6]

8.3.3. TEMPORAL STABILITY

The power supply for the magnet should provide a current which ideally keeps the magnetic field constant. This can be done by employing a field-

[6] J. M. S. Hutchison, R. J. Sutherland, and J. R. Mallard, J. Phys. E Sci. Instrum. **11**, 217 (1978).

controlled feedback loop to regulate the current and may be achieved by using a Hall probe field sensor or a separate NMR spectrometer dedicated to field control. Hall probes are not very stable because of rf pickup and thermal drift problems. NMR probe stabilization is probably the best method of controlling magnetic fields at the present, but here too, the rf fields used in the imaging experiments should not interfere. These may be difficult to screen out. The effects of gradients, especially when time dependent, also create problems for magnetic field stabilization. We are thus inclined toward the use of current-stabilized power supplies which should suffice under proper conditions. Power supplies stabilized to a few parts in 10^6 are commercially available.

The fractional short-term stability of a current-stabilized supply should be better than the static homogeneity and in this case will not be detectable in terms of resonance shift during one experiment. By one experiment we mean typically the expected time of an imaging experiment or set of experiments. This is important if one hopes to use the imaging system to produce a set of correlated cross-sectional images. Thus the time scale of any permissible drift should be an hour or two at a minimum. More stringent drift conditions are obviously desirable, for example, over an 8-hr period. It is therefore essential that the conditions of drift and short-term stability should apply for permissible mains voltage variations. In the U.K. these can be up to 5% without warning.

8.3.4. Thermal Stability

It is clear that for a current-regulated magnet system, the coil structure should be stable against thermal variations. Otherwise, small changes in the structure will cause inevitable variations in the magnetic field, even though the current supply may be rock steady.

Thermal variations may arise from a number of causes. Lack of proper heat dissipation from the coils via water cooling is one important cause. General variations in ambient temperature could also be important, especially for very large structures in which coil supports are not cooled or thermally stabilized. The power supply itself may be sensitive to changes in ambient temperature and typically the design should accommodate temperature changes of $\pm 5°C$.

The whole problem of thermal stability is a complex one, since practically all magnet parameters are affected by temperature changes and it would be difficult, if not impossible, to design against all such variations. In the long run it may be safer to regulate the air temperature.[6a]

[6a] Modern supercons largely overcome both temporal and thermal stability problems.

8.3.5. COUPLING OF MAGNETIC FIELDS

The use of modulated or switched-gradient coils within or in the vicinity of the magnet creates problems through inductive coupling. In order to minimize this it is necessary to ensure that no continuous conducting loop is present within the magnet structure unless specifically introduced for safety reasons (*vide infra*). This means that the metal coil formers supporting the various coils should be split and a plastic spacer inserted to prevent circulating induced currents. Of course, whether the gradient coils enclose the magnet or the magnet the gradient coils, there will be induced voltages in the magnet coils whenever the gradients are switched or modulated in some manner. For balanced gradient coils, this voltage should, of course, be zero, provided all the magnet coils are in series. This is the case since the net flux change through the gradient coils is zero.

8.3.6. SAFETY PRECAUTIONS

The split-coil formers may carry high induced voltages during gradient switching, so precautions should be taken to insulate the patient against a chance electric shock in medical imaging systems. This also applies, of course, to the gradient coils themselves, as well as the rf transmitter and receiver coils.

The most important consideration must be given to patient protection in the event of magnet power failure or, worse still, the breakage or loss of a supply conductor causing the magnetic field to collapse abruptly. The total flux change through a patient in such a circumstance, when the main field drops from 1 kG to 0 in a fraction of a microsecond could result in sizeable induced voltages causing currents to flow around the circumference of the patient. If this were to happen, it could result in serious injury or even death. Three simple methods may be used to give protection in the eventuality of a power-lead failure.

A low-resistance shorted copper coil is enclosed within (or just outside) the main magnet but outside the gradient coils. The protection coil may be a single coil located at the symmetry plane of the main magnet, or a number of separate coils distributed throughout the volume of the magnet. In the latter case, it is essential that the separate sections of the protection coil are wired in series. If now the main field collapses a current will be induced in the shorted protection coil, which will tend to sustain the main field thus reducing its rate of collapse. The field collapse rate will depend on the resistance of the protection coil and can be slowed down to a safe level. The symmetry of the protection coil with respect to the gradient coils should prevent any net coupling between the two circuits.

An alternative method is to include a large capacitance and series damping resistor directly across the magnet input terminals. The magnet is tuned to a low and safe frequency and is ideally critically damped.

A third method[7] is simply to use a stack of diodes in reverse across the magnet input. If the supply leads are broken, a reverse voltage is produced causing the magnetic energy to be absorbed in the diode stack.

The latter two methods give protection for broken supply leads to the magnet, however, only the first precautionary method gives patient protection in the case of intercoil lead breaks. It might, therefore, be wise to routinely employ two safety precautions.

8.3.7. Manufacturing Technique

A number of different construction methods may be employed in the manufacture of the magnet coils. The actual winding geometry can affect the rate at which the heat generated in the coils can be dissipated to a water-cooled jacket. In one method, the coil is wound from a single continuous wide strip of aluminum foil suitably insulated with a thin plastic sheet. The whole coil assembly is set in epoxy resin and both edges of the coil are then machined parallel. The finished coil may then be parted into two separate coils, each of which is cemented in good thermal contact to one water-cooled aluminum plate through a thin plastic sheet. In this arrangement, each turn of the coil is edge cooled, which is fairly efficient. However, there will be thermal gradients across the coil thickness which could be further reduced by cooling from both sides of each coil. A disadvantage of this constructional technique is the possibility of electrical breakdown at the machined edges of the coil. Under normal working conditions, the inter-turn voltage is likely to be quite small in this design, but in the event of power failure large voltages may be developed causing interturn arcing. The switching of adjacent gradient coils as used in many imaging schemes, is also likely to give trouble.

Another method uses anodized aluminum foil directly wound between pairs of edge-cooling plates. No machining subsequent to winding the coil is necessary, but the winding operation must be accurately done.

An advantage of the foil-wound construction method is that the current distribution throughout the coil cross section is very uniform, thus conforming more closely to the theoretical design. The performance of this type of coil is therefore more likely to approach the designed homogeneity in practice.

Wire-wound coils are likely to be better insulated but could suffer from cooling problems. Winding pitch may also cause deviations in coil performance from that predicted for uniform current distributions. Thus in order to

---

[7] W. Edelstein, (private communication).

achieve the desired performance from such a magnet system it may be necessary to incorporate additional field shimming coils.

Finally, we consider what is probably the most efficient coil-cooling system of all, that of direct as opposed to indirect cooling as described above. For direct cooling, the coils are wound from copper tubing through which coolant is directly circulated. The same arguments apply regarding winding pitch, but this problem could be avoided by using rectangular cross-section tubing wound to form a "pancake" coil. To increase the coil cross-sectional thickness several such coils could be placed side by side. A disadvantage of direct cooling when water is used, is that special low-voltage power supplies must be used to prevent chemical breakdown of the specially purified water coolant. Of course, paraffin or some other nonconducting coolant can be employed in a closed-circuit system which is buffered via a heat exchanger to a water-cooling system.

A significant consideration is the cost of the magnet system. For a given size of magnet and power consumption, the cost, which is essentially the cost of the metal windings, goes up as the square of the magnetic field, to a first approximation. In a practical design, however, the cost is likely to rise more sharply, since it will not always be practicable to simply increase the winding cross section, and a point will be reached when the coil diameter has to be increased as well to accommodate the extra windings. Alternatively, a factor of three or four gain may be obtained by reducing the winding resistivity by cooling the whole coil in liquid nitrogen. In this case the square law on cost would clearly no longer apply. Of course, a detailed costing would have to be undertaken, but for a large enough magnet system it is conceivable that a nitrogen-cooled resistive magnet could become an economical proposition.

8.4. Radio-Frequency and Gradient Coil Design

8.4.1. INTRODUCTION

The S/N of an NMR signal varies with static magnetic field according to a seven-fourths power law (see Section 6.3.2). Ostensibly it would appear, therefore, that the higher the polarizing static magnetic field used the better. In fact there are a number of other considerations which vitiate the simple S/N argument, as already discussed in Chapter 6.

In many of the imaging techniques, spatially uniform rf power must be delivered to the specimen over a prescribed volume. To achieve this, the rf coil must itself be specially shaped to achieve maximum homogeneity. This usually means having a coil somewhat larger than the volume to be irradiated. It is a simple matter to convince oneself experimentally that rf coils with more than a couple of turns and large enough to take a human body with some slack cannot be tuned with a nominal tuning capacitance of

50 pF above about 5.0 MHz. A discussion of both transmitter and receiver coils has been given by Hoult and Lauterbur and by others.[8-10]

Of course, for a nonresonant coil, one could go somewhat higher in frequency, although the self-capacitance of the coil and also the coil capacitance to ground will eventually limit the practical achievable frequency. The driving power required from the transmitter in this circumstance could be substantially higher than that required in the case of a tuned resonant circuit.

The human body is an inhomogeneous conductor. The rf fields generated throughout the body will induce currents to flow and depending on their size, will tend to screen the interior of the body. Put differently, the rf field does not completely penetrate the subject. As we have seen already, the penetration depth depends on the tissue resistivity and the rf frequency. We recall that low frequencies and high resistivities are required to obtain full rf penetration. At 4.0 MHz, for example, the penetration depth for an assumed average tissue resistivity of 500 Ω cm^{-1} is approximately 50.0 cm. Thus full penetration of a cylindrical or elliptical object of diameter 36 cm may be assumed in this case. Of course, the assumption implicit in the average calculation is that of a homogeneous distribution of tissue and this is clearly invalid in many parts of the body. Penetration problems will vitiate the rf spatial uniformity of the transmitter coil and are thus better avoided. The nuclear signals from the body interior must also penetrate the body tissue to produce a received signal. As discussed and demonstrated in Chapter 6, penetration problems can be largely avoided by working at rf below 5.0 MHz. At 10.0 MHz the deleterious effects of rf penetration are likely to be too severe for a viable NMR imaging system without phase and amplitude correction of the data.

8.4.2 GENERAL THEORY

In the design of magnetic field coils, whether for transmitter probes, gradient assemblies, or the static magnet field itself, we start with the Biot–Savart expression. This gives the total magnetic field **B** produced at a point P distance r from a current I flowing in a particular wire configuration as

$$\mathbf{B} = \frac{\mu_0 I}{4\pi} \int \frac{d\mathbf{l} \times \mathbf{r}}{r^3}, \tag{8.8}$$

where **r** is the unit vector between the conductor line element $d\mathbf{l}$ and the point P, and μ_0 $(=4\pi \times 10^{-7})$ is the permeability of free space.

[8] D. I. Hoult and P. C. Lauterbur, *J. Magn. Reson.* **34,** 425 (1979).

[9] F. E. Terman, "Radio Engineers' Handbook." McGraw-Hill, New York, 1943.

[10] J. M. Libove and J. R. Singer, *J. Phys. E. Sci. Instrum.* **13,** 38 (1979).

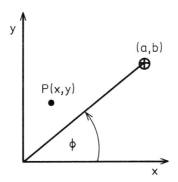

FIG. 8.8. Sketch showing coordinates and axes for a conductor at (a, b) and the point $P(x, y)$. (See text for details.)

Equation (8.8) may be simply integrated to obtain the field at the center of a circular conducting loop. Two such loops of radius a, spaced a apart in a parallel concentric configuration form the well-known Helmholtz coil giving a volume of uniform field about the coil center. This common arrangement, used for both transmitter and receiver coils does find application in NMR equipment, but for imaging applications it is generally uneconomic in terms of the space required.

Another straightforward case where Eq. (8.8) may be integrated is that of a straight wire. The field at a point distance r from the middle of a straight wire of length $2l$ along the z axis is from Eq. (8.8)

$$B = \frac{\mu_0 I}{2\pi r} \frac{l}{(l^2 + r^2)^{1/2}}, \tag{8.9}$$

and is a field tangential to a circle coaxial with z. In the limit $l \rightarrow \infty$ we obtain the well-known result

$$B = \mu_0 I / 2\pi r. \tag{8.10}$$

Consider now this wire displaced from the origin to coordinates a, b in the x, y plane (Fig. 8.8). The component of field at the point $P(x, y)$ along the x direction is from Eq. (8.10)

$$B_x(x, y) = \frac{\mu_0 I}{2\pi} \frac{a - y}{(a - y)^2 + (b - x)^2}. \tag{8.11}$$

Following Zupancic and Pirs,[11] we write Eq. (8.11) as the real part of the complex function

$$B_x(x, y) = (\mu_0 I / 2\pi) \, \mathrm{Re}\{[(a + ib) - (x + iy)]^{-1}\}. \tag{8.12}$$

[11] I. Zupancic and J. Pirs, *J. Phys. E Sci. Instrum.* **9**, 79 (1976).

Let $x + iy = \xi$ and $a + ib = ie^{i\phi}$, then Eq. (8.12) may be expanded as a Taylor series, provided $|\xi| < r$. In this case we may rewrite Eq. (8.12) as

$$B_x(x, y) = (\mu_0 I/2\pi) \operatorname{Re} \sum_{n=0}^{\infty} (\xi/r)^n e^{-i(n+1)\phi}. \qquad (8.13)$$

This rapidly converging expansion is particularly useful for quickly estimating the main field contributions for a particular set of straight conductors and current directions.

8.4.3. rf RECEIVER AND TRANSMITTER COILS

The design aim in both receiver and transmitter coils is to produce uniform magnet fields over as large a fraction of the coil volume as possible. The degree to which rf inhomogeneities can be tolerated depends on the NMR imaging technique and also on the use of the coil. For example, rf inhomogeneity in the transmitter coil may be more critical than nonuniform reception in the receiver coil, since the latter, though undesirable, may be compensated to some degree. Separate orthogonal transmitter and receiver coil probes have the advantage that both geometries may be individually optimized.

Using Eq. (8.13), we can easily estimate the optimum geometry for a four-wire transmitter coil with wires displaced as in Fig. 8.9. From the symmetry of the conductors, the field produced by wire pairs 1, 3 and 2, 4 contain only even orders of the expansion parameter ξ/r. Adding all components we obtain to fourth order in the expansion

$$B_x = (2\mu_0 I/r) \operatorname{Re}[1 - i(\xi/r)^2 \sin 3\phi - i(\xi/r)^4 \sin 5\phi + \ldots]. \qquad (8.14)$$

If $\phi = 60°$, the third-order correction vanishes and gives for the field in teslas per unit ampere turn (the field multiplier) at the center of the conductor assembly,

$$B_x{}^0 = B_x/nI = 8/10^7 r \qquad (8.15)$$

where r is in meters.

For simplicity, we have considered infinitely long conductors. An actual coil system is, of course, finite. The arrangement of currents in Fig. 8.9 suggests a number of possibilities for the current-return paths for finite coil systems. For example, if wires 1, 2 and 3, 4 are connected by return arcs, we have the well-known saddle geometry of Fig. 8.10. This geometry has been considered by Ginsberg and Melchner,[12] who suggest that for optimum homogeneity over the coil volume, the length-to-radius ratio should be

[12] D. M. Ginsberg and M. I. Melchner, *Rev. Sci. Instrum.* **41**, 122 (1970).

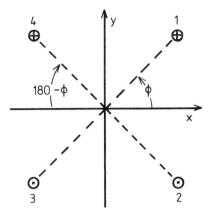

FIG. 8.9. Positions of the four infinite wires 1–4 and the current directions necessary to produce a uniform field B_1 along x. (N.B.: $\phi = 60°$ in this case. See text for further details.)

$l/a = 3.322$ with the field multiplier

$$B_x{}^0 = 7.53/10^7 r. \qquad (8.16)$$

The factor 7.53 is only slightly smaller than that of Eq. (8.15), which suggests that the return arcs and/or the infinite conductors are perhaps not so important. The field distribution plot for this coil configuration has already been discussed in Section 6.1.6.

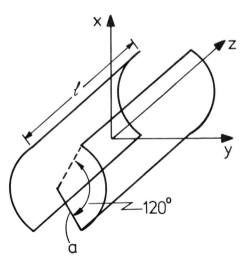

FIG. 8.10. Four wires as in Fig. 8.9 of finite length l with current return paths formed into arcs to give the standard saddle-coil geometry with radius a.

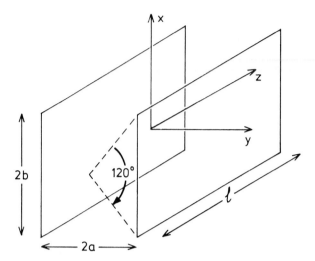

FIG. 8.11.   Four wires as in Fig. 8.9 of finite length l with straight current return paths, forming a rectangular coil pair of spacing $2a$. Note that field is produced along y axis, i.e., B_y.

The return paths could, of course, be straight as in Fig. 8.11. We have calculated the field distribution for this case which shows that the best homogeneity is obtained when $l/b = 2$. The excellent field uniformity obtained is shown in the field distribution plot of Fig. 8.12 for one quadrant of the mid-z, y plane. This result shows that B_y is uniform to better than 5% over most of the usable cross section. The field multiplier value of 54.36 mG/A turn is remarkably high in this case.

Another interesting return-path geometry is shown in Fig. 8.13. Both forward and return paths lie in the same optimum plane for each quadrant and end effects are minimized. Unlike the rectangular coil of Fig. 8.11, this coil allows full axial access. However, the field multiplier reduces and from Eq. (8.14) and (8.15) for infinite wires, becomes to third order,

$$B_x{}^0 = \frac{8}{10^7}\left(\frac{1}{r_1} - \frac{1}{r_2}\right). \tag{8.17}$$

Although we have stressed the field distributions for coils comprising a number of straight wires, it is of course possible to integrate Eq. (8.8) to obtain field distributions for curved geometries. Examples where we have performed this calculation were given for the flat circular coil in Section 4.3.2 and the standard saddle coil in Section 6.1.6. We have also considered other saddle geometries where the ratio l/a is varied. For example, Fig. 8.14a shows one quadrant of the midplane field distribution for $l/a = 1$. The field multiplier here is 22.37 mG/A turn. Figure 8.14b shows the improvement in field distribution when $l/a = 2$. The field multiplier increases to 29.45 mG/A turn.

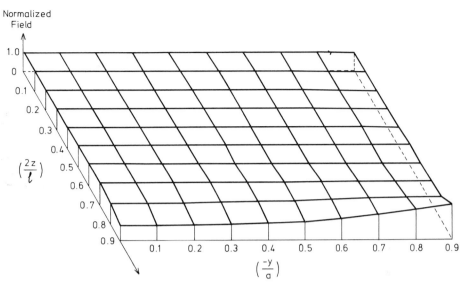

FIG. 8.12. Normalized field plot for one quadrant in the z,y plane for the rectangular coil ($l/a = 2\sqrt{3}$) of Fig. 8.11 above. The field is constant to better than 5% over an elliptic region of major axis $0.9l$ and minor axis $1.8a$. (See text for field multiplier.)

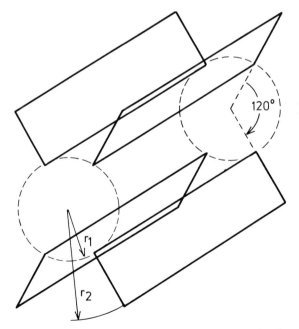

FIG. 8.13. Four wires as in Fig. 8.9 of finite length on a cylinder of radius r_1. The return paths have the same angular displacements but lie on a second cylinder of radius r_2.

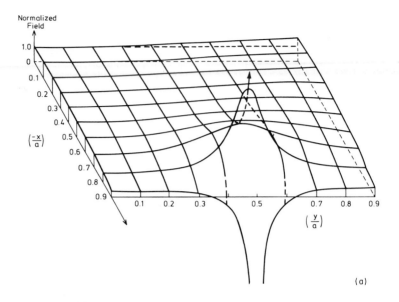

(a)

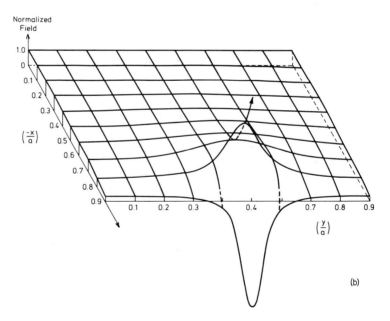

(b)

FIG. 8.14.   (a) Normalized field distribution in one quadrant of the mid-x,y plane for the saddle coil of Fig. 8.10. The ratio $l/a = 1$. This coil gives uniform fields to 10% within a circular region of radius $0.55a$. (N.B.: The field divergence occurs at a wire position.) (b) As in (a) above but with $l/a = 2$. The uniformity is much improved and is constant to 10% within a circular region of radius $0.7a$. (See text for field multipliers.)

Yet another interesting receiver or transmitter structure, in principle similar to the short saddle coil, is the crossed elliptical coil arrangement shown in Fig. 8.15. For an included angle $\theta = 90°$ between coil planes, as indicated, we obtain the field distribution plot for one quadrant in the mid-plane shown in Fig. 8.16. The ratio $l/a = 2.0$ for this coil and is thus comparable to the saddle-coil case with $l/a = 2$. The field multiplier for the cross-elliptical assembly is 25.80 mG/A turn. Both the saddle and cross-elliptical coils produce variations of up to 10% in the field distribution over cross sections out to $r = a/2$ and there seems little to choose between them on this score. The cross-elliptical configuration, however, does give a significantly lower field multiplier, suggesting that the saddle arrangement is to be preferred.

8.4.4. FIELD GRADIENT COILS

As with transmitter coil arrangements, there are a number of conductor configurations useful for the generation of magnetic field gradients. Perhaps the simplest arrangement is the Maxwell pair.[13] This is a pair of circular coils each with radius a, rather like a Helmholtz system, but with opposing currents giving zero field in the midplane. The optimum field gradient linearity is obtained when the coils are spaced $a\sqrt{3}$ apart. The Maxwell coil will produce a gradient $G_z = \partial B_z / \partial z$. In NMR imaging, however, we require gradients $G_x = \partial B_z / \partial x$ and $G_y = \partial B_z / \partial y$ in addition to G_z. From the symmetry argument it is clear that G_y can be produced from the same coil assembly as that used to generate G_x if the structure is physically rotated through 90°. We shall therefore consider ways of producing G_x.

From the theory of infinite straight conductors developed in Section 8.4.2, we see that four appropriately spaced wires all with currents in the same direction will produce the required gradient. Figure 8.17 shows a symmetric arrangement with wires lying on a circle of radius r with coordinates of the positive quadrant wire 1, for example, at $y = a$, $z = b$ (note the coordinate change). Since wires 1,3 carry current in the same direction, the sum of the corresponding field expressions, Eq. (8.13) for a point $P(y, z)$ contains only odd orders in the expansion parameter ξ/r. Adding the effects of all four conductors we obtain the expression

$$B(y, z) = (2\mu_0 I / \pi r) \, \text{Re}[(\xi/r) \cos 2\phi + (\xi/r)^3 \cos 4\phi$$

$$+ (\xi/r)^5 \cos 5\phi + \ldots] \tag{8.18}$$

where now $\xi = y + iz$. We notice that the third-order term vanishes in general when $\phi = (\pi/2 + n\pi)/4$, n integer. Two angles of particular interest are $\phi = 22.5°$ and $\phi = 67.5°$. When $\phi = 22.5°$, Eq. (8.18) gives the field

[13] I. E. Tanner, *Rev. Sci. Instrum.* **36**, 1086 (1965).

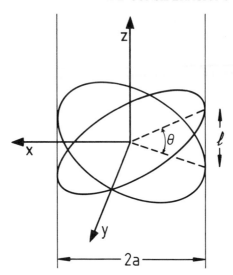

FIG. 8.15.   Crossed-ellipse transmitter/receiver coil wound on a cylinder of radius a. Included angle between elliptic planes is θ.

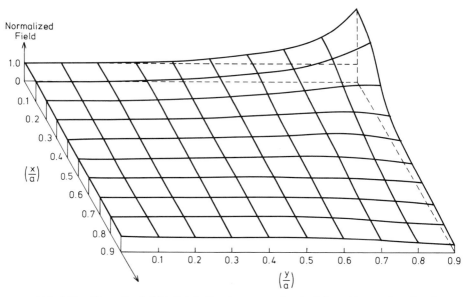

FIG. 8.16.   Normalized field distribution in one quadrant of the mid-x, y plane for the crossed-ellipse coil of Fig. 8.15 when $\theta = 90°$. The ratio $l/a = 2$ for this coil and the field uniformity is constant to 10% within a radius of $0.5a$. See text for field multiplier.

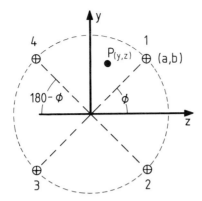

FIG. 8.17. Positions of the four infinite wires 1–4 and the current directions (all the same) necessary to produce a linear field gradient along y. (N.B.: $\phi = 67.5°$ in this case.)

gradient up to the fourth order in the expansion coefficient as

$$G_y = \partial B(y, z)/\partial y = 1.414\mu_0 I/\pi r^2,$$ (8.19a)

whereas for $\phi = 67.5°$,

$$G_y = -1.414\mu_0 I/\pi r^2.$$ (8.19b)

Equations (8.19a,b) suggest two interesting possibilities for current-return paths for the basic coil structure, both sketched in Fig. 8.18. The first case, Figure 8.18a is a fairly compact design but has the disadvantage that the return paths reduce the gradient-per-ampere turn produced. Equation (8.19b) becomes

$$G_y = -\frac{1.414 I}{\pi}\left(\frac{1}{r_1^2} - \frac{1}{r_2^2}\right).$$ (8.20)

The second gradient coil design employs two optimum angles, 22.5° and 67.5°, resulting in the arrangement sketched in Fig. 8.18b. Although a much more open design, and therefore of higher inductance, the forward and return paths give gradients which add. In this case G_y is given by

$$G_y = -\frac{1.414 I}{\pi}\left(\frac{1}{r_1^2} + \frac{1}{r_2^2}\right).$$ (8.21)

If, in addition, all four wires lie on parallel planes, as drawn, the fields from connecting wires which lie along the z axis, will not contribute to G_y. Unfortunately, for an axial access of $2a$, the overall length of this coil is $2l = 4.828a$, which becomes somewhat unwieldy for practical whole-body imaging systems.

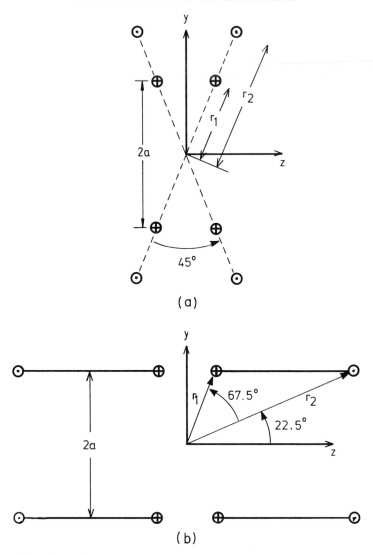

FIG. 8.18.   (a) Infinite-conductor arrangement for a compact coil to produce a gradient G_y and access diameter $2a$ along z. (b) Alternative return conductor path for coil (a) above.

We emphasize that the expressions given do not apply to finite wire lengths. The results given should therefore be treated as a guide to optimum coil geometries. For exact performance of coils with practical dimensions, one must resort to detailed computer integration of Eq. (8.1). This has been done for a number of the gradient coils mentioned and full details will be reported elsewhere.[14,14a]

Another class of gradient coil suitable for the generation of large transverse gradients comprises variants of the quadrupole structure in which both current density and angular displacement of the wires are varied.[15,16] Since many more conductors are generally used, the inductance of these variants is likely to be quite high, although the region of uniform gradient should be substantially larger than that achievable using just four wires.

8.5. Imaging-System Requirements

8.5.1. GENERAL SYSTEM OVERVIEW

An NMR imaging apparatus is basically an rf spectrometer, but with special additional items such as modulated rf carrier frequency, switched or modulated field gradient coils and even static-field sweep control under the direct command of the spectrometer. Because of the coherent nature of both the rf carrier wave and the various switched or modulated magnetic field gradient states and also through its sheer complexity, it is desirable to have the whole system under central computer control.[17]

Signal reception, digitization, and data handling are also integral parts of such a system and commercial signal averagers are available to perform this task if desired. The time-varying signals often take the form of fast, transient decays in NMR imaging schemes. Having obtained the averaged time signal, it will at some stage be necessary to operate on this using an algorithm from which the actual picture matrix may be calculated or extracted. This can be performed on- or off-line as desired. Finally, the data matrix must be displayed in some form, usually as a TV picture in which intensity or color variations or both are made proportional to signal size and hence mobile-proton content.[18] Other NMR parameters may also be displayed in picture form such as T_1 maps and flow data.

Obviously, for rapid alignment of the apparatus, it is better to have all facilities "on-line" and located in control and spectrometer consoles. This includes, ideally, the color TV display system. However, we find experimentally that the stray field from the main static magnet is too strong for a normal color TV set and upsets the color balance. It is therefore better to have this particular facility mobile so that it can be moved out of range of the magnet when running.

---

[14] V. Bangert, P. Mansfield, and P. G. Morris (to be published).

[14a] V. Bangert and P. Mansfield, *J. Phys. E Sci. Instrum.* (in press).

[15] G. Ödberg and L. Ödberg, *J. Magn. Reson.* **16**, 342 (1974).

[16] H. Brener, *Rev. Sci. Instrum.* **36**, 1666 (1965).

[17] P. Mansfield, A. A. Maudsley, and T. Baines, *J. Phys. E Sci. Instrum.* **9**, 27 (1976).

[18] I. L. Pykett and P. Mansfield, *Phys. Med. Biol.* **23**, 961 (1978).

Static-Field Coil
Transmitter Coil
G_z Gradient Coil
Patient-Handling Table
G_x Gradient Coil
Receiver Coil

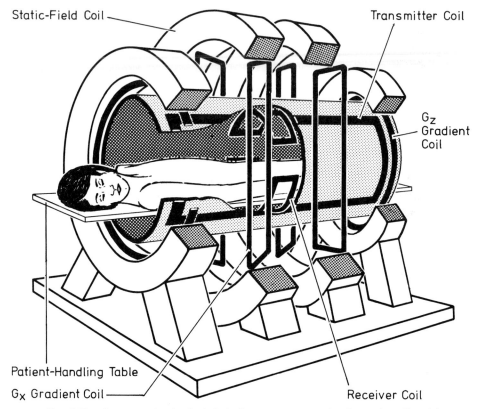

FIG. 8.19.   Cut away sketch of whole-body magnet system showing patient disposition. The various coil arrangements indicated are considered separately and shown in previous figures. The G_z coil is a straightforward Maxwell pair.

Black-and-white cathode-ray tube monitor displays in which electrostatic deflection of the beam is employed, do not seem so vulnerable to the stray magnetic field, and are therefore to be preferred as an "on-line" monitoring device while imaging.

A typical magnet and probe assembly including field gradient coils for whole-body imaging is sketched in Fig. 8.19. Axial disposition of the patient is indicated here, but transverse dispositions are also possible although they require a large gap between the central coils. Figure 8.20 shows schematically a complete NMR imaging system (*sans* magnet) for performing line-scanning experiments and planar imaging using either the echo-planar or the projection reconstruction technique. In the latter, FIDs following an initial selective pulse are examined in various fixed field gradients which are stepped through a series of projection angles from 0–180°. We shall take the various

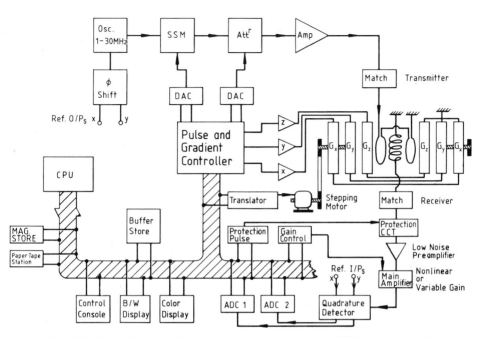

FIG. 8.20. Block diagram of the electronic control system for an NMR imaging machine. [See Fig. 8.19 for the various coil positions and main static magnet (not shown on this diagram).]

items referred to in Fig. 8.20 and describe them in turn, trying where possible to indicate the necessary specifications for such a working system. Although the system described incorporates many of the design features of our equipment, it is sufficiently flexible to perform many of the imaging techniques discussed in Chapter 4.

8.5.2 OSCILLATOR

This is a standard crystal oscillator operating at fixed frequency in the range of 1.0–30 MHz with or without thermal control. Such crystal oscillators are usually capable of producing a frequency with an accuracy of $1:10^6$ and thermal stabilization can improve this to $1:10^7$ or better. There seems little point in having a variable-frequency source, such as a stable frequency synthesizer. Once the system is designed and set up, that is to say, a center frequency chosen to correspond with the available static magnetic field, there is little likelihood that the frequency will need to be varied. If it does turn out to be necessary, for example, a 1.0-kG magnet will actually work at 4.26 MHz for protons or 1.72 MHz for phosphorus nuclei, then it is simple enough to have a replacement crystal specially cut to operate at the new frequency. Of course, implicit in this statement is the assumption that the bandwidth of the subsequent electronics is sufficient to handle the new frequency, or alternatively, that the rf electronics are retuneable.

The output from the crystal oscillator should have a power level corresponding to about 1–2-V peak into a 50 Ω load. The whole system should be matched to 50 Ω as this makes the task of design of subsequent rf stages so much easier, because many firms now offer 50 Ω matched components which are useful for quickly building up the rf system as a set of "plug-in" modules.

8.5.3. SINGLE-SIDEBAND MODULATOR

We have found it generally much more convenient to modulate the rf carrier when scanning specimens under fixed-field gradients, than to sweep the static magnetic field. Both methods are commonly used, but we restrict discussion to our own preferred method here.

In selective excitation a long, shaped 90° rf pulse is applied to the specimen which is also subjected to a linear magnetic field gradient. The rf pulse may be modulated to give a particular frequency offset. This modulation is best achieved by computing the Fourier transform of the stick spectrum with the required offset. The time-domain data are then transferred to a buffer store in the pulse controller. (The pulse controller performs more than one function and is fully discussed in Section 8.5.7.) Each time-domain word is strobed consecutively from the buffer store into the amplitude modulation unit. This may be either a carrier-suppressed double- or single-sideband modulator (SSM). In selective excitation only one of the sidebands is used, the other represents wasted rf power, and if too close to the carrier frequency, can give NMR interference with the other sideband. Thus, although all our work has used double-sideband modulators, we recommend SSM.

The SSM may be a digital-controlled attenuator system[17,19,20] or an analog-controlled system as described here. In this case the n-bit time data word is converted to an analog signal via the DAC[20a] (Fig. 8.21). This modulation signal is fed to phase splitter 2, and the equal-amplitude quadrature output signals are each fed to the X ports of two double-balanced mixers (for example, the HP10514A or the Merrimac DMF-2A-500). The carrier signal is split into two equal-amplitude quadrature phase signals (Merrimac OH-5-30) and fed to the L ports of the double-balanced mixers. The R port outputs are then combined in a 50-Ω combiner (Merrimac PDS-20-110). The output is amplified using an Avantek GPD402 amplifier, for example, which

[19] P. Mansfield and U. Haeberlen, *Z. Naturforsch.* **7**, 1082 (1973).
[20] T. Baines and P. Mansfield, *J. Phys. E Sci. Instrum.* **6**, 459 (1973).
[20a] Digital-to-analog converter (DAC).

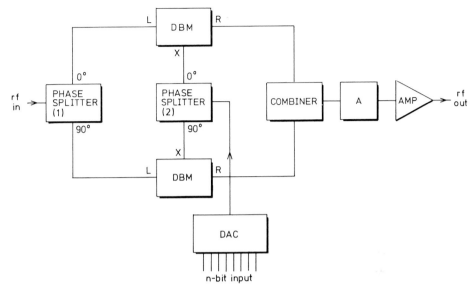

FIG. 8.21. Block diagram of a carrier-suppressed single-sideband modulator. Attenuator A is included to adjust the rf output level.

is preceded by an attenuator A_1 so that the SSM output is of the order of 1.0-V peak. This signal is then fed to the rf power amplifier stages. Although the SSM circuit has a fairly high carrier suppression, there may be still sufficient leakage to saturate the sensitive receiver system. It is therefore advisable to follow the SSM by a high-attenuation transmission gate, as indicated in Fig. 8.20. The gate may comprise several balanced modulator units (30-dB attenuation per stage), or a specially designed and shielded rf gate (130-dB attenuation).[20] (See Notes Added in Proof, Note 8.1.)

8.5.4. rf AMPLIFIER

Although indicated here as a single amplifier, it may be desirable to use a low-power broad-band amplifier which then drives a higher power narrower band system. The bandwidth needed is in fact quite narrow. The broadest bandwidth required is about 200 kHz and is set by the shortest pulse length that might be required, here assumed to be about 10.0 μsec.

The bandwidth required through gradient broadening of the specimen naturally depends on the total field variation across the specimen. In our case, we tend to use gradients giving a total field spread of about 1.0 G or about 4.2 kHz frequency variation in the slice selection phase. Thus a tuned rf amplifier with a center frequency of 4.0 MHz and a bandwidth of 0.2 MHz is more than adequate to produce a nonselective pulse with the selection gradient on. In fact, it is probably easier to buy a broadband amplifier of the requisite power covering the frequency range 0–32 MHz and not bother with tuned circuits until the probe stage. If a single amplifier unit is used, it will be necessary in general to precede it with an attenuator A as in Fig. 8.20 so that proper signal levels may be adjusted experimentally.

The rf power required depends very much on the imaging method used and also on the transmitter coil design and tuning arrangements. For whole-body imaging using selective excitation, we are currently using only between 4–6 W for a 7.4-msec 90° rf pulse. However, we have the immediate possibility of up to 300 W (attenuator A set to zero). The transmitter matching arrangement is a series tuned circuit,[17] and this presently limits the maximum transmitter coil turns to two (Fig. 8.22a). Parallel tuning allows more turns but requires a more complicated matching arrangement (Fig. 8.22b.) It should be stressed that power and field requirements are really no great problem for selective rf schemes. For SSFP and DEFT schemes, how-ever, much higher powers are required, typically 4–5 kW for 10-μsec 90° pulses.[21-24]

8.5.5. Receiver System

We have used an orthogonal receiver coil arrangement for two reasons: It gives some receiver protection and it is much easier to noise match. Also, the lower coupling to the transmitter gives less received transmitter noise than is picked up in a single-coil system. Of course, both single- and cross-coil probes require series noise blocking diodes in the transmitter output.

The actual circuit used in our system is a double-coil version of the circuit described elsewhere.[17] The series tuned receiver coil is broadband matched into a 50 Ω line with a crossed-diode termination before entering the low-noise receiver preamplifier. This is based on a low-noise commercial design. An interesting low-noise preamplifier circuit designed to operate at 4.0 MHz has recently been described.[25,26] This has a noise figure of 1.0 dB.

The main amplifier and phase-sensitive detector (PSD) are home built using Avantek GPD 402's as the broadband amplifying elements, while a

[21] W. S. Hinshaw, *J. Appl. Phys.* **47**, 3709 (1976).
[22] P. Rohan, M. Viola, E. Templar, and J. Wilcox, *Electron. Lett.* **4**, 442 (1968).
[22a] M. R. Osborne, *Electron. Eng.* **40**, 436 (1968).
[23] C. L. Ruthroff, *Proc. IRE* **47**, 1337 (1959).
[24] G. N. Holland and E. Heysmond, *J. Phys. E Sci. Instrum.* **12**, 480 (1979).
[25] D. I. Hoult, *Rev. Sci. Instrum.* **50**, 193 (1979).
[26] D. I. Hoult, *in* "Progress in NMR Spectroscopy" (J. W. Emsley, J. Feeney, and L. H. Sutcliffe, eds.), Vol. 12, p. 41. Pergamon, Oxford, 1978.

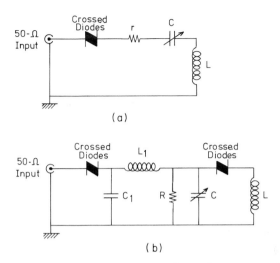

FIG. 8.22. (a) A 50-Ω matched series-tuned transmitter coil arrangement. The series crossed diodes improve the ring down following the pulse and also serve to reduce transmitter noise during signal reception. (b) An alternative 50-Ω matched parallel tuned transmitter. Again, the cross diodes are necessary for noise suppression and improved transient response. (N. B.: Both circuits, but with *parallel* crossed-diode arrangements, can be used for signal reception.)

double balanced mixer like the Merrimac DMF-2A-500 may be used for phase detection. The total voltage gain of the system, including the pre-amplifier and main amplifier is approximately 10^4 but can be reduced by introducing attenuation within the main amplifier.

Preamplifier and main amplifier protection are allowed for, but as yet we have found their use unnecessary. So far we have used only single-channel phase-sensitive detection. But quadrature detection has advantages, not the least being a $\sqrt{2}$ improvement in the S/N.

The reference signal for the PSD is taken from the 4.0-MHz source and fed via a limiter amplifier through a phase shifter. In our case, this is a 0.3-μsec 10 turn helical delay line made commercially by AD-YU and ensures a full 360° phase shift. While discussing phase shifting, we should point out a serious difficulty already encountered with whole-body line-scan imaging. This is the problem of rf phase shifts especially in the case of received signals. It is almost impossible to manually correct for phase shifts while scanning and, in general, it saves considerable time and effort to simply record and store all time data for subsequent Fourier transform rotation, i.e., phase correction via software at a later stage. By this means apparent picture degradation while scanning can easily be rectified provided the time data or the full frequency data are saved.

There are a number of analog-to-digital converters (ADCs) commercially available. We have used a 10-bit converter from Datalabs, but this is rather slow. Again, the conversion speed required of the ADC depends on the type of imaging experiment planned, but we do not envisage conversion rates faster than about 1.0 μsec. This figure includes data transfer time to either the computer core or a fast buffer store. If more than eight bits are required, then additional time and software might be needed (depending on the type of computer available) to transfer the contents of ADC1 and ADC2 sequentially following simultaneous signal sampling of two quadrature channels.

8.5.6. DYNAMIC RANGE

An important point which arises in the receiver system as the input data rate is increased is its dynamic range. Put in its simplest terms, consider an image comprising a rectangular array of m^2 pixels. Suppose each pixel produces a voltage v_p at the receiver input. The receiver will therefore be presented, at least initially, with a total input voltage of $m^2 v_p$. Taking $m = 128$, this voltage is $16,384 v_p$. If we further assume that $v_p = 0.1 \mu V$, the initial signal to be amplified is ~ 1.6 mV. The preamplifier therefore must be able to handle linear signals from 0.1 μV to 1.6 mV. Taking a modest gain for the low-noise preamplifier as 50, the output signal will range from 5 μV to 80 mV. This may be further amplified in a postamplifier, but we soon see that maintaining the dynamic range becomes a problem. For example, to maintain one-pixel resolution, the lowest signal must be amplified up to the digitization threshold of the ADC. If we take this as 1.0 mV per bit resolution, the largest signal now becomes 16.384 V and the ADC capacity must be $(7 + d)$ bits, where d is the final picture depth. Taking a 16-level gray scale ($d = 4$) necessitates an 11-bit ADC. Up to 12-bit ADCs of sufficient speed are now commercially available, falling comfortably within the central processor unit (CPU) word length for the machines quoted (see Section 8.5.8). The situation described is true even for noiseless input signals, in which case the 1-bit uncertainty per pixel in the final image arises from the quantization error in sampling and performing the FT. From this standpoint, the system is just determined. More bits in the ADC allow greater picture depth for a given matrix size. The same reasoning applies when the input signal is noisy. In this case the initial S/N of the FID must be at least 2^{n+d} for single-bit resolution in the final picture where $m = 2^n$. If this condition is not met, then some signal averaging is required. In this case, the bit size of the ADC can be less than $n + d$. For example, with a 9-bit converter, the requisite picture depth may be built up by a 16-shot average. The final data is thus, again, 11 bits.

Our discussion so far has been centered on single-slice planar imaging.

The dynamic range problem can be expected to worsen with three-dimensional imaging or multiplanar imaging.[27]

One solution to this dilemma is to employ data compression in the post-amplifier or preamplifier or both. Logarithmic amplifiers for this purpose are available, although we emphasize that logarithmic compression is not necessarily the best route. The procedure is therefore to compress the dynamic range of the detected nuclear signal so that it spans the ADC bit range. The linear data are recovered by stretching numerically in the CPU.

If the input data were noiseless, the procedure outlined would produce an acceptable reproduction of the original spin distribution, subject only to rounding errors. In fact the noise inherent in all imaging schemes exacerbates the rounding problem resulting in distortions in the final picture. For data compression to work well therefore, it is essential to have the best S/N possible. Signal averaging should therefore be performed in the time domain before data stretching.

Fortunately, in NMR imaging experiments, the full dynamic range of the receiver system is required only for the first few points of the transient signal. The majority of the long-time signal behavior, which contains the high-resolution detail, can usually be expected to fall well within the capacity of a linear 10-bit ADC.

In the case of projection reconstruction methods,[28] where signal averaging is implicit, the problem is eased somewhat, since in order to image a cubic lattice of $m^3 = 2^{3n}$ points, the dynamic range of the ADC does not have to equal $d + 3n/2$, but $d + n$, and for two-dimensional imaging, $d + n/2$ for fully determined matrices. A 128^3 three-dimensional array, for example, requires an ADC capacity of only $(7 + d)$ bits.[28a] If greater picture depth is required, therefore, exceeding the ADC capacity, this may be achieved at the expense of further signal averaging, for example, coadding several signals for a given projection direction. (See Notes Added in Proof, Note 8.2.)

8.5.7. Pulse Controller

The pulse controller, as its name implies, produces pulses and shaped waveforms which directly control the various parts of the imaging system. We have already discussed one function of the pulse controller in Section 8.5.3 in connection with single-sideband modulation of the rf carrier. Another important function is the control of the field gradient modulation unit, *vide infra*.

---

[27] P. Mansfield and I. L. Pykett, *J. Magn. Reson.* **29**, 355 (1978).
[28] P. C. Lauterbur and C.-M. Lai *IEEE Trans.* **27**, 1227 (1980).
[28a] We are grateful to Dr. M. Burl for clarifying discussions regarding dynamic range.

Most of our imaging to date has been performed by combining the function of the pulse controller with the computer itself. Initially this seemed the most economical way of using a computer to "drive" an NMR spectrometer. However, a disadvantage of that approach is that the CPU is permanently tied up either in control, data acquisition, computation, or display. Of course, that is precisely what is required of a computer in terms of cost effectiveness. But as the type of experiment becomes more complicated, as with NMR imaging, we often arrive at a situation where it is desirable to change the order of things or even to use the CPU in a multitask mode. For example, it is convenient to be able to continuously display the picture data while it is being accumulated; and even better, to be able to manipulate the data while running. Dedicated sequential operation makes this kind of use impossible and that is why we now discuss multitasking systems. Figure 8.20 is essentially a multitask system and allows control, data acquisition, and display to proceed independently. Of course, the computer has general overview of the complete system, but can be branched or interrupted via the Control Console.

Our first planar images[27,29] were produced using a home-built pulse controller operating in closed-loop synchronism with the CPU. But even this arrangement is not flexible enough, and a more versatile solution lies in using localized microprocessors for the pulse controller, gradient-coil rotation (as discussed), and possibly as a buffer between the ADCs and the CPU.

8.5.8. THE CENTRAL PROCESSOR UNIT

A number of suitable CPUs are currently available which have the requisite number of bits and a fast enough memory or machine cycle time. Two computers with which we have had experience and which appear to adequately fill the role are the Honeywell 716 (an updated and faster version of the H316) and the Data General Nova 3, both with hard-wired multiply and direct-memory access (DMA). Many other computer systems are available, for example, the Data General Eclipse, often used with X-ray CT imaging systems. A memory size of $32 \text{ K} \times 16$ bits seems adequate for most situations, but for large-picture manipulation an additional $32 \text{ K} \times 16$ bits is often a necessity.

Magnetic core memory is to be preferred, but solid-state volatile memory is cheaper. Most current computers offering solid-state memory supply refreshable memory and, in the Nova 3, for example, this has to be refreshed every 25 μsec. An automatic machine interrupt does this, but it means that

[29] P. Mansfield, P. G. Morris, R. J. Ordidge, I. L. Pykett, V. Bangert, and R. E. Coupland, *Phil. Trans. R. Soc. London Ser. B* **289** (1980).

the CPU cannot be used for sequential timed operations as we currently do with the H316. However, as stated previously, the present control philosophy would not require the CPU to be used in this manner.

Data storage capacity can be extended with magnetic tape, disk, diskette, or even paper tape. But we stress that for manipulation of picture data, and by this we mean spatial averaging, Fourier transformation, windowing, rotation, interpolation, etc., a large capacity dynamic memory is really a prerequisite. For example, a picture comprising 256^2 data points with 8-bit depth will require a memory of $2^{15} \times 16$ bits or 32 K words of memory for the picture data alone. For 512^2 data points the picture storage memory required increases to 128 K. We point out that many NMR imagers work with 128^2 real data arrays and rely on interpolation[30] to increase the picture size. In this case, interpolation may be performed on-line, thus occupying little over 16 K of computer memory. Naturally, a larger data store is required ultimately for the final interpolated data, but this can be treated externally to the CPU and may well constitute a buffer memory for the picture display system. (See Notes Added in Proof; Note 8.3.)

8.5.9. FIELD GRADIENT MODULATION UNIT

An essential part of the imaging schemes concentrated on here is the provision of linear magnetic field gradients along the three principal coordinate axes of the system. We have already discussed the principles of coil design. We now discuss the current drive unit.

A difficulty with any imaging technique requiring fairly rapid gradient switching, especially with apparatus of whole-body proportions, is the provision of large modulated currents. Full-scale line-scan imaging has already been achieved and in our early experiments,[31,32] for example, the largest gradient used was 0.04 G cm^{-1}. The gradient coil required a drive current of 25 A and was switched in 50 μsec.

The modulation unit comprises essentially three balanced on/off switches. Each switch consists of complementary driven pairs of power Darlingtons, the gradient coil in each switch being placed in series with one of the collector leads through a suitable current-limiting resistor. This particular approach, though unsophisticated, maintains the total current to each switching-unit constant, thereby minimizing power supply surges. The current rise time is controlled simply by the ratio of inductance to series resistance of each gradient coil. A Zener diode stack directly across the gradient coil helps to

[30] R. N. Bracewell, *Aust. J. Phys.* **9,** 297 (1956).
[31] P. Mansfield, I. L. Pykett, P. G. Morris, and R. E. Coupland, *Br. J. Radiol.* **51,** 921 (1978).
[32] P. G. Morris, P. Mansfield, I. L. Pykett, R. J. Ordidge, and R. E. Coupland, *IEEE Trans.* **26,** 2817 (1979).

speed up the trailing edge of the current pulse when switched off. Each switch is driven by complementary signals derived from low-level TTL circuitry.

For echo-planar imaging, a more sophisticated switching system is required, since it is necessary to periodically reverse at least one of the gradient currents. This kind of power-switching circuitry works well for a fixed-imaging scheme, but in general it is too inflexible as part of a development system.

A more versatile but much more costly approach is to use wide-band bipolar current amplifiers driven from a TTL waveform-shaping circuit. In a more ambitious arrangement, the shaping would be done using a dedicated microprocessor slaved to the CPU.

A low-power gradient coil driver which uses a microprocessor has recently been described by Lai et al.[33] This circuit has been specially developed to generate orthogonal current drives to two gradient coils G_x and G_y. With this arrangement, a rotating vector gradient can be electronically stepped by small angular displacements to produce a series of image projections as required in the projection reconstruction method of imaging. An alternative approach to generating a rotating vector gradient, which uses multiplying DACs or MDACs, has recently been described.[33a]

A simpler approach which we have tried uses one gradient coil which is mechanically stepped through the desired angular displacement under local microprocessor control.[34] Audiofrequency driver amplifiers may also be employed in the modulation of field gradients which are used in some techniques to define the imaging plane and a rotating vector gradient.

8.5.10. PICTURE DISPLAY

The spin parameter distribution is, of course, a set of numbers with a data-point depth of d bits. Early images were either direct printouts of the number matrix or utilized some crude gray-scale approximation by selecting particular teletype characters or groups of characters to represent the gray tone.[35,36] Some image presentations employ a graph plotter.[21] In one version the plot of consecutive data lines is laterally displaced to give the appearance of a third-angle projection (see Fig. 6.25). Another method uses a graph plotter in which the pen is caused to oscillate while traversing one picture line. The data are used to amplitude modulate the pen oscillation to produce an effective gray scale. An alternative to this is frequency modulation of the

[33] C.-M. Lai, J. W. Shook, and P. C. Lauterbur, *Chem. Biomed. Environ. Instrument.* **9,** 1 (1979).
[33a] P. A. Bottomley, *J. Phys. E.* **14,** 1052 (1981).
[34] V. Bangert and P. Mansfield, (to be published).
[35] P. C. Lauterbur, *Pure Appl. Chem.* **40,** 149 (1974).
[36] A. Kumar, D. Welti, and R. R. Ernst, *J. Magn. Reson.* **18,** 69 (1975).

pen at constant amplitude to produce a variable density shading. It becomes somewhat difficult by any means to present the data distribution as a gray-scale modulation when d is greater than 4, corresponding to 16 levels. However, in this case, selected parts of the distribution may be examined by compressing the desired data slice into a 4-bit window. This process effectively increases the picture contrast. Each data point may be transferred from the CPU, windowed as required and stored in a buffer memory.

In a simple electronic video display system, the 4-bit data are converted to an analog signal which is then used to directly modulate the beam intensity of a monochrome TV display or CRT. In practice, this method does not work too well, since the screen intensity is a nonlinear function of the luminance drive voltage. Hardware or software compensation can, of course, be done, and this is precisely the method used for standard TV systems. However, our experience is that standard TV systems tend to compress the intensity range from 16 levels to around 8–10 levels covering black to white.

An alternative to amplitude modulation of the luminance input is what we have called exposure-time modulation (ETM). In the system described,[37] a CRT beam is switched on for various time intervals t_{on} given by

$$t_{on} = (\sqrt{2})^{n-1} \times 0.1 \ \mu sec, \qquad (8.22)$$

for $n = 1, \ldots, 15$. The beam on time is therefore varied from 0.1 to 12.4 μsec and produces a very linear intensity step-wedge density distinguishing 14–15 levels. In this system, the luminance voltage is constant. Subsequent photography usually reduces the distinguishable levels to about eight or so at the printing stage. Nevertheless, for direct screen viewing, there seems to be some distinct advantage using this method.

Image contrast may be artificially enhanced by using color display systems. Here, each picture point is converted to arbitrary color and intensity by encoding the data point into a prescribed red, green, blue, and luminance signal combination. This can be done by modifying a domestic TV set,[38] however, very sophisticated systems are commercially available which can present data arrays in excess of 512^2, thereby producing very high-resolution pictures.[18]

A mechanical system for producing color pictures has been described,[39] which uses a modified graph plotter and three ETM colored lights focused onto a photographically sensitive sheet. The modulated light spot is scanned across and down the photographic sheet in the manner of a very slowly scanned TV spot.

[37] T. Baines and P. Mansfield, *J. Phys. E. Sci. Instrum.* **9**, 809 (1976).
[38] A. K. Boardman, *Phys. Med. Biol.* **21**, 289 (1976).
[39] P. A. Bottomley, W. S. Hinshaw, and G. N. Holland, *Phys. Med. Biol.* **23**, 309 (1978).

An advantage of TV displays is the immediate readout of the data. However, for hard copy, either black and white or color, mechanical scanning can produce excellent results.[40] Commercial photowriters working on a similar principle to the above-described mechanical system, have been used for several years in the recording and analysis of monochrome and colored satellite pictures of the earth for geographical and military use. The resolution and contrast obtainable are exceedingly good and surpass any NMR imaging requirement at present. The picture is usually produced directly on a transparency, which photographically has a much larger linear intensity range. The problem of photographic printing and loss of contrast still remains, especially in monochrome.

A way of artificially increasing contrast in black-and-white pictures is by using a multiple-cycle display. This means that of the full number of data levels to be displayed, for example, 256 in a two-cycle display, the first 128 levels in the cycle are arranged to span the full contrast range, zero corresponding to black and 127 corresponding to white. Level 128, although of higher intensity, is reset to zero and further levels in the second cycle in the range 128–255 take on the same set of gray levels starting at black and proceeding through to white. Three-cycle displays may be similarly produced. Examples of pictures using multicycle displays[29] are given in Chapter 7.

It is perhaps worth pointing out that having produced an image, there are generally a variety of possibilities for manipulating the data in order to improve its appearance.[41] These manipulations are currently termed data "massaging," a somewhat unfortunate expression. However, information theory fixes certain constraints so that, for example, spatial S/N enhancement may be performed at the expense of point resolution. Edge enhancement is another technique which may be important if a small signal is to be detected on a larger "baseline" signal, i.e., if the picture contrast is poor. These manipulations are concerned with the picture data once obtained. It is emphasized, however, that there is no substitute for good initial data, We have mentioned already coaddition of data or signal summing which gives $\sqrt{N}$ improvement in S/N, where N is the number of signals summed. However, exponential averaging can be advantageous for optimization of an experiment while running continuously. In this method the current data average A_n is constantly updated relative to its previous value A_{n-1} using the expression

$$A_n = A_{n-1} + (S_n - A_{n-1})/M, \qquad (8.23)$$

where S_n is the current signal in a stream of measurements 1, 2, ..., n. The weighting or integration factor M ($=2^l$, l integer) effectively controls the

[40] E. R. Andrew, P. A. Bottomley, W. S. Hinshaw, G. N. Holland, W. S. Moore, and C. Simaroj, *Phys. Med. Biol.* **22**, 971 (1977).
[41] W. K. Pratt, "Digital Image Processing." Wiley, New York, 1978.

response time of A_n to current changes through S_n. The S/N improvement is $(2M - 1)^{1/2}$. Small values of M give fast response but poor S/N, and vice versa.

8.6. Optimization of Signal-to-Noise Ratio

8.6.1. INTRODUCTION

Important considerations in the discussion of signal-to-noise ratios are the optimization of the coil geometry, receiver system, and operating frequency. These considerations will apply to all imaging methods. As we have already seen in Chapter 6, the range of operating frequency is set by rf penetration problems, which in turn revolve around object size. The receiver system has already been discussed in Section 8.5.5. We here concern ourselves, therefore, more with the probe and its effect on signal reception, than with receiver electronics.

For good signal reception a high quality factor or Q is required for the receiver coil, in addition to a low-noise preamplifier. For traditional NMR experiments a high coil filling factor is usually advocated. The implication is that a tightly wrapped coil about the specimen should be best, and indeed that would be the case if the specimen did not alter the electrical properties of the coil. In fact, as we have seen, for biological imaging, the body is a conducting mass with a resistivity ρ of between 200–500 Ω cm. This means that a coil closely wrapped about the body when energized, will induce circulating currents which naturally absorb power from the coil. At high frequencies, the induced current penetration depth is small but the current is high, thereby shielding the interior of the body from the rf field. At lower frequencies, the rf field B_1 can reach inside the body to excite the spins. If there is any attenuation of B_1 going into the subject, one can expect a signal attenuation from the interior regions of the body as the nuclear signals try to radiate out from the conducting mass.

8.6.2. EFFECTIVE Q

The effect of induced currents in the body produces a marked change on the Q values of tightly wound coils. In particular, we find it difficult to obtain receiver Q values in excess of 20 for the thoracic region and about 30 for the abdominal region for a body-sized flat five-turn elliptical coil.[32] Whatever wire gauge is used, and even with coil cooling, one cannot expect to obtain significantly higher Q values. However, if the coil is made larger the Q value will increase, but the filling factor decreases. What is of importance is the resultant received signal which depends on Q and filling factor. An empty coil will have a high Q, but produces no signal. A full coil on the other hand has a large filling factor but low Q. The question, therefore, is whether a geometrical optimization of sample/coil diameter can be used to maximize the received signal. To take the discussion further we must consider the factors which affect both the effective coil quality factor Q_{eff} and the S/N for the receiver coil.

In the following analysis, we consider for simplicity the case of a solenoidal receiver coil. However, the general argument developed there may be shown by simple dimensional analysis arguments to be valid for any receiver geometry, in particular, the saddle geometry, or its crossed elliptical coil variant, *vide supra*.

The effect of a long specimen in a coil of inductance L may be represented by an equivalent series resistance r_e in addition to the intrinsic coil resistance r_i. The equivalent circuit is shown in Fig. 8.23a. The effective quality factor for this circuit is given by

$$Q_{eff} = \omega L/(r_i + r_e) = Q_i Q_s/(Q_i + Q_s), \tag{8.24}$$

where

$$Q_i = \omega L/r_i, \tag{8.25}$$

$$Q_s = \omega L/r_e. \tag{8.26}$$

An assumption implicit in Eq. (8.24) is that the induced loss does not change the coil inductance L. Strictly speaking, this is not true in general. However, for relatively high resistivity specimens, we completely ignore this change.

We point out that Q_s as defined in Eq. (8.26) is not exclusively concerned with the specimen. A more general definition of Q for a system is the ratio of stored energy to mean energy dissipation per cycle. Clearly such a definition

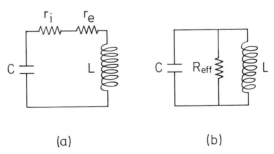

(a)   (b)

FIG. 8.23.   (a) Simple tuned circuit with series damping resistors. The intrinsic coil resistance is r_i and the equivalent sample loading resistance is r_e. (b) As in (a) above, but with r_i and r_e combined to form a single, parallel effective damping resistor R_{eff}.

could apply specifically to the stored magnetic energy and dissipation within the specimen itself. This could be evaluated by integrating the known magnetic energy distribution within the sample using Eqs. (6.26) and (6.27). The dissipation could similarly be evaluated by integrating the dissipation per unit volume σE^2 where σ is the sample conductivity and E the electric field [Eq. (6.25); see also Chapter 9]. In our view, this approach though rigorous is not as useful as one based on easily measured Q values.

8.6.3. SPECIMEN LOADING

Two energy-loss mechanisms are responsible for Q changes in a coil, magnetically induced losses and dielectric losses. In a solenoid containing a specimen as illustrated in Fig. 8.24, the dielectric losses in the specimen will be quite small, especially at low frequencies, since the highest electric field exists *between* coil turns. Only stray applied E fields will couple with the specimen. The total stray capacitance C_s across the coil is related to the interturn distributed capacitance C_d by the approximate expression $C_s = C_d/n$, which for a practical probe circuit should be less than the tuning capacitor. This

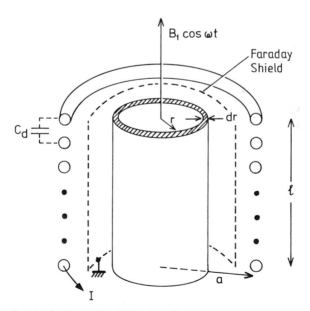

FIG. 8.24.   Sketch of a short solenoidal coil, radius a, and length l containing a cylindrical annulus of conducting sample, radius r and thickness dr. The fluctuating current I is taken to produce a uniform field $B_1 \cos \omega t$ throughout the coil. The interturn distributed capacitance of the coil is C_d. The Faraday shield reduces dielectric absorption in the specimen.

amounts to requiring the self-resonant frequency of the coil to be much higher than the desired operating frequency.

For a conducting sample, therefore, we shall take the principal loss mechanism as that magnetically induced, and we ignore entirely dielectric losses.[42] This assumption can be validated if necessary by using a Faraday screen. The case for spherical samples has been treated by Hoult and Lauterbur,[8] but for medical imaging one is generally more interested in extended samples placed in relatively short-probe assemblies. As a first approximation to this case, therefore, we consider a homogeneous conducting cylindrical specimen in a solenoidal coil. Let the specimen conductivity be σ. For a cylindrical element of radius r, thickness dr, and length l, the elementary conductance is from Fig. 8.24,

$$dg = \sigma l \, dr / 2\pi r. \tag{8.27}$$

If we further assume that an alternating magnetic field $B_1 \cos \omega t$ couples uniformly with all regions of the specimen, then the peak induced-loop voltage in the annular specimen is given simply by

$$\mathscr{E} = \pi r^2 \omega B_1. \tag{8.28}$$

Equation (8.28) is only approximately true for low frequencies in biological tissue. Exact expressions are developed and discussed in Chapter 9. The total power absorbed in a cylindrical specimen is therefore from Eqs. (8.27) and (8.28),

$$P = \int_{r_1}^{r_2} \mathscr{E}^2 \, dg/2 = \pi \sigma \omega^2 B_1^2 l (r_2^4 - r_1^4)/10. \tag{8.29}$$

This power is dissipated in the equivalent series resistor r_e (Fig. 8.23), that is to say

$$P = \tfrac{1}{2} I^2 r_e. \tag{8.30}$$

The B_1 field produced at the center of a solenoid of length l and radius a is[43]

$$B_1 = \mu_0 N_t I \, \frac{l/2}{[(l/2)^2 + a^2]^{1/2}} \tag{8.31}$$

where N_t is the number of coil turns and I the current flowing. For a coil with $l/2 > a$, we may assume that B_1 is fairly uniform over the whole of the coil volume. In this circumstance, from Eqs. (8.29)–(8.31), we obtain

$$r_e = (\pi \omega^2 N_t^2 \mu_0^2 \sigma / 8l)[r_2^4 - r_1^4]. \tag{8.32}$$

---

[42] For a full discussion of dielectric losses see D. G. Gadian and F. N. H. Robinson, *J. Magn. Reson.* **34**, 449 (1979); also Hoult and Lauterbur,[8] and Terman.[9]

[43] C. A. Coulson, "Electricity." Oliver & Boyd, London, 1953.

For a solid cylinder of material $r_1 = 0$, and when the coil is completely filled $r_2 = a$, giving from Eq. (8.32) the maximum equivalent resistance

$$r_{\text{emax}} = Ka^4. \tag{8.33}$$

where the constant K is the first term in parentheses in Eq. (8.32). For $r_2 < a$ we obtain from Eq. (8.33),

$$r_e/r_{\text{emax}} = (r/a)^4. \tag{8.34}$$

It is also clear from Eq. (8.29) that if the B_1 field penetrates only part of the specimen with annular thickness approximately equal to the rf penetration depth δ, then we obtain the alternative higher frequency relationship

$$r_e/r_{\text{emax}} = (r/a)^3. \tag{8.35}$$

We stress that the assumption that L is constant begins to break down at higher frequencies, so that Eq. (8.35) should be regarded only as a guide, and perhaps one should adopt a more general functional relationship of the kind

$$r_e/r_{\text{emax}} = f(r/a), \tag{8.36}$$

where $f(r/a)$ can be determined experimentally.

8.6.4. SIGNAL-TO-NOISE RATIO

The signal-to-noise ratio produced by a sample of volume V_s in a solenoidal coil of volume V_c (see Fig. 8.23b) is given by[44]

$$\frac{S}{N} = \frac{K_0 N_t \omega_0{}^2 A Q_{\text{eff}}(V_s/V_c)}{(4kT \, \Delta v R_{\text{eff}})^{1/2}}, \tag{8.37}$$

where k is Boltzmann's constant and $K_0 = 2\pi\chi_0(T)/\gamma$ in which T is the absolute temperature of the specimen (here also assumed to be the coil temperature), $\chi_0(T)$ the specimen susceptibility, and γ the magnetogyric ratio.

The noise bandwidth $\Delta v = \Delta\omega/2\pi$; Q_{eff} and the effective noise resistance R_{eff} are of course interrelated. For example, we may write R_{eff} as

$$R_{\text{eff}} = Q_{\text{eff}}\omega_0 L. \tag{8.38}$$

The coil inductance L is given by[9]

$$L = \alpha_c N_t{}^2 A^2/V_c, \tag{8.39}$$

where A is the coil cross-sectional area and α_c a harmless constant. By substituting Eqs. (8.38) and (8.39) into Eq. (8.37) and introducing the pre-

[44] A. Abragam, "The Principles of Nuclear Magnetism." Oxford Univ. Press (Clarendon), London and New York, 1961.

amplifier noise factor F_n defined by

$$F_n = [(V_{ns}^2 + V_{na}^2)^{1/2}]/V_{ns},\tag{8.40}$$

in which V_{ns} and V_{na} are the source and amplifier noise voltages respectively, we obtain finally

$$S/N = (S/N)_0 Q_{eff}(r/a)^2,\tag{8.41}$$

where the specific signal-to-noise ratio $(S/N)_0$ is given by

$$(S/N)_0 = K_0\omega_0 \left/ \left[\frac{2kT\alpha_c}{\pi V_c}\right]^{1/2} F_n,\right.\tag{8.42}$$

and is fixed for constant temperature, frequency, and coil volume. Our formulation differs from that of Hoult and Richards[45] in that we have explicitly sought to retain the Q dependence, which is probably the easiest coil parameter to measure. We note that $(S/N)_0$ is proportional to ω_0; rf skin-depth factors in the coil wire itself, which give rise to a frequency dependent r_i, are completely accounted for in Q_{eff}.

Equation (8.41) allows us to seek an optimization using the previously derived expression for Q_{eff} [Eq. (8.24)], together with the equivalent resistance Eq. (8.34). For example, the specimen Q from Eqs. (8.26) and (8.34) may be written as

$$Q_s = Q_{min}(a/r)^4,\tag{8.43}$$

where Q_{min} is the value of Q_s when $r = a$, that is for a completely filled coil. We define a S/N optimization factor F_o from Eq. (8.41), given by

$$F_o = \frac{S/N}{(S/N)_0} = \frac{Q_i Q_{min}(a/r)^2}{Q_i + Q_{min}(a/r)^4}.\tag{8.44}$$

This expression gives an optimum value F_{omax} when

$$(a/r)^4 = Q_i/Q_{min}.\tag{8.45}$$

As an example, consider a coil which when unloaded gives $Q_i = 200$ and when loaded gives $Q_{min} = 20$; F_{omax} occurs when $r = 0.562a$ and the factor $F_{omax} = 31.62$, compared with $F_o(a) = 18$ for a full coil. In this particular example there is nearly a factor of two in S/N ratio to be gained. Clearly, the greater the ratio Q_i/Q_{min}, the more is to be gained by such an optimization procedure. The parametric plot of Eq. (8.44) in normalized form, i.e., F_o/Q_{min} is shown in Fig. 8.25 where the parameter $\alpha = Q_i/Q_{min}$.

Finally, we remark that the more general formulation

$$Q_s = Q_{min} f(a/r)\tag{8.46}$$

[45] D. I. Hoult and R. E. Richards, *J. Magn. Reson.* **24,** 71 (1976).

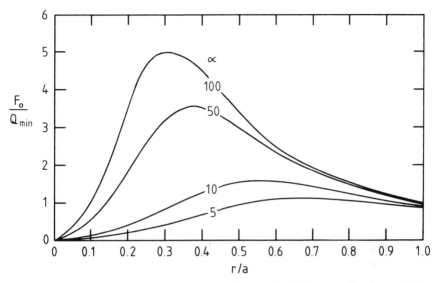

FIG. 8.25. Graph of the noise optimization factor/unit minimum quality factor F_o/Q_{min} versus r/a for various values of the parameter $\alpha = Q_i/Q_{min}$.

may also be used to optimize S/N empirically using an experimental determination of $f(a/r)$. (See Notes Added in Proof, Note 8.4.)

8.6.5. NOISE ALIASING IN SAMPLING

In our signal-to-noise analysis in Section 8.6.4 we have assumed implicitly that the noise arises from thermal sources limited by a bandwidth Δv. The transient nuclear signals are digitized by an ADC at a sampling frequency f_S. The sampling process introduces additional frequencies in the data in the form of sidebands given by

$$f_{SB} = nf_S \pm f_B, \qquad (8.47)$$

where f_{SB} is the sideband frequency, n an integer, and f_B the detector bandwidth. If the sampling frequency is not equal to twice the receiver bandwidth, noise aliasing will introduce additional low-frequency noise. For example, if we take $f_B = 30$ kHz and $f_S = 50$ kHz corresponding to a 20-μsec sample rate, $f_{SB} = 20, 70, 80, 130, \ldots$ kHz. Thus undesirable noise in the range 20–30 kHz is enhanced.

Of course, slow roll-off low-pass filters, like the simple and often-used RC filter may introduce even higher sideband frequencies in the example quoted. For this reason, it is really essential in S/N optimization to use a high-performance low-pass filter. Three commonly used filters are the Butterworth,

Bessel, and Chebyschev low-pass filters. The Butterworth filter has the flattest possible amplitude response, but the overshoot in its pulse response can rise to 16% in an eight-pole filter. Nevertheless, this filter is widely used for prevention of aliasing errors in data acquisition. The Bessel filter is not quite so flat up to the cutoff frequency or 3-dB point, but its pulse response is excellent, giving only a fraction of 1% overshoot. Because of the quite linear phase relationship for frequencies up to twice the cutoff frequency, Bessel filters are sometimes used to provide time delays. The Chebyschev filter has very fast roll-off characteristics, but introduces minor ripple in the amplitude of the passband response. As with the Butterworth filter, the pulse response contains transient overshoot leading to the longest settling time of all three filters.

The phase characteristic of these filters, if used in NMR imaging experiments, will lead to frequency-dependent distortions. Phase shifts which are linear with frequency are relatively easy to correct for in the Fourier transform software. Although not as linear as the Bessel filter, the Butterworth filter has a fairly linear phase shift covering the range $0 \rightarrow -p\pi/4$ radians over the passband where p is the number of poles in the filter. Thus, for an eight-pole filter the total phase shift is -2π radians. The most flexible filters are active devices in which the bandwidth is variable to suit the experimental situation. However, before leaving the subject of filters, it should be pointed out that the most flexible arrangement employs the so-called digital filtering technique.[46,47] In this method, the precise characteristics of a given filter, including the number of poles, may be simulated mathematically. Partially filtered time domain data are thus digitally processed to yield the optimum response from an idealized filter design. This type of filtering has found little application so far in NMR imaging because of the increased time penalty in data handling and the higher input data sampling rates required.

[46] M. R. Smith and S. Cohn-Sfetcu, *J. Phys. E Sci. Instrum.* **8**, 515 (1975).
[47] L. R. Rabiner and C. M. Rader (eds.), "Digital Signal Processing." IEEE, New York, 1972.

9. Biomagnetic Effects

9.1. Introduction

In this chapter we wish to consider the known and possible effects of magnetic fields on living biological systems. In NMR imaging, these magnetic fields can be static or dynamic. In both cases, the fields may be spatially uniform or deliberately inhomogeneous. For medical imaging purposes, however, we shall distinguish three important classes of magnetic field related to the above-mentioned categories. These are static magnetic fields (including static gradients), rf fields, and time-dependent gradient fields.

There is a considerable and growing literature concerning both static and rf magnetic fields, but there is a paucity of information on the effects of time-dependent uniform fields and even less on time-dependent gradients. With this in mind, we attempt to review the current state of our knowledge of static-field biomagnetics with special concern for the conceivable physical mechanisms of interaction. However, our major interest will be potential biomagnetic effects in man, and apart from a brief mention here, we do not intend to pursue the effects of magnetic fields on birds, bees, fish, snails, and other organisms.[1] However interesting they may be, our feeling is that these established effects are probably irrelevant to the main issues relating to man. Having said that, we briefly mention what is currently known about some of these effects.

According to Gould[2] and others,[3,4] bees have a permanent magnetism

---

[1] See articles in "Proceedings of the Biomagnetic Effects Workshop" (T. S. Tenforde, ed.). Laurence Berkeley Lab., Div. of Biology and Medicine, Berkeley, California, 1978.

[2] J. Gould, Princeton University.

[3] D. R. Griffin, "Bird Migration, the Biology and Physics of Orientation Behaviour." Heinemann, London, 1965.

[4] B. Greenberg, in Ref. 1, p. 14.

associated with them which is not apparent in the pupal stage, but develops rapidly in the adult. The source of the magnetic moment is a collection of small granules of magnetite in the abdominal region. It is not believed, at the moment, that bees navigate using their built-in compass. The magnetic moment decays to zero on death of the bee, but can be restored by repolarization in a magnet. This suggests that the 10-μm granules are bound together by a biodegradable agent, which, when absent, allows a random orientation of the magnetite particles giving a zero net moment.

Magnetite deposits have also been found in the heads of pigeons,[2,5] lying between the pia and dura mater membranes surrounding the brain. Initially it was thought that these were excretion dumping sites for unwanted magnetite, but it now appears that the sites are innovatory in that nerves are present leading from the magnetic deposits. This suggests that an adaptive mechanism has been evolved which explains in a straightforward manner how pigeons could navigate using the earth's field. It should be pointed out, however, that even when magnetically perturbed by affixing small magnets to their heads, pigeons still manage to home under the most adverse conditions.[5]

Other simple organisms living in seawater have been found to contain magnetite granules, and it is thought that polarization of the magnetite is used in an innovatory manner to orientate the creature below deep mud deposits where natural light levels are zero. The adaptive mechanism is further supported by the fact that creatures found in the northern hemisphere are north-seeking, whereas those found in the southern hemisphere are south-seeking.[6,7]

So far as is known, magnetite deposits do not occur naturally in humans. Furthermore, the presence of a magnetic field is not sensed using any normal mechanism by man.[7a] The question is, therefore, whether at a more fundamental level, the body responds to static magnetic fields.

With time-dependent fields, the situation is quite different. Their effects if sufficiently strong can be detected through the central nervous system (CNS) either as a sensation of heat, or depending on the details of the waveform and

[5] W. T. Keeton, in Ref. 1, p. 11.

[6] R. Blakemore, R. Frankel, and F. Wolfe, *Science* **203,** 1355 (1979).

[7] R. Blakemore, in Ref. 1, p. 6.

[7a] A recent report in *New Scientist*, Sept. 18 (1980) claims to have detected a magnetic sense in the human brain. However, the nature of the report suggests that a great deal more work needs to be done before the claimed effect can be substantiated (p. 844). Another recent report, by P. Semm, T. Schneider, and L. Vollrath, *Nature* (*London*) **288,** 607 (1980), claims to have observed changes in the electrical activity of pineal cells when magnetic fields of around 0.5 G are reversed. The electrical activity decreases on reversal of the field, but can be restored on further reversal of the field. Since the time scale between field reversals is quite long, it is hard to understand how a nonmagnetic cell can "remember" its previous magnetic history. Clearly much more work is required to substantiate this claim.

its duration, a tingling sensation, not unlike that experienced in electric shock. Such tangible and reproduceable effects make the dynamic field situation much clearer from a theoretical standpoint and safe limits to exposure can readily be ascertained. Even in areas of uncertainty, hypotheses can be tested by modeling the biological system and by observation *in vivo*.

With this strong contrast in sensibility of static vis-à-vis dynamic magnetic fields in mind, we next wish to review some of the biomagnetic experiments performed in static fields and consider possible theoretical mechanisms which might lead to static biomagnetic interactions.

9.2. Static Magnetic Fields

Because of the lack of a clear effect of static magnetic fields on the body, a mystique has evolved over the last 100 years or so with claims for biomagnetic effects ranging from therapeutic at one extreme to damaging at the other. The subject is receiving increasing attention resulting in a proliferation of publications too numerous to report here. A useful bibliography of some of the more serious attempts to study biomagnetic effects up to 1962 is given by Davis et al.,[8] although it should be emphasized that much of the work is superficial and lacks scientific rigor. More recent reviews and discussions of the subject are given by Barnothy,[9] Beischer,[10] Budinger,[11] Hauf,[12] Llaurado et al.,[13] Mahlum,[14] Saunders,[15] and Tenforde.[1] One of the problems which soon becomes apparent to anyone trying to investigate biomagnetics *in vivo* is the very large number of degrees of freedom present in an integrated biological system. Extreme care must be taken in handling animals and providing proper habitation. Proper controls are, of course, essential, and

[8] L. D. Davis, K. Pappajohn, and I. M. Plavnieks, *Fed. Proc. Fed. Am. Soc. Exp. Biol.* **21** (Suppl. 12), No. 5, Pt. II (1962).

[9] J. M. Barnothy, *in* "Medical Physics" (O. Glasser, ed.), Vol. 3. Yearbook Publ., Chicago, Illinois, 1960.

[10] D. E. Beischer, *Astronautics* **7**, 24 (1962).

[11] T. F. Budinger, *IEEE Trans. Nuclear Sci.* **NS-26**, 2821 (1979).

[12] R. Hauf, in "Manual on Health Aspects of Exposure to Non-ionizing Radiation." World Health Organization Regional Office for Europe, 1978.

[13] J. G. Llaurado, A. Sances, and J. H. Battocletti, "Biological and Clinical Effects of Low Frequency Magnetic and Electric Fields." Thomas, Springfield, Illinois, 1975; see also J. H. Battocletti, "Electromagnetism, Man and the Environment" in the series "Environmental Studies" (J. Rose and E. W. Weidner, eds.). Paul Elek, London, 1976.

[14] D. D. Mahlum, *Environ. Health Perspect.* **20**, 131 (1977).

[15] R. D. Saunders, Interim Review on Safety Aspects of NMR Whole Body Imaging. Private communication, (1979).

above all, large numbers of animals are required in order to produce statistically meaningful results. In many cases where positive results are reported, no controls were used and poor statistics exist. Indeed, in some cases, even field strengths and exposure times are unrecorded. It is therefore our view that such reports should be entirely disregarded. Some reports were made by biologists or zoologists who reveal in their work a sad lack of knowledge and competence in physics. Other reports by physicists reveal a lack of knowledge and understanding of the particular biological system considered. Most of the work in biomagnetics reported to date must therefore be regarded with considerable scepticism. Nevertheless, with the prospect of a magnetic imaging modality, it is important to carefully study this area. It is our strong view that interdisciplinary teams of physicists, biochemists, and biologists are an essential prerequisite to meaningful and worthwhile studies. Having said that, we now wish to consider in a selective manner, some of the work reported in the literature. We restrict our discussion to high fields, that is to say, fields in excess of several hundred gauss.

The experiments range from objective organism and organ or tissue specific tests *in vivo* and *in vitro* to what might be termed subjective behavioral studies on animals and observations of an epidemiological nature on humans. However, before looking into these complicated systems, it seems worth considering the possible interaction mechanisms.

9.2.1. STATIC-FIELD BIOMAGNETIC MECHANISMS

Several biomagnetic interaction mechanisms with static magnetic fields could be of importance potentially and in this section we wish to list these and consider each one in turn. We rule out the possibility of interaction with distributed magnetite, since, as stated previously, there is no evidence that such deposits exist in humans. If there is an effect, it will be most subtle and not detectable by the body's own senses. The effects that spring to mind are:

molecular polarization effects;
fluid and gaseous diffusion effects;
static hemomagnetic effects;
magnetohemodynamic effects;
CNS effects;
genetic and enzymic effects;
kinetic effects;
tunneling and related effects;
chemically induced magnetic polarization effects; and
metabolic effects.

9.2.1.1. Molecular Polarization Effects

Certain long-chain molecules which form liquid-crystal phases are well known to be magnetically polarizable.[16] Among these are the amphiphillic soap solution derivatives of perfluorinated octanoic acid; the so-called fluorooctanoates.[17,18] The arrangement of these surfactants in the lamellar phase, for example, is not unlike the biological lipid bilayer membrane structure. The magnetic alignment of these fluorooctanoates depends simply on the anisotropy of the molecular susceptibility. For ^{19}F long-chain molecules, this can be quite high, leading to a dominance of the magnetic energy over the thermal reorientation energy $\frac{1}{2}kT$, which is typically 0.6 kcal mol^{-1}. Certain hydrocarbon surfactants, in particular, cesium octanoate also align spontaneously in a magnetic field. However, the physical conditions for spontaneous alignment, especially density, are very critical. There is no evidence for polarization effects in phospholipids themselves, at least for field strengths of the order of 1.0 T. As with organic liquid crystals, magnetically polarizable lamellar-phase liquid biocrystals would easily form miscelle structures. If such materials existed in living organisms, they would quickly fall apart and die through their lack of cohesive strength. In other words, it would seem that biostructures which are both living and magnetically polarizable in fields of the order of 1.0 T are mutually incompatible.[18a] However, it is suggested that were it possible to polarize such biological membranes, for example, in extremely high magnetic fields, translational diffusion through the membrane should be affected.[19]

Natural surfactants do occur in the body as lubricants in, for example, the alveoli of the thorax, but these free fluids are unlikely to be troublesome in magnetic fields.

9.2.1.2. Fluid and Gaseous Diffusion Effects

The physiological functions of the body depend on a constant supply of nutrient and oxygen carried via the vascular system to the cells of the various organs. Waste materials are likewise removed by the same vascular system and normal function depends on differences of osmotic and hydrostatic pressures across tissue membranes. Oxygen molecules are, of course, paramagnetic. Dissolved protein and sugar nutrients are likely to be weakly

[16] P. G. de Gennes, "The Physics of Liquid Crystals." Oxford Univ. Press (Clarendon), London and New York, 1974.

[17] A. Jasinsky, P. G. Morris, and P. Mansfield, Mol. Cryst. Liq. Cryst. **45**, 183 (1978).

[18] P. G. Morris, P. Mansfield, and G. J. T. Tiddy, Faraday Disc. Chem. Soc. Symp. **13**, 37 (1979).

[18a] We are grateful to Dr. G. J. T. Tiddy for clarifying discussions on this point.

[19] M. M. Labes, in Ref. 1, p. 84.

diamagnetic. But in either case, a small magnetic force could exist if the external static magnetic field were nonuniform. The question is, therefore, whether such an additional molecular force would interfere significantly with the diffusion transport process. Another related point, is that even if there were a significant effect across a single membrane in an ideal experiment, presumably it would not have the same effect in a biological system. This is because increased diffusion into a cell on one side would be largely balanced by decreased diffusion into the cell from the other side, due to the isotropic nature of the transport processes servicing a cell. Nevertheless, it seems worth considering the somewhat artificial undirectional flow situation. Although we have considered NMR diffusion effects previously in Chapters 2 and 7, we make the point here that in these cases the applied gradients were used simply as measuring devices, and it was assumed implicitly that they did not affect the diffusion dynamics.

The transport of oxygen, for example, is controlled by the diffusion equation and flows from high to low concentrations. Since matter flows, a molecular force exists, which in the steady-state case is simply balanced by momentum changes through the viscosity η. We therefore wish to calculate the Stokes force on a single O_2 molecule, here assumed to be spherical and of radius a. When the viscous force balances the magnetic force we obtain

$$6\pi a\eta v = B_0 V_{mol}\chi G_x = (vkT)/D, \qquad (9.1a)$$

where v is the molecular terminal velocity, χ the magnetic susceptibility of O_2, G_x the external field gradient, k is Boltzmann's constant, T the absolute temperature, D the diffusion coefficient, and V_{mol} the O_2 molecular volume. For simplicity, we shall assume that the diffusion coefficient for O_2 molecules is not too different from that of water molecules. Taking $G_x = 100$ G cm^{-1}, $\chi = 10^{-4}$ cgs units, $D = 2 \times 10^{-5}$ cm^2 sec^{-1}, and $V_{mol} = 3.2 \times 10^{-20}$ cm^3, we obtain from Eq. (9.1a) the oxygen terminal velocity $v = 5 \times 10^{-12}$ cm sec^{-1}. Under these circumstances the molecular force is given approximately by

$$vkT/D \simeq mg/10 = V_{mol}\rho g/10, \qquad (9.1b)$$

where m is the molecular mass, g the gravitational acceleration, and ρ the O_2 density.

The drift velocity v corresponds to a momentum transfer per second or pressure excess given by

$$\Delta P = nmv^2 = \rho v^2, \qquad (9.2a)$$

where n is the O_2 concentration.

From simple kinetic theory, the ratio of this excess pressure to the atmospheric oxygen partial pressure is simply

$$\Delta P/P = (3v^2)/\overline{c^2}, \tag{9.2b}$$

where P is the atmospheric partial pressure and $\overline{c^2}$ the mean squared molecular velocity of oxygen at NTP. Taking $(\overline{c^2})^{1/2} = 5 \times 10^4$ cm sec^{-1} for O_2 in Eq. (9.2b) yields an excess pressure $\Delta P = 2.3 \times 10^{-29}$ mm Hg. The above calculation assumes that the partial densities of both dissolved and atmospheric oxygen are equal and is therefore a worst-case approximation.

This excess pressure should be compared with the typical pressure difference range of between 10 and 40 mm Hg across a membrane. It is clear, even from this simple viewpoint, that typical magnetic field gradients of 100 G cm^{-1} make no significant difference to the membrane transport processes.

9.2.1.3. Static Hemomagnetic Effects

Oxygen transport in the body occurs via the erythrocytes or red corpuscles of the blood. The active constituent is the heme group of hemoglobin. This contains an Fe ion and in deoxygenated blood is paramagnetic. However, in the oxyhemoglobin composite, the molecule is strongly diamagnetic. The leucocytes or white blood cells comprising the plasma are nonmagnetic, or at best weakly diamagnetic. Application of strong magnetic field gradients, therefore, might be expected to enhance the sedimentation rate of red cells. Indeed, the idea of using field-gradient blood separators has been suggested and successfully demonstrated.[20] In this case, the force balance equation involves the gravitational acceleration g. Artificially increasing g by spinning is, of course, the basis of the ultracentrifuge. The effective acceleration g_{eff} of deoxyhemoglobin molecules is given by

$$g_{\mathrm{eff}} = B_0 \chi V G_x/m, \tag{9.3}$$

where m is the erythrocyte mass, χ its magnetic susceptibility, and V its volume. The other terms have been defined previously. Melville et al.[20] have evaluated Eq. (9.3) and show that for a gradient $G_x = 10$ G cm^{-1}, $g_{\mathrm{eff}} \simeq g/100$. Such a small g_{eff} would not therefore be expected to affect the normal sedimentation rate. In their magnetic separator, they estimate, using Eq. (9.3), that gradients in excess of $G_x = 10^5$ G cm^{-1} are produced, thereby obtaining substantial separation at relatively low flow rates.

Bearing in mind that the body as a whole can withstand total accelerations of several g without damage, and that normal blood circulation would strongly discourage sedimentation, our conclusion is, therefore, that with

[20] D. Melville, F. Paul, and S. Raoth, *Nature (London)* **255**, 706 (1975).

typical static-field gradients contemplated in NMR imaging experiments, static hemomagnetic effects are negligible.

9.2.1.4. *Magnetohemodynamic Effects*

Of particular interest is the interaction of static magnetic fields with the cardiovascular system. The effect of high fields (7 and 10 T) on ECGs of squirrel monkeys[21,22] was studied and showed no arrhythmia, but there was an additional potential superimposed which is generated by the blood flow in the field. The additional voltage was 70 μV T^{-1} and is around two orders of magnitude *lower* than the expected electromotive force (emf) generated across the aortic wall due to the magnetohemodynamic effect. The emf (V) can be estimated from the expression[23]

$$\mathscr{E} = \mu_0 dB_0 v \sin \theta \tag{9.4}$$

where $\mu\mu_0$ is the permeability of blood (H m^{-1}), d the diameter of the aorta (m), B_0 is the magnetic field strength (A m^{-1}), v the fluid velocity (msec^{-1}), and θ is the angle between the magnetic field and flow vectors. Since the generated emf depends on blood-flow velocity and vessel diameter, the major effects are expected in the pulmonary artery and ascending aorta. Some effect may also be seen within the heart itself.

In the case of man,[24] with a peak blood flow of 0.63 msec^{-1} and an aortic diameter of 0.025 m, Eq. (9.4) predicts a flow potential of 16 mV T^{-1} across the aorta. The actual potential across the muscle fibers in the vessel wall would be much smaller. The threshold change in cardiac potential required to initiate depolarization of cardiac muscle is about 40 mV and this could be exceeded in a 3-T field if the equivalent total aortic flow potential appeared across cardiac muscle. The corresponding calculation for squirrel monkeys[22] predicts a peak flow potential of 5 mV T^{-1}. Experiments on monkeys exposed to 10 T for 15 min, where the flow potential exceeds the depolarization potential threshold, showed no physiological effect. One-hour exposure to 7 T did produce arrhythmia and a decreased heart rate. However, a meaningful base heart rate is notoriously difficult to measure and is most susceptible to psychological factors which can increase as well as decrease the rate. In magnetic fields of 1–2 kG, we do not consider blood-flow potentials to be a problem. No evidence of ECG abnormalities for humans in 1.0-kG fields has been reported.

[21] D. E. Beischer and J. C. Knepton, *Aerospace Med.* **35**, 939 (1964).

[22] D. E. Beischer, *in* "Biological Effects of Magnet Fields" (M. F. Barnothy, ed.), Vol. 2, p. 241. Plenum, New York, 1969.

[23] A. Kolin, *Rev. Sci. Instrum.* **23**, 235 (1952).

[24] C. H. Best and N. B. Taylor, "Physiological Basis of Medical Practice" (J. R. Brobeck, ed.). Williams & Wilkins, Baltimore, Maryland, 1973.

9.2.1.5. *CNS Effects*

In integrated animal systems with a highly developed and sophisticated CNS, it has been suggested that since there is clear measurable electrical activity associated with the passage of information along neurons, magnetic fields might be expected to have some effect on the charge carriers and conduction process. Indeed, the Hall effect, strong in semiconducting materials is the kind of effect one might look for. We may postulate that were such an effect large enough it might lead to cross talk between axons and thus interfere with neural transmission in some measurable manner.

Surprisingly, perhaps, experiments to determine the effects of magnetic fields on dogs and man were performed as early as 1892, all with negative results.[25] Some 50 years ago, careful studies of isolated nerve–muscle preparations were made in uniform and inhomogeneous static magnetic fields by Drinker and Thompson.[26] They measured the gastrocnemius muscle fatigue curves with the sciatic nerve and muscle in high fields. Using suitable controls, they found no effects with fields up to about 1.8 T and in field gradients of around 0.36 T cm^{-1}. Measurements of nerve action potential on popliteal nerve sections from a cat also showed no difference with controls. Live dogs were placed in the magnetic field and revealed no changes in behavior. Finally, the experimenters placed their own heads in the magnetic field and reported no effects.

More recently Becker[27] has reviewed, somewhat uncritically, the reported effects of magnetic fields on the CNS. The fact remains, however, that there is a singular lack of direct evidence of a significant effect of magnetic fields on the CNS. Other postulated interactions are via the synaptic junction, where the neural electrical impulse is converted to chemical transmitters, e.g., acetylcholine and norepinephrine, which diffuse across the synaptic junction. The release of the chemical agent and its subsequent transfer are areas of possible interaction. However, bearing in mind our comments on membrane diffusion, this seems a most unlikely interaction mechanism, and again is not supported by any experimental evidence.

9.2.1.6. *Genetic and Enzymic Effects*

A number of reports have appeared suggesting that magnetic fields have an effect on growth-rate[28,29] genetic mutations, leucocyte count,[9] and sterility.[29] Growth rate and mutant effects have been reported on animals

[25] F. Peterson and A. E. Kennally, *Med. J.* **56**, 729 (1892).

[26] C. K. Drinker and R. M. Thompson, *J. Indust. Hyg.* **3**, 117 (1921).

[27] R. O. Becker, *in* "Biological Effects of Magnetic Fields" (M. F. Barnothy, ed.), Vol. 2. Plenum, New York, 1969.

[28] G. C. Kimbol. *J. Bacteriol.* **35**, 109 (1937). See also Notes Added in Proof, Note 9.1.

[29] H. B. Brewer, *Biophys. J.* **28**, 305 (1979).

and also in some early work on yeast spores.[28] Careful efforts to reproduce some of the claimed effects have produced negative results.[30,31] In integrated biological systems like the higher order vertebrates, the number of parameters affecting the animal are very large, thus making it very difficult to design proper control experiments. In order to eliminate many of these ill-defined parameters we have recently completed a study of *E. coli* subjected to both static and combined static, switched-gradient and rf fields.[32] Using the sophisticated Ames tests specially adapted for this study, we have demonstrated conclusively that there is no significant effect on the survival rate or mutagenicity of these bacteria over many generations. Other tests on the bacterial enzyme β-galactosidase again show no significant effect on the enzyme's ability to break down lactose into dextrose and galactose in static fields up to 1.0 T.

Other experiments which we have recently performed[33] on the leucocytes of fresh whole blood exposed to static fields of up to 1.0 T and combined static fields, switched-gradients, and rf fields showed no effects on the frequency of gross lesions, sister chromatid exchanges, and on the proportion of amodal cells exposed prior to culture. Cultured cells have also been submitted to extended NMR imaging exposure during active growth and induced mitotic division. In all cases no significant effects over the controls were observed.

This most recent work strongly suggests that genetic effects and enzyme effects do not occur at the field strengths contemplated by NMR imagers. Similar conclusions are drawn by Baum *et al.*[34] for the plant *tradescantia*, and by Wolff *et al.*[34a] for tests on DNA for chromosomal damage in cultured Chinese hamster ovary cells and in HeLa cells.

The recent work of Brewer[29] on guppy fish, where increased size and eventual sterility are reported in uniform static magnetic fields over three generations of progeny suggest that the effect might be an induced change in hormonal secretion. The experiments lasted for two years, during which time the "magnetic" fish lived in a 500-G environment. However, supposedly sterile third generation "magnetic" fish of both sexes were subsequently mated with nonmagnetic control fish of the same generation and produced normal offspring. This observation would seem to suggest that genetic effects are not responsible.

The hypothalamus is the most sensitive hormonal regulatory gland and it

[30] B. S. Eiselein, H. M. Boutell, and M. W. Biggs, *Aerospace Med.* **32,** 383 (1961).

[31] M. W. Biggs, in Ref. 1, p. 36.

[32] A. Thomas and P. G. Morris, *Br. J. Radiol.* **54,** 615 (1981).

[33] P. Cook and P. G. Morris, *Br. J. Radiol.* **54,** 622 (1981).

[34] J. W. Baum *et al.*, in Ref. 1, p. 15.

[34a] S. Wolff, L. E. Crooks, P. Brown, R. Howard, and R. B. Painter, *Radiology* **136,** 707 (1980).

is conceivable that magnetic fields in some way interact with its secretions. An alternative, and in our view more likely, explanation is that induced electric currents caused by the fish's motion interfere with the hormonal secretions. Fish are known to be extremely sensitive[35] to induced emfs.

Unlike the guppy fish, experiments with *Drosophila melanogaster* exposed continuously to fields up to 0.4 T over one to three generations produced no visible effects.[36] However, further experiments performed later by another group with *Drosophila* in fields of up to 520 G for 24 hr report mutations and deviations in sex ratio.[37] This sort of contradiction pervades much of the earlier work on biomagnetics.

Our conclusions based on the most recent available data are that mutagenic effects are not induced by magnetic fields up to 1.0 T, nor is there any evidence that magnetic fields up to 1.0 T affect enzyme action. (See Notes Added in Proof, Note 9.2.)

9.2.1.7. Kinetic Effects

An interaction related to our discussion of blood flow is that of body movement in static magnetic fields. For sufficiently high fields around 20 kG, a mild tingling sensation in metallic tooth fillings has been reported.[10,38] This effect is presumably simple emf induction in the filling during the transit of the head through the inhomogeneous fringe field. Of course, for sufficiently rapid motions, or alternatively for rapidly alternating fields, one can expect to experience other effects. For example, in time-dependent fields, stimulation of the optic nerve or the retina or both occurs producing a flashing sensation in the eyes.[25] This is often referred to as the magnetic phosphene effect and has been extensively experienced and reported for many years and is not thought to be hazardous at the stimulation threshold. However, we shall return to the subject of nerve stimulation later when discussing time-dependent field effects.

9.2.1.8. Tunneling and Related Effects

In the review by Saunders[15] the suggestion by Barnothy is quoted that magnetic fields may influence the proton tunneling rate in DNA molecules, thereby altering the base pairing and so causing point mutations. However, the previously described work on genetic effects does not substantiate this proposal.

[35] Ad. J. Kalmijn, in Ref. 1, p. 8.

[36] I. L. Mulay and L. N. Mulay, *in* "Biological Effects of Magnetic Fields" (M. F. Barnothy, ed.), Vol.1. Plenum, New York, 1964.

[37] T. R. Tegenkamp, *in* "Biological Effects of Magnetic Fields" (M. F. Barnothy, ed.), Vol. 2. Plenum, New York, 1969.

[38] E. E. Ketchen, W. E. Porter, and N. E. Bolton, *Am. Indust. Hyg. Assoc. J.* **39,** 1 (1978).

Another mechanism, somewhat related to this suggestion, is that magnetic fields, though not generally strong enough to affect molecular structures, may do so in the presence of some additional radiation field. Conceivably, molecular conformational changes could occur which might be magnetically more favorable when a field is present. Such a situation might obtain, for example, in the presence of natural background or even high rf radiation.

We hasten to emphasize that there is no evidence to support this idea. Indeed, the epidemiological studies and industrial reports of workers exposed to static magnetic fields, though of an anecdotal nature, strongly suggest that no such mutagenic problem exists. Reports of human whole-body exposure in fields as high as 20 kG for integrated periods up to 3 days/year[10] indicate no effects, harmful or otherwise. Finally, experiments to test cell viability and repair of X-ray damage in a uniform 2.05-T field and an inhomogeneous field of 1.74 T with a gradient of 2.3 kG cm^{-1} gave negative results.[39]

9.2.1.9. *Chemically Induced Magnetic Polarization Effects*

Another possibility for static field interactions is the modification of chemical reactions observable by chemically induced dynamic electron polarization (CIDEP) or chemically induced dynamic nuclear polarization (CIDNP).[39a]

These effects are brought about by minor modifications of the singlet and triplet electronic energy levels of free radicals caused by the static field. The main effects are generally to change recombination rates, but it is also possible to create different recombination products when energetically favorable.

Radical generation is currently accomplished either photolytically or with high-energy electron beams, and applications limited mainly to reactions of small organic molecules. However, photo-CIDNP has been observed more recently in protein solutions.[39b]

It seems clear that normal background radiation could provide a singlet–triplet excitation mechanism in bioradicals, thus giving a potential biomagnetic interaction. Such interactions, if present would not necessarily represent cell damage or change viability, but perhaps could introduce subtle changes in the cell biochemistry either in terms of different radical recombination products, or simply through changes of reaction rate.

There is no evidence for such effects in NMR imaging, but it is also fair to say that such effects have not been looked for systematically.

---

[39] R. Nath, S. Rockwell, P. Bongiorni, and R. J. Schulz, in Ref. 1, p. 48.
[39a] R. Kapstein, *Adv. Free-Radical Chem.* **5**, 319 (1975).
[39b] R. Kapstein, K. Dijkstra, and K. Nicolay, *Nature (London)* **274**, 293 (1978).

9.2.1.10. *Metabolic Effects*

We have already discussed the important studies *in vivo* of phosphorus metabolites by NMR. Typical field strengths for these studies are around 10 T. The metabolic levels, as measured by NMR agree remarkably well with direct chemical assays.[40] This taken together with the fact that the heart and kidneys continue to function perfectly *in vivo* or when excised, and operated independently, strongly suggests that static magnetic fields do not upset the body chemistry or physiology. Admittedly, these experiments last only for an hour or so and subtle behavioral changes are not being monitored (see Section 9.2.1.9). On the other hand, one would expect such effects if present and permanent to be the result of detectable chemical changes.

9.2.2. SUBJECTIVE AND CLAIMED EFFECTS

We have so far considered a number of physical mechanisms that might produce biomagnetic effects, and as we have stressed, firm evidence for any reproducible effect is lacking. Nevertheless, in some epidemiological studies,[12] certain effects are claimed to be produced by static magnetic fields. This type of survey is by its nature anecdotal, and it is important to note that to a large extent, for every story claiming a positive effect one can find another denying any effect at all under similar or more stringent conditions. We therefore classify these reported positive effects as subjective.

The effects claimed include: general symptoms comprising headaches, weakness, tiredness, attacks of dizziness, an inflated feeling in the head, difficulty in concentration, sleep disturbances, nausea, and impotence. Claims that are not adequately controlled or otherwise substantiated include electroencephalographic modifications, unstable pulse and blood pressure, decreased sensitivity and articulation of the hand. Metabolic changes include an increase in the triglyceride level and a decrease in the cholesterol level. Cardiovascular disorders include hypotension, hypertension, bradycardia, tachycardia, arrhythmia, and ECG modifications. Blood changes include an increase in the leucocyte count. Other claimed effects include enhanced growth, diminished growth, longevity, shortened life span, and desquamation of the hands.

We again emphasize that in addition to the sometimes contradictory claims, many of the effects mentioned here, which have been looked for specifically in carefully controlled experiments, are nonreproducible, the experimenters reporting no effects, adverse or therapeutic. Later in this

---

[40] R. Richards and G. Radda (1979), private communication.

chapter we consider the various unofficial guidelines currently used to define safe limits for static and time-dependent magnetic fields.

9.3. Radio-Frequency Field Biomagnetic Effects

9.3.1. INTRODUCTION

Many attempts to calculate power absorption in biological tissues and to generally assess the effects of rf are to be found in the literature.[41-45] Most of the earlier attempts to calculate power deposition are based on simple models consisting of homogeneous tissue cylinders or spheres.[46,47] Since much interest has been concerned with power absorption of personnel in the vicinity of radio or radar antennas, more recent calculations[48] are concerned with the so called far-field case, namely plane-wave absorption by the subject. This is undoubtedly the simplest case to consider, and solutions for transverse plane magnetic waves incident on an extended cylindrical object have been fully discussed in Chapter 6. However, a more realistic model is that of a cylinder of finite length or better still, a prolate spheroid. Naturally, non-homogeneous models can be constructed and are considered elsewhere.[48] When finite-length models are used,[49,50] absorption resonances occur when the object length is equal to $\sim 0.4\lambda$, where λ is the incident wavelength. Durney et al.[48] have developed empirical expressions predicting the specific absorption power (SAP) for plane-polarized EM waves incident on a prolate spheroid of major axis 1.75 m and minor axis 0.276 m, and when the electric vector lies along the major axis. The incident power flux (IPF) taken in their calculations is 1.0 mW cm^{-2}. Their results are reproduced in Fig. 9.1.

[41] H. P. Schwan and G. M. Piersol, *Am. J. Phys. Med.* **33,** 371 (1954).

[42] H. P. Schwan and G. M. Piersol, *Am. J. Phys. Med.* **34,** 425 (1955).

[43] C. C. Johnson and A. W. Guy, *Proc. IEEE* **60,** 692 (1972).

[44] S. M. Michaelson, *Proc. IEEE* **60,** 389 (1972).

[45] H. P. Schwan, *in* "Fundamental and Applied Aspects of Non-Ionizing Radiation" (S. M. Michaelson, M. W. Miller, R. Magin, and E. L. Carstensen, eds.), p. 3. Plenum, New York, 1975.

[46] A. S. Presman, "Electromagnetic Fields and Life," (F. L. Sinclair, transl.; F. A. Brown, ed.). Plenum, New York, 1970.

[47] J. C. Lin, *IEEE Trans. Microwave Theory Techniques* **MTT-21,** 791 (1973).

[48] C. H. Durney, C. C. Johnson, P. W. Barber, H. Massoudi, M. F. Iskander, J. L. Lords, D. K. Ryser, S. J. Allen, and J. C. Mitchell, "Radiofrequency Radiation Dosimetry Handbook" (2nd Ed.), 1978. (USAF School of Aerospace Medicine (RZP), Aerospace Medical Division, Brooks Air Force Base, Texas, 78235.

[49] H. R. Kucia, *IEEE Trans. Instrum. Measurement* **IM-21,** 412 (1972).

[50] O. P. Gandhi and M. J. Hagmann, in *Abstr. URSI Int. Symp. Biol. Effects E.M. Waves, Airlie, Va., Oct.* page 51 (1977).

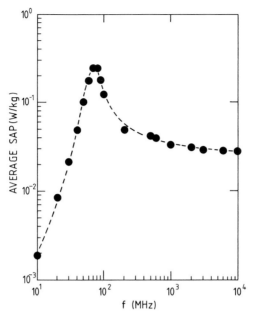

FIG. 9.1.   SAP (W/kG) versus frequency for an oblate spheroid tissue model of a man. An IPF of 1.0 mW cm^{-2} is assumed for a plane-polarized wave with E parallel to the major axis. [From Durney *et al.*, "Radiofrequency Radiation Dosimetry Handbook." USAF School of Aerospace Medicine, Brooks Air Force Base, Texas, USA, 1978.]

Although not specifically relevant to the NMR imaging situation, the graph emphasizes the important resonance effect at around 70 MHz and strongly suggests that this frequency should be avoided in any contemplated imaging or related experiments. In fact, the curve predicts that about 2.5 times the IPF is absorbed at 70 MHz over that at 50 MHz whereas at 10.0 MHz only 0.02 times the IPF is absorbed. An IPF of 1.0 mW cm^{-2} corresponds to an SAP of 10^{-1} W kg^{-1} at 50 MHz in Fig. 9.1.

Of great value in estimating SAP values are the electrical characteristics of the various biological tissues. A fairly comprehensive set of data are contained in Durney *et al.*,[48] and a more recent but limited set of measurements for excised rat tissue are given by Bottomley and Andrew.[51] Apart from the shortcomings of a particular model calculation, it is not clear whether published data on tissue resistivity measured *in vitro* is very reliable, bearing in mind the known changes of other tissue parameters that occur during the rigor process. With these points in mind we could not expect to predict

[51] P. A. Bottomley and E. R. Andrew, *Phys. Med. Biol.* **23**, 630 (1978).

effects to better than about an order of magnitude. For most purposes, therefore, this should serve as a guide only.

9.3.2. Power Absorption in NMR Imaging

An altogether more practical approach to the complicated problem of calculating absorbed power in a subject will be taken in this section. The reasons for this are simply that none of the manageable approximations we have discussed already really applies. As discussed in Chapter 8, the real problem is that for NMR imaging applications we generally have a short coil producing a transverse magnetic field B_1 in an extended object. Of course, when no other reasonable course presents itself, we are forced to adopt a simplified model *vide infra*.

In the present case, however, it is a simple matter to develop the formulation of Chapter 8 to calculate exactly the total absorbed power over a specific region of the subject. This calculation is based on Q changes of the transmitter coil with and without the subject. The SAP will, of course, depend on the rf penetration depth. If we are considering a saddle coil, then the curves of Figs. 6.10 and 6.11 can be used to estimate the SAP at different points of the specimen. At 4.0 MHz, however, the rf penetration may be considered complete.

Let the transmitter coil inductance be L_t. We shall assume a Faraday screened coil so that only magnetically induced losses are of importance.[52] We further assume that L_t remains constant, independent of the specimen characteristics and size.

Let the transmitter coil quality factor with no specimen be

$$Q_t = R_t/\omega L_t, \tag{9.5}$$

where R_t is the transmitter-coil parallel damping resistor. Similarly, the coil Q value when the subject is in the coil is

$$Q_{eff} = R_{eff}/\omega L_t, \tag{9.6}$$

where

$$R_{eff} = R_t R_s/(R_t + R_s), \tag{9.7}$$

in which R_s is the equivalent parallel damping resistor of the subject.

Let P be the incident power which is matched into the transmitter probe. Then

$$P = V_1^2/2R_t = V_2^2/2R_{eff}, \tag{9.8}$$

---

[52] D. G. Gadian and F. N. H. Robinson, *J. Magn. Reson.* **34**, 449 (1979).

TABLE 9.1

EFFECTIVE Q VALUES FOR TRANSMITTER COIL
PLACED AT VARIOUS POSITIONS
OVER A PATIENT[a]

	Q_{eff}	
Body region	No Faraday shield	With Faraday shield
Head	27.86	34.52
Thorax	28.57	31.55
Abdomen	26.19	31.00
Thighs	29.76	32.62
Empty coil, $Q_{eff} = Q_t$	39.28	35.00

[a] See text for coil details. The coil impedance ωL_t at 4.0 MHz is 420 Ω. (N.B.: Actually measured at 1.0 MHz.) Note the generally lower absorbed power in the specimen when using the Faraday shield.

where V_1 and V_2 are the voltages developed across the transmitter inductance without and with the subject, respectively.

The power absorbed by the subject is

$$P_s = V_2^2 / 2R_s. \tag{9.9}$$

From Eqs. (9.5)–(9.8) we may eliminate V_2 and R_s from Eq. (9.9) to obtain the particularly simple expression

$$P_s = P(1 - Q_{eff}/Q_t). \tag{9.10}$$

The fraction f_s of incident power absorbed by the subject is from Eq. (9.10),

$$f_s = 1 - Q_{eff}/Q_t. \tag{9.11}$$

Typical Q values at 4.0 MHz for whole-body imaging of the head, thorax, abdomen, and legs are given in Table 9.1 for a two-turn saddle coil with $l/a = 1.75$ and $a = 26.0$ cm. The subject has an approximately elliptical cross section through the trunk region with a major axis of 32 cm and a minor axis of 24 cm.

From Eq. (9.10) and Table 9.1, together with the coil and subject dimensions, we can straightforwardly evaluate the SAP. Peak transmitted power levels for slice selection in selective irradiation imaging experiments in our laboratory vary from $P \simeq 10 - 250$ W, depending on the imaging technique and slice width. The total shaped pulse length is 7.4 msec and appears rather like a nonlinear ramp with offset modulation, rising from 0 V at $t_w = 0$ to its peak voltage at $t_w = 7.4$ msec. With our measured values, and for $P = 250$ W, we estimate a maximum SAP in the thoracic/abdominal region of 0.72 W kg^{-1} and an average SAP during the pulse of 0.36 W kg^{-1}. The duty cycle in our experiments is approximately 1/40. The mean SAP over a 10-min period is therefore 9.0 mW kg^{-1}. These values are based on an estimate of the weight of body tissue enclosed by the transmitter coil of 30 kg.

9.3.3. rf EFFECTS ON THE CNS

Some speculation exists concerning rf effects other than ohmic heating.[53–55] In particular, some concern is expressed that rf fields (both electric and magnetic) can somehow affect the CNS.[56] As a result of this speculation, the maximum recommended IPF is likely to be reduced to 1.0 mW/cm^2. The physical basis of this supposed effect is not entirely clear, nor is it clear whether the CNS effect, if any, is produced at the rather low frequencies contemplated by NMR imagers. Most of the discussion centers around uhf and microwave frequencies. Certainly at radar frequencies, some individuals claim to be able to hear the radar pulse repetition frequency.[57] However, this has been shown to be a heating effect on the head. The very small but rapid expansions produce a pressure wave which acts on the cochlea. Most of the current data on this effect have been obtained with 10-μsec pulses in the frequency range 200–3000 MHz. The threshold energy detectable is about 40 μJ cm^{-2}.[57a] No effects of this kind have been reported at 4.0 MHz, even when transmitted powers are used as high as 7 kW and with pulse lengths of around 20 μsec. In the next section we shall discuss nerve stimulation thresholds and responses and suggest reasons why we believe rf interactions with the CNS are unlikely.

9.4. Time-Dependent Field Biomagnetic Effects

9.4.1. INTRODUCTION

In this section we shall concern ourselves with the biological effects of time-dependent magnetic fields and gradients. It is true that rf fields are time

[53] H. P. Schwan, *IEEE Trans. Microwave Theory Techniques* **MTT-19**, 146 (1971).

[54] H. P. Schwan, *IEEE Trans. Biomed. Eng.* **BME-19**, 304 (1972).

[55] S. F. Cleary, *Health Phys.* **25**, 387 (1973).

[56] E. R. Adair and B. W. Adams, *Science* **207**, 1381 (1980).

[57] A. H. Frey, *Aerospace Med.* **32**, 1140 (1961).

[57a] J. C. Lin, "Microwave Auditory Effects and Applications." Thomas, Springfield, Illinois, 1978.

dependent, and the distinction between what we have already discussed and the present section is really only a matter of frequency and waveform shape. As already discussed, the effect of rf is thought to be purely bulk ohmic heating of the body tissues. However, when the frequencies are lowered and the wave shapes changed to ramped or perhaps square waveforms, a different regime obtains, with potential hazards to the CNS, and other excitable tissue (for example, the sinoatrial node and possibly the autonomic nervous system). This arises principally through the stimulation time–current product discussed below.

The basis of the concern is that since the body is a conductor, Faraday induced currents are produced when the body is subjected to a changing magnetic flux. The subject is further confused by mention of the fact that dc currents as low as 17 μA have been known to produce ventricular fibrillation when passed directly across the heart itself. The difficulty with evaluating such statements is that contact area is not quoted so that the current density is not known. Also, it is notoriously difficult, even with flat electrodes, to be absolutely sure what the *actual* contact area is from metal conductor to tissue.

Hasty and unsubstantiated statements concerning safety can cause unnecessary harm to a new and potentially useful technique. Certainly it is prudent to be cautious, but caution should be exercised by performing carefully controlled experiments on animals. Experiments of this sort are under way in our laboratory and elsewhere.[57b] Preliminary results already suggest that the dangers of field switching have been exaggerated. In this section, therefore, we wish to develop a theoretical analysis of the effect of time-dependent fields and gradients on homogeneous tissue models. Naturally, such models are not exact but allow estimates of the maximum induced currents. These currents may then be compared with the nerve stimulation sensory threshold known from other work. However, before embarking on this, we next consider the action of a neuron and the model for electrical stimulation of nerves.

9.4.2. Nerve Action

Neural transmission along an axon is electrical and is caused by an imbalance of Na^+ and K^+ ions across the axon membrane. The conduction process is thus ionic. Compared with electrical transmission along cables, nerve conduction velocities are rather slow, typically in the range of 10–100 msec^{-1}, depending on the axon diameter. The relationship between the nerve conduction velocity u (m sec^{-1}) and axon diameter d (μm) is given by the empirical expression $u \simeq 2.5d$. The relatively massive charge carriers also

[57b] M. J. R. Polson, A. T. Barker, and S. Gardiner. *Clin. Phys. Physiol. Measurement* (to be published).

means that it is unlikely that high frequencies would interact (see also stimulation strength–duration product discussed below).

Neural impulses are triggered by a multiredundant coupling called a synapse. This is a connection between the signal receptors, or dendrites, and the signal transmitters, or axon feet, of an adjoining neuron. The synaptic junction does not actually touch and the signals are passed across synapses by release of various chemical transmitters including acetylcholine, which is rapidly hydrolyzed by the subsequent release and action of the enzyme cholinesterase. Many axons are surrounded by an insulating myelin sheath enclosed by the neurilemma. Along the sheath are Ranvier nodes regularly spaced about 2 mm apart. These are circular grooves in the myelin and serve to periodically expose the axon core through a fine membrane. It is at these points that external electrical currents may be introduced. Other axons are free from myelin and the whole fiber responds to ionic and electrical changes.

The nodal membrane is selectively diffusable. In the nerve resting state, there is a preponderance of K^+ ions in the axoplasm and Na^+ ions outside the axon. This ionic imbalance creates a rest potential for the K^+ ions within the axon of about -70 mV with respect to outside.[58,59] During neural propagation Na^+ ions flow rapidly through the Ranvier nodes sending the axon potential to about $+50$ mV. Although this process, brought about by a modification of the nodal membrane properties, is not fully understood, measurements show that the Na^+ ion diffusion rate is much faster than that of K^+, at the leading edge of the propagating neural wave.

The maximum voltage difference or transmembrane potential during passage of a neural impulse is thus about 120 mV and is called the nerve action potential. The action potential rise time is typically 300 μsec falling with a light undershoot back to the resting state in about 600 μsec. This is followed by a refractory period of around 1.0 msec during which time the neuron cannot be retriggered.

In addition to the natural process of neural stimulation, external electrical currents introduced though the Ranvier nodes or directly to the axon can also trigger neural pulse propagation. The current density stimulation threshold level which just initiates neural propagation follows a stimulation strength–duration curve which may be calculated knowing the electrical properties of the nodal membrane. In the resting state, the nodal resistance–area product is typically 1 kΩ cm^2, but drops drastically during the passage of the neural impulse to a minimum value corresponding to the peak value of the action potential.

Let the minimum membrane depolarization potential difference required

[58] R. Plonsey, "Bioelectric Phenomena." McGraw-Hill, New York, 1969.
[59] A. L. Hodgkin, "The Conduction of the Nervous Impulse," p. 60. Liverpool Univ. Press, 1967.

to stimulate an excitable membrane be $\Delta v = v_T - v_R$ in which v_T is the transmembrane stimulation potential and v_R the axoplasm resting potential. Let us take $\Delta v = f_t v_a$, where v_a is the action potential and f_t the threshold fraction having a value in the range $0 < f_t < 1$. During conduction the nodal membrane behaves like a linear electrical circuit with a membrane resistance–area product r_m and a membrane capacitance/area of C_m. In order to trigger the nerve, it is necessary to depolarize the membrane by discharging the nodal capacitance, thereby creating a transmembrane potential equal to v_a. For square wave stimulation, this simple electrical discharge process of the nodal membrane is well approximated by the expression

$$\Delta v = j_s r_m [1 - \exp(-t/\tau_m)], \qquad (9.12a)$$

where

$$\tau_m = r_m C_m, \qquad (9.12b)$$

and j_s is the stimulation current density applied. Clearly when $t \gg \tau$, Eq. (9.12a) gives

$$f_t v_a = j_{s0} r_m, \qquad (9.12c)$$

where j_{s0} is the stimulation threshold current density. Rearranging Eq. (9.12a) using Eq. (9.12c), we arrive at the stimulation strength–duration expression

$$j_s/j_{s0} = [1 - \exp(-t/\tau_m)]^{-1}, \qquad (9.13a)$$

which for $t < \tau_m$ may be approximated by

$$j_s t \simeq j_{s0} \tau_m. \qquad (9.13b)$$

Using the following typical values for nodal membranes[59] of $r_m = 20\ \Omega\ cm^2$ and $C_m = 7.5\ \mu F\ cm^{-2}$, and taking $f_t = 0.25$, we find that $j_{s0} = 1.5\ mA\ cm^{-2}$ and $\tau_m = 150\ \mu sec$. Thus for square wave current stimulation lasting 50 μsec, for example, $j_s \simeq 4.5\ mA\ cm^{-2}$ using Eq. (9.13b) or from the exact expression [Eq. (9.13a)], $j_s = 5.2\ mA\ cm^{-2}$. We emphasize that this calculation refers to one type of excitable tissue. Many neurons are unmyelinated. Electrically excitable tissue also includes muscle fibers and sensory receptors, all of which will have different properties. Nevertheless, our value of $j_{s0} = 1.5\ mA\ cm^{-2}$ agrees well with Schwan's estimate[54] for stimulation threshold current density of 1–10 $mA\ cm^{-2}$ for ventricular fibrillation.

Equations (9.13a) and (9.13b) clearly demonstrate that for long stimulation times the neural stimulation threshold current density is lower. Of course, the converse is also true and this shows why rf currents, with their very short periodicity, would not be expected to induce nerve stimulation, except, perhaps, when enormous current densities were applied greatly exceeding j_{s0}. On receipt of this current, a nerve action potential is created and propagates along the axon to the axon feet. An important point to remember is that induced electrical stimulation around a Ranvier node is generally isotropic. Because of this simple fact, it is conceptually difficult to envisage excitation of currents into and out of Ranvier nodes with spatially uniform induced current distributions. Of course, all nerves do not run parallel *or* normal to the body axis, and we presume that this fact, together with the general deviations or spatial anisotropy caused either by coil design or by tissue inhomogeneities, is the reason why induced currents interact with the CNS at all. With this in mind we introduce an anisotropy factor α_a which varies from $0 \rightarrow 1$. Electromagnetically induced current densities, discussed later, must therefore be multiplied by this factor. That is to say, the actual induced current density entering a node is $j\alpha_a$ where j is our computed current density based on a homogeneous tissue model. Clearly, $\alpha_a = 1$ is the worst case, but from our discussions above we would expect α_a in general to be less than unity.

With direct contact to a nerve axon external stimulation is easier to understand. There is a considerable literature on the effects of direct contact dc and ac electric shock[60,61] and from this it is clear that certain frequencies are particularly hazardous. For example, the range 30–150 Hz should be avoided.

In designing experiments to produce neuron stimulation, one can work quite satisfactorily with stimulation times up to 100–150 μsec, a time range which is fairly easy to achieve experimentally.

The problem of calculating induced electrical effects on the CNS is thus complicated by several major obstacles; general tissue inhomogeneity, the geometrical irregularity of the CNS, and the electromagnetic field calculations themselves, which require simplified models in order to obtain analytical solutions.

In the next section, we shall therefore look for analytical solutions for the EM wave equations with simplified tissue models, bearing in mind that we are not expecting precise answers but only order-of-magnitude results.

9.4.3. ELECTROMAGNETIC FIELD EQUATIONS

We have already considered particular solutions of Maxwell's electromagnetic (EM) field equations for rf fields in Chapter 6. The results obtained there refer to the particular transverse geometry of the rf coil with respect to

[60] F. Christopher, (ed.), "A Textbook of Surgery," p. 72. Saunders, Philadelphia, Pennsylvania, 1944.
[61] C. H. Best and N. B. Taylor, "The Physiological Basis of Medical Practice." Bailliere, London, 1945.

the specimen. In addition, the results apply for uniform rf fields and are not valid for the special situations encountered with field gradients. In this Section we consider more general solutions of the source free field equations derived with the minimum of assumptions.

Figure 9.2 shows the coordinate system for a point $P(r, \theta, z)$ within a cylindrical specimen of radius a. In all instances, we assume that $E_z = 0$, in which case Maxwell's equations reduce to

$$k_1 H_r = \partial E_\theta / \partial z, \tag{9.14}$$

$$k_1 H_\theta = -\partial E_r / \partial z, \tag{9.15}$$

$$\frac{\partial^2 H_z}{\partial r^2} + \frac{1}{r} \frac{\partial H_z}{\partial r} + \frac{1}{r^2} \frac{\partial^2 H_z}{\partial \theta^2} + \frac{\partial^2 H_z}{\partial z^2} - k_1 k_2 H_z = 0, \tag{9.16}$$

where $\mathbf{E}$ and $\mathbf{H}$ are the electric and magnetic field vectors respectively, and

$$k_1 = i\omega\mu\mu_0, \tag{9.17a}$$

$$k_2 = \sigma + i\omega\epsilon\epsilon_0, \tag{9.17b}$$

in which ω is the field fluctuation angular frequency, μ and ϵ the relative permeability and permittivity respectively, and σ the conductivity of the medium. The free-space permeability $\mu_0 = 4\pi 10^{-7}$ sec^2 F^{-1} m^{-1} and the free-space permittivity $\epsilon_0 = 8.419 \times 10^{-12}$ F m^{-1}. In evaluating expressions developed later, it will be useful to note that

$$c^2 = (\mu_0 \epsilon_0)^{-1}, \tag{9.18}$$

where the free-space velocity of light $c = 3 \times 10^8$ m sec^{-1} and more importantly, the free-space impedance

$$[\mu_0/\epsilon_0]^{1/2} = 376.8 \ \Omega. \tag{9.19}$$

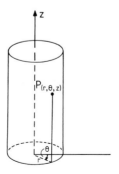

FIG. 9.2.   Coordinate system for a point $P(r, \theta, z)$ within a tissue cylinder of radius a.

We shall also assume our media are linear and obey Ohm's law, namely,

$$\mathbf{j} = \sigma\mathbf{E}, \tag{9.20}$$

where $\mathbf{j}$ is the current density vector.

Equation (9.16) may be solved by separation of variables giving a general solution

$$H_z = R(r)\Theta(\theta)Z(z), \tag{9.21}$$

in which $Z(z)$ is the solution of

$$\frac{1}{Z(z)}\frac{\partial^2 Z}{\partial z^2} = -m^2\left(\frac{2\pi}{L}\right)^2, \tag{9.22}$$

where m is an integer and L a length. The periodic solutions of Eq. (9.22) are of the form

$$Z(z) = a_m \cos(2\pi mz/L) + b_m \sin(2\pi mz/L). \tag{9.23}$$

Similarly, $\Theta(\theta)$ is the solution of the equation

$$\frac{1}{\Theta(\theta)}\frac{\partial^2 \Theta}{\partial \theta^2} = -l^2, \tag{9.24}$$

where l is an integer. The single-valued solutions of Eq. (9.24) are of the form

$$\Theta = A_l \cos l\theta + B_l \sin l\theta. \tag{9.25}$$

Using Eqs. (9.22) and (9.24) in Eq. (9.16), we find that $R(r)$ satisfies Bessel's equation

$$\frac{\partial^2 R}{\partial r^2} + \frac{1}{r}\frac{\partial R}{\partial r} + \left(K^2 - \frac{l^2}{r^2}\right)R = 0, \tag{9.26}$$

where

$$K^2 = k^2 - m^2(2\pi/L)^2, \tag{9.27}$$

$$k^2 = k_1 k_2. \tag{9.28}$$

In solving Eq. (9.26) it is convenient to transform the variable from $R(r)$ to $R(Kr)$. Solutions of the transformed version of Eq. (9.26) are given by

$$R_l(Kr) = J_l(Kr), \tag{9.29}$$

where J_l are Bessel functions of the first kind. Although the argument Kr is in general complex, it is just as convenient for our purposes to work with the unmodified Bessel functions.

Our general solution for the magnetic field H_z is

$$H_z = J_l(Kr)[A_l \cos l\theta + B_l \sin l\theta]$$
$$\times [a_m \cos(2\pi mz/L) + b_m \sin(2\pi mz/L)]. \tag{9.30}$$

This solution is valid both inside and outside the specimen, where we may approximate $k^2 = 0$, provided that $m \neq 0$ and L is finite. However, when $2\pi r/L \ll 1, K^2 r^2 \simeq 0$, in which case the external solution for $R(r)$ is of the form

$$R(r) = (P_l/r^l) + Q_l r^l. \tag{9.31}$$

We shall require this solution later when dealing with transverse gradients. For the moment, we consider gradients along the z axis.

9.4.4. FIELD GRADIENTS ALONG THE z AXIS

Equation (9.30) describes both uniform and gradient fields along the z axis. The uniform field case, discussed elsewhere,[51] corresponds to $b_m = 0$ and $x/L \ll 1$ with $l = 0$ and $B_0 = 0$, giving an internal field

$$H_z{}^i = A_0 J_0(k^i r), \tag{9.32}$$

where superscripts "i", used above, and "e", used later, refer to the internal and external value of the indexed variable or parameter.

The boundary conditions that must necessarily be satisfied simultaneously at the medium interface are

$$H_z{}^i = H_z{}^e, \tag{9.33a}$$

$$E_\theta{}^i = E_\theta{}^e. \tag{9.33b}$$

In the case of Eq. (9.32), this means that

$$A_0 = H_1 J_0(k^e a)/J_0(k^i a), \tag{9.34}$$

where H_1 is the externally applied fluctuating field. There is a small but important difference between the solution given here and that quoted in Chapter 6 [Eq. (6.7)] which does not satisfy the second boundary condition Eq. (9.33b). However, since $J_0(k^e r) \simeq 1$ for all practical frequencies, we see from the Maxwell expressions

$$k_2 E_\theta = (\partial H_r/\partial z) - (\partial H_z/\partial r), \tag{9.35}$$

$$k_2 E_r = (1/r)(\partial H_z/\partial \theta) - (\partial H_\theta/\partial z), \tag{9.36}$$

that $E_r = 0$, and since $\partial H_r/\partial z = 0$, the tangential E field approximates to

$$E_\theta{}^i \simeq k^i H_1[J_1(k^i r)/k_2{}^i J_0(k^i a)], \tag{9.37}$$

and therefore,

$$j_\theta = \sigma E_\theta{}^i. \tag{9.38}$$

If $|k^i r| < |k^i a| < 1$, Eq. (9.38) reduces to

$$j_\theta \simeq i\omega B_1 r/2, \tag{9.39}$$

where

$$B_1 = \mu\mu_0 H_1. \tag{9.40}$$

Equation (9.39), which has been derived by simpler arguments elsewhere,[11] is valid for typical tissue conductances (0.2–0.5 Ω^{-1} m^{-1}) and frequencies up to about 1.0 MHz according to Eq. (9.37).

Of particular interest in NMR imaging is the establishment of a linear field gradient $G_z = \partial H_z/\partial z$ along the z axis. In this case we require $l = 0$, $B_0 = 0$, $m = 1$, and $a_1 = 0$ in Eq. (9.30), giving

$$H_z = A_0 b_1 J_0(Kr) \sin(2\pi z/L). \tag{9.41}$$

The Maxwell pair gradient coil spacing is chosen to satisfy the conditions

$$\rho_0 \sqrt{3} = L/2, \tag{9.42}$$

where ρ_0 is the coil radius. Figure 9.3 shows how the periodic field [Eq. (9.41)] approximately satisfies the generated field over the restricted region of interest.

For $z/L < 1$, the gradient is linear. When the boundary conditions [Eqs. (9.33a) and (9.33b)] are satisfied at $r = a$, we obtain, using Eq. (9.14), the following field-component solutions:

$$H_r{}^i = G_z J_1\left(\frac{2\pi a}{L}\right) \frac{J_1(K^i r)}{\Delta(a) K^i} \cos\left(\frac{2\pi z}{L}\right), \tag{9.43}$$

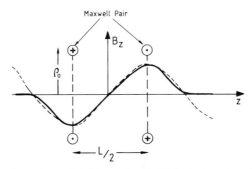

FIG. 9.3.   Maxwell gradient-coil field variation (solid line) along the axis and approximate fit to the periodic gradient field G_z [Eq. (9.41)] over the restricted region of interest.

$$H_z^i = \frac{G_z L}{2\pi} J_1\left(\frac{2\pi a}{L}\right) \frac{J_0(K^i r)}{\Delta(a)} \sin\left(\frac{2\pi z}{L}\right), \tag{9.44}$$

$$E_\theta^i = \frac{G_z L}{2\pi} k_1^i J_1\left(\frac{2\pi a}{L}\right) \frac{J_1(K^i r)}{\Delta(a) K^i} \sin\left(\frac{2\pi z}{L}\right), \tag{9.45}$$

$$H_r^e = G_z J_1\left(\frac{2\pi a}{L}\right) \frac{J_1(K^i r)}{\Delta(a) K^i} \cos\left(\frac{2\pi z}{L}\right), \tag{9.46}$$

$$H_z^e = \left[\frac{J_1(K^i a)}{D(k)\Delta(a)} J_0\left(\frac{2\pi z}{L}\right) + 1\right] \frac{G_z L}{2\pi} \sin\left(\frac{2\pi z}{L}\right), \tag{9.47}$$

$$E_\theta^e = \frac{G_z L k_1^e}{2\pi} J_1\left(\frac{2\pi a}{L}\right) \frac{J_1(K^e a)}{\Delta(a) K^e} \sin\left(\frac{2\pi z}{L}\right), \tag{9.48}$$

where

$$\Delta(a) = J_0(K^i a) J_1(2\pi a/L) - J_0(2\pi a/L) J_1(K^i a)/D(k), \tag{9.49}$$

$$D(k) = 2(K^i/k^e)^2. \tag{9.50}$$

We note from Eq. (9.45) that when $|K^i r| < 1$, E_θ^i reduces to a form similar to that of Eq. (9.39) producing induced ring currents in the specimen. An interesting result is the interchange of the H_r^i and H_z^i field components as z varies. Near the z origin the gradient G_z is linear and the field $H_z^i \rightarrow 0$. In this region, the radial field component H_r^i maximizes but though dominant, fortunately does not interfere with the NMR imaging experiment, since it is orthogonal to the static field B_0.

In the low-frequency limit, the induced current density with the specimen is from Eq. (9.45),

$$j_\theta \simeq (G_z L/2\pi) i\omega\mu\mu_0 r \sin(2\pi z/L). \tag{9.51}$$

9.4.5. TRANSVERSE FIELD GRADIENTS

We now consider the case when a linear gradient of the form $G_x = \partial H_z/\partial x$ is applied to a cylindrical specimen. The coordinate system is shown in Fig. 9.4. This type of gradient is conceptually more difficult to deal with as we shall see in the detailed solutions of Maxwell's equations, mainly because simple derivations of induced currents are not so obvious as in the previously discussed case of induced ring currents.

The field outside the specimen at a distant point $P(r, \theta)$ and at zero frequency is given by

$$H_z(r, \theta) = G_x r \sin\theta. \tag{9.52}$$

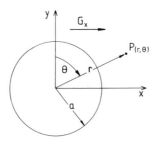

FIG. 9.4. Coordinate system for a point $P(r, \theta)$ outside an infinite tissue cylinder of radius a subjected to a transverse magnetic field gradient G_x.

This means automatically that Bessel function solutions to our problem must involve the lowest order angular functions, that is to say, $l = 1$, in the general solution equation [Eq. (9.30)]. In addition, we require that $m = 1$, with $A_1 = b_1 = 0$ and L large. In this case Eq. (9.30) gives for the field inside the specimen,

$$H_z(r, \theta) = a_1 B_1 J_1(kr) \sin \theta. \tag{9.53}$$

The E-field components [Eqs. (9.36) and (9.37)] inside the specimen, with $\partial H_\theta/\partial z = \partial H_r/\partial z = 0$, together with Eq. (9.53) reduce to

$$E_r = \frac{a_1 B_1}{k_2 r} J_1(kr) \cos \theta, \tag{9.54}$$

$$E_\theta = -\frac{a_1 B_1}{k_2 r} \frac{\partial}{\partial r} J_1(kr) \sin \theta. \tag{9.55}$$

Using the external solutions for $R(r)$ [Eq. (9.31)], together with the boundary conditions [Eq. (9.33)] evaluated at the cylinder surface $r = a$, we obtain the following field-component solutions

$$H_z^i = \frac{2aG_x}{[1 + a/\Gamma(a)]} \frac{J_1(k^i r)}{J_1(k^i a)} \sin \theta, \tag{9.56}$$

$$E_r^i = \frac{2aG_x J_1(k^i r)}{[1 + a/\Gamma(a)]k_2{}^i r J_1(k^i a)} \cos \theta, \tag{9.57}$$

$$E_\theta^i = -\frac{2aG_x J_1{}'(k^i r)}{[1 + a/\Gamma(a)]k_2{}^i J_1(k^i a)} \sin \theta, \tag{9.58}$$

$$H_z^e = \left[\frac{a^2 G(a)}{r} + r\right] G_x \sin \theta, \tag{9.59}$$

$$E_r^e = \left[\frac{a^2 G(a)}{r} + r\right] \frac{G_x}{r k_2{}^e} \cos \theta, \tag{9.60}$$

$$E_\theta^{\,e} = \left[\frac{a^2 G(a)}{r^2} - 1\right] \frac{G_x}{k_2^{\,e}} \sin\theta, \qquad (9.61)$$

where

$$G(a) = [\Gamma(a) - a]/[\Gamma(a) + a], \qquad (9.62)$$

$$\Gamma(a) = \frac{k_2^{\,i} J_1(k^i a)}{k_2^{\,e} J_1'(ka)}. \qquad (9.63)$$

The initial far-field condition [Eq. (9.52)] implies a zero frequency current distribution in a conducting medium of conductance σ^e. This pseudocurrent is necessarily introduced in order that the field equations be consistent. Naturally, no current flows inside or outside the specimen when $\omega = 0$ (only in the gradient coils), and we shall subtract this pseudostatic contribution from our solutions in order to obtain the true induced currents. This means that

$$k_2^{\,i} = \sigma^i + i\omega\epsilon\epsilon_0, \qquad (9.64a)$$

$$k_2^{\,e} = \sigma^e + i\omega\epsilon_0. \qquad (9.64b)$$

From the definitions of Eqs. (9.62) and (9.63), we obtain in the case $|k^i a| < 1$,

$$\Gamma(a) = ak_2^{\,i}/k_2^{\,e}, \qquad (9.65)$$

$$G(a) = (k_2^{\,i} - k_2^{\,e})/(k_2^{\,i} + k_2^{\,e}). \qquad (9.66)$$

Using Eq. (9.65), in Eqs. (9.57) and (9.58) we obtain the approximate E-field components

$$E_r^{\,i} = [2G_x/(k_2^{\,i} + k_2^{\,e})] \cos\theta, \qquad (9.67)$$

$$E_\theta^{\,i} = -[2G_x/(k_2^{\,i} + k_2^{\,e})] \sin\theta, \qquad (9.68)$$

which together produce a resultant E field,

$$E^i = 2G_x/(k_2^{\,i} + k_2^{\,e}). \qquad (9.69)$$

The field is *constant* throughout the specimen and is *parallel* to the y axis (Fig. 9.5). The initial conditions and behavior at $\omega = 0$ discussed above dictate that $\sigma^i = \sigma^e = \sigma$. Clearly, since no current flows at $\omega = 0$, the induced frequency-dependent field $E_1^{\,i}$ is obtained from Eq. (9.69) by subtracting the background field at $\omega = 0$. The induced current density is then given by

$$j = -\frac{i\omega G_x \epsilon_0(1 + \epsilon)}{2\sigma} \left\{ 1 + \left[\frac{\omega\epsilon_0(1 + \epsilon)}{2\sigma}\right]^2 \right\}^{-1/2} e^{i\theta}, \qquad (9.70a)$$

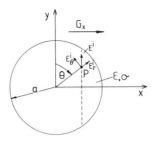

FIG. 9.5.   Induced E-field components within an infinite tissue cylinder of radius a.

where

$$\theta = \tan^{-1}[\omega\epsilon_0(1 + \epsilon)/2\sigma]. \qquad (9.70b)$$

For all frequencies up to 1.0 MHz and for $\sigma = 0.2 \; \Omega^{-1} \; m^{-1}$ and $\epsilon = 2000$ at 1.0 MHz, the quantity $[\omega\epsilon_0(1 + \epsilon)/2\sigma]^2 \ll 1$. Rearranging Eq. (9.70a) in this approximation, we obtain the induced current density

$$j = -i\omega\mu_0 a^2 \, \frac{G_x\sigma}{2} \, F(a, \sigma, \epsilon), \qquad (9.71)$$

where we introduce the dimensionless parameter

$$F = \epsilon_0\epsilon/\mu_0 a^2\sigma^2, \qquad (9.72)$$

to stress the similarity between Eq. (9.71) and previously derived results.

The induced external fields are given approximately by

$$E_{r1}^e = [a^2 G(a)G_x/r^2 k_2^e] \cos\theta, \qquad (9.73)$$

$$E_{\theta1}^e = [a^2 G(a)G_x/r^2 k_2^e] \sin\theta. \qquad (9.74)$$

Unlike the internal field component, there is no sign difference between the angular and radial components.

Equations (9.73) and (9.74), together with Eqs. (9.67) and (9.68) predict an induced field and current distribution indicated in Fig. 9.6. Since div $D = \epsilon$ div $E = 0$, there are no net charge sources in the system. The induced currents, therefore, produce a balanced positive and negative surface-charge distribution as indicated. In a body, as opposed to a homogeneous tissue cylinder, the outer epithelial layer or stratum corneum is likely to approximate reasonably well to an insulating layer. In this case, the E- and D-field boundary conditions may be suitably modified by allowing induced external static charge on the cylinder surface and, at the same time, allowing some tangential charge flow or current within the tissue cylinder. We would expect the modified current path near the cylinder surface to follow the dotted curves in Fig. 9.6. (See Notes Added in Proof, Note 9.3.)

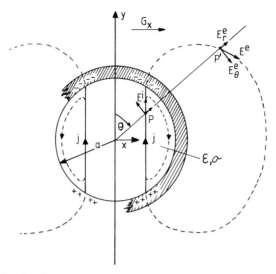

FIG. 9.6.   Induced E-fields and current density j in an infinite tissue cylinder of radius a subjected to a time-dependent linear transverse gradient G_x. Note the surface-charge accumulation and the parallel current flow within the cylinder. Current path deviations (dotted) would be expected within and near the cylinder surface for a conducting cylinder surrounded by a nonconducting tissue layer.

9.4.6. EVALUATION OF CURRENT DENSITIES

We have cast the current density distribution for the time-dependent gradient fields in a form, in common with that for spatially uniform fluctuating fields, valid at low frequencies. For precise results, linearly ramped fields may be Fourier analyzed into their frequency components and the current density contributions for each frequency summed to give the total ramped response. However, a simpler approach, which we shall adopt, particularly since the expressions derived are all valid for typical tissues up to about 1.0 MHz, is to replace $i\omega$ in our expressions by the operator d/dt. This simple artifice allows direct estimation of ramped gradients and will be valid for ramp rise times down to about 1.0 μsec. A complication in this simplified approach is the experimentally observed frequency dependence of both ϵ and σ. Measurements of both $\epsilon(v)$ and $\sigma(v)$ above 1.0 MHz[48,51] and limited tissue measurements down to 10 Hz[48] suggest that we will incur small error by ignoring the frequency dependence of σ. The limited data available[48] suggest that $\epsilon(v)$ is inversely proportioned to frequency over at least two decades below 1.0 MHz. That is to say,

$$\epsilon(v) = \epsilon_{v0}/v = \epsilon_{r0}\tau, \tag{9.75}$$

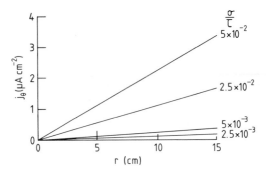

FIG. 9.7. Parametric plot of maximum induced circular current density j_θ versus r [Eq. (9.76)] in a tissue cylinder of radius $a = 15$ cm subjected to an axial time-ramped gradient $G_z = 1.0$ G cm^{-1}. The gradient is produced by a Maxwell coil pair with radius $\rho_0 = 30$ cm. The parameter σ/τ has units of Ω^{-1} m^{-1} μsec^{-1} and corresponds to typical ranges of tissue conductivities (0.25–0.5 Ω^{-1} m^{-1}) and ramp times 10.0–100.0 μsec.

where v is the frequency in MHz and v_0 the base reference frequency. Equation (9.75) may also be carried over to the time domain for short pulse intervals τ measured in microseconds. At $v_0 = 1.0$ MHz, corresponding to $\tau_0 = 1.0$ μsec, the relative permittivity of tissue is typically $\epsilon_1 = 2 \times 10^3$. The maximum induced current density for a linearly ramped gradient $G_z = 1.0$ G cm^{-1} applied to a cylindrical specimen of radius $a = 15$ cm and Maxwell coil radius $\rho_0 = 30$ cm is from Eq. (9.51),

$$j_\theta = (\sqrt{3}/\pi)(\sigma r/\tau) \quad (\mu\text{A cm}^{-2}), \tag{9.76}$$

where r is in cm, σ in Ω^{-1} m^{-1}, and τ in μsec. Equation (9.76) is plotted versus r with (σ/τ) as a parameter in Fig. 9.7. The typical range of ramp times is $\tau = 10$–100 μsec. The range of conductivities used is $\sigma = 0.5$–0.25 Ω^{-1} m^{-1}.

The induced current density expression [Eq. (9.71)] for a linearly ramped gradient $G_x = 0.1$ G cm^{-1} applied to a cylindrical specimen, together with Eq. (9.75) reduces to

$$j = 1.1738(\epsilon_1/\sigma)10^{-3} \quad (\mu\text{A cm}^{-2}). \tag{9.77}$$

Equation (9.77) is plotted versus ϵ_1/σ in Fig. 9.8. We note that over the range of ramp times $\tau = 1.0$–100 μsec, j is independent of τ in this approximation. A full Fourier analytical treatment, together with measured $\epsilon(v)$ and $\sigma(v)$ values would no doubt reintroduce some mild frequency and hence time dependence, especially for very low frequencies corresponding to long ramp times. The expressions derived for current density are directly proportional to gradient strength and may therefore be easily scaled.

We note from both Figs. 9.7 and 9.8 that current densities never exceed 6.0 μA cm^{-2}. Even if both gradients were increased by a factor of ten, the

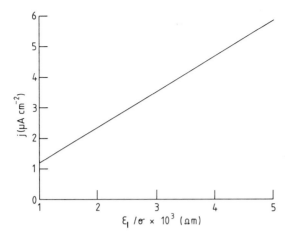

FIG. 9.8.   Plot of induced-current density j versus ϵ_1/σ [Eq. (9.77)] for an infinite tissue cylinder subjected to a time-ramped transverse gradient $G_x = 0.1$ G cm^{-1}. The range of relative permittivity ϵ_1 at 1.0 MHz and the tissue conductivity are chosen to correspond to typical values for body tissues.

maximum induced current density would still be less than 60 μA cm^{-2} for the x gradient. Bearing in mind that our estimate of the dc neural stimulation threshold current density is around 1.5 mA cm^{-2}, it seems clear that use of ramped gradients as high as $G_x = 1.0$ G cm^{-1} and $G_z = 10.0$ G cm^{-1} would produce current densities which would be still two orders of magnitude lower than the dc sensory threshold taking the anisotropy factor $\alpha_a = 1$. Such high gradients have not been used, to our knowledge, in NMR imaging experiments.

We conclude from this section, therefore, that the ramped field gradients currently being used in NMR imaging are well below any possible danger level. The margin is sufficiently large that one could envisage increasing gradient strengths by an order of magnitude without cause for concern. Indeed, even at this level, the margin of safety is so large that possible current focusing effects of up to an order of magnitude could be tolerated without danger. Such effects might be caused by tissue inhomogeneities, although we hasten to add that there is absolutely no evidence for such a focusing mechanism.

The results and discussion of Section 9.4.2 also suggest that if induced current density levels approach the dc sensory threshold level, faster rather than slower gradient switching times are to be preferred, at least for currents induced by time-dependent x or y gradients. For here, the actual magnitude of j is relatively insensitive to the ramp switching time τ. However, for the circular induced current density j_θ, we see from Eq. (9.76) that $j_\theta\tau$ is a constant for fixed r. From the graph of Fig. 9.7, the maximum value of this product is 33.7 μA cm^{-2} μsec when $r = 15$ cm. From Eq. (9.13b) we see that the product $j_{s0}\tau_m = 22.5 \times 10^4$ μA cm^{-2} μsec. If we again assume $\alpha_a = 1$, there is thus a safety factor of around 10^4 below the neural stimulation level when switching a z gradient of 1.0 G cm^{-1}.

It is useful to convert the calculated field-gradient switching rates into the units of T m^{-1} sec^{-1} for ease of comparison with the suggested maximum values of other workers.[11] For both G_x and G_z our maximum switching rate used in Figs. 9.7 and 9.8 is $\partial G_{x,z}/\partial t = 10^3$ T m^{-1} sec^{-1}. On the basis of the above discussion regarding sensitivity threshold levels, we would estimate that effects would become noticeable in the case of G_z at around 10^7 T m^{-1} sec^{-1}. A similar sensory threshold point might be expected for G_x. Bearing in mind that the above calculation assumes $\alpha_a = 1$, which is almost certainly an overestimate, and allowing a further two orders of magnitude safety margin, switched or ramped field gradients up to 10^5 T m^{-1} sec^{-1} should be perfectly safe to work with in human imaging systems.

9.5. Safety Recommendations in Biomagnetics

We finish this Chapter with a brief overview of the various unofficial guidelines and recommendations that have evolved independently in association with the use of magnetic fields in other disciplines and in NMR imaging.

The most established guidelines have evolved principally with the use of high static magnetic fields in particle accelerator machines where personnel are variously subjected to a range of magnetic fields in the course of operating or servicing the machines.[10,62] In this respect it would appear that the fields and exposures quoted have been pretty well tailored to the existing magnetic environment and in that sense are purely arbitrary.

In the case of the Brookhaven National Accelerator Laboratory, the guidelines listed in Table 9.2 permit personnel to make film changes near the bubble chamber in fields greater than 10 kG for up to 15 min. At the Stanford linear accelerator, the unofficial guidelines for static magnetic field exposure, Table 9.3, are more conservative and specify different exposures for limbs and body. Here again, the short time exposures in high fields are essentially tailored to allow film changes in the bubble chamber area. Similar practices are observed at CERN's Geneva laboratory.

Other unofficial recommendations for static magnetic field exposure have

---

[62] E. L. Alpen, in Ref. 1, p. 19.

TABLE 9.2

UNOFFICIAL STATIC MAGNETIC-FIELD EXPOSURE LEVELS AT THE
BROOKHAVEN NATIONAL ACCELERATOR LABORATORY[a]

Field	Exposure time	Body region	Comments
10 kG	Unlimited	Whole	Only with permission
5–10 kG	< 1 hr	Whole	
100 G to 5 kG		Whole	Exposure to be minimized

[a] Brookhaven, New York.

TABLE 9.3

STANFORD LINEAR ACCELERATOR UNOFFICIAL EXPOSURE LEVELS TO
STATIC MAGNETIC FIELDS[a]

Field	Exposure time	Body region	Comments
200 G		Whole	No precise exposure time quoted, only extended periods
2 kG	Several minutes	Whole	
2 kG		Arms and hands	Extended periods
20 kG	Several minutes	Arms and hands	

[a] Stanford, California.

TABLE 9.4

UNOFFICIAL RECOMMENDATION OF VYALOV[63,64] FOR UNIFORM AND
GRADIENT MAGNETIC-FIELD EXPOSURE[a]

Field	Body region	Comment
300 G	Whole ⎱	Intended for continuous
700 G	Hands ⎰	occupational exposure
5–20 G cm^{-1}	Whole ⎱	Field gradients in
10–20 G cm^{-1}	Hands ⎰	occupational exposure

[a] For the U.S.S.R.; exposure time, 8 hr.

TABLE 9.5

N.R.P.B.[a] GUIDELINES FOR EXPOSURE TO NMR CLINICAL IMAGING[b]

Field	Comments
<2.5 T	Exposure should be minimized
<20 T sec^{-1}	Maximum rate of change of a uniform or gradient field for pulses of duration ≥ 10 msec
rf fields up to 15 MHz	< 1 W/kG absorbed power. Exposure should be minimized

[a] National Radiological Protection Board, United Kingdom.
[b] Whole-body region; no limit given for exposure time.

been suggested by Vyalov[63,64] and include values for both uniform and gradient magnetic fields. These are listed in Table 9.4. The uniform field limits are about an order of magnitude lower than those of Brookhaven and accord more with the Stanford recommendations.

With the advent of NMR imaging and designed exposure, as opposed to occupational exposure to magnetic fields, the National Radiological Protection Board in Britain has set up an *ad hoc* advisory committee which offers interim guidelines for NMR in clinical imaging. Their recommendations, which cover static fields, switched gradient fields, and rf fields are listed in Table 9.5. The rf-level recommendation is consistent with the United States standard for IPF of 10 mW cm^{-2}, although as mentioned in Section 9.3.4, this could well be reduced to 1.0 mW cm^{-2}, particularly for vhf through to microwaves. If this were the case, such a change may not be relevant for NMR imaging. (See Notes Added in Proof, Note 9.4.)

[63] A. M. Vyalov, *Vestnik* **8**, 52 (1967).
[64] A. M. Vyalov, *in* "Influence of Magnetic Fields on Biological Objects" (Y. Kholodov, ed.). National Technical Information Service, Report JPRS 63038.
[65] N. R. P. B. Announcement, *Lancet* **2**, 103 (1980).

10. Conclusion

10.1. Choice of Technique

Nuclear magnetic resonance imaging has the advantage over X-ray and radioisotope-based systems in that both phase and intensity information can be studied. In this respect, it therefore resembles ultrasonography. This possibility of encoding information in the phase of an NMR signal is the reason for the large number of different NMR imaging techniques currently under investigation (see Chapter 4).

Although it is difficult to be completely objective about the choice of "the best technique," we have seen in Chapter 5 that, in general, it is those methods which simultaneously gather information from a complete object or an isolated slice within it which will be the most efficient in terms of the image S/N per unit time. Those people considering entry into the field would therefore do well to restrict their attention to techniques such as projection reconstruction (Section 4.5.4), echo-planar imaging (Section 4.5.5), or Fourier zeugmatography and its variants (Sections 4.5.1 and 4.5.2). In principle, there is little to choose between the efficiencies of any of these methods. However, as we have seen in Section 5.3.6, echo-planar imaging does have the advantage of a very short minimum performance time (typically, ~ 40 msec). Unfortunately, it is by no means a simple technique to implement, particularly for matrix sizes in excess of 32×32.

The only advantage of the point and line methods (Sections 4.3 and 4.4, respectively) lies in their comparative simplicity. Thus, for example, the sensitive-point method of Hinshaw does not require any form of computer or microprocessor for control. This simplicity often carries with it much less stringent requirements on the homogeneities of static and rf fields, a feature which may prove attractive if imaging time is not an important consideration or if more complex experiments such as the measurement of chemical shifts are envisaged.

Of course, it is not hard to devise new variants of the basic methods, and radically different approaches may still await discovery. We can say with confidence however that although such techniques may have their own particular advantages, they will not exceed the theoretical performance limits calculated in Section 6.3.5 which are fast being approached by a number of the methods presently being developed.

Commercially, we may expect that the accumulation of experience with X-ray CT, combined with a high efficiency, will ensure that the projection reconstruction method will find application in at least some of the first-generation NMR scanners.

10.2. Imaging Times and Measurement of NMR Parameters

The calculation of an absolute imaging time for a specified resolution and S/N ratio given an optimum technique is a difficult problem which we have addressed in Chapter 6. For typical NMR parameters, Fig. 6.13 indicates that whole-body proton imaging times of 10 sec for a resolution of 2 mm and S/N of 10:1 are the best that can be hoped for. This poor sensitivity is a fundamental problem arising from the small value of the Zeeman splitting, which in turn implies a very small excess of spins in the low-energy state at normal temperatures. This is the reason why, to date, the bulk of NMR imaging studies in biological subjects have concentrated on the measurement of the mobile-proton density of water, fats, and oils.

The philosophy adopted in Chapter 6 was to assume that one works at the maximum possible frequency as dictated by the skin-depth effect and to estimate imaging times based on the assumption that all noise originates in the resistance of the receiver coil. However, as we have seen in Section 6.4.2, biological samples are lossy and can themselves be an important source of noise. The imaging-time estimates of Figs. 6.13–6.16 must therefore be regarded as lower bounds.

It is possible to optimize the receiver coil geometry for this effect using a theoretical method. However, in Section 8.6 we have adopted a new and more practical approach to this problem based on measurements of the quality factor Q. In addition to designing for maximum sensitivity, it is also important to obtain a good uniformity of response. This aspect is dealt with in Section 8.4. It may therefore be appreciated that the art of probe design, which at first sight seems a relatively straightforward business, is in reality one of the most important and difficult aspects in the construction of an NMR imaging system.

As mentioned above, proton density and T_1 images have accounted for virtually all NMR imaging studies so far undertaken. However, it is possible to retain the traditional analytic role of NMR by measurement of chemical shift distributions. There is particular interest in the phosphorus nucleus, which, although much less sensitive and abundant than the proton, offers the chance of performing biochemistry *in vivo* through determination of the spatial distribution of the various phosphorus-containing metabolites such as ATP (see Section 7.3.6).

It is also possible to observe other nuclei which may occur naturally, for example, sodium, or which are introduced as contrast agents, for example, the fluorine-containing group of blood-substitute compounds.

A number of techniques are currently being developed for the imaging of flow, and these are likely to have their major impact in the study of the vascular system. The measurement of diffusion coefficients is another interesting possibility which has yet to be explored. In short, one can envisage many of the standard NMR experiments becoming associated with imaging studies.

10.3. Hardware Considerations

One feature of NMR imaging which is particularly attractive is the small number of moving parts. Indeed, some systems have been constructed without any.

As is to be expected, the various imaging techniques have their own peculiar experimental requirements, but nevertheless, a large proportion of the hardware is common to all systems. It has been the custom with most research instruments to perform the experiments under software control using a modest minicomputer such as a Data General Nova 3. This approach allows great flexibility. For instance, our system, with relatively minor modifications, has been used to perform line-scan, projection reconstruction, and echo-planar imaging experiments. The same philosophy permits future improvements in technique and hardware to be easily incorporated into the system. It could be, however, that some manufacturers may choose to adopt the cheaper alternative of a dedicated microprocessor-based system.

One area where improvement is likely is in data handling where the use of array processors and/or hard wired devices will greatly reduce image processing times.

The most expensive piece of NMR imaging hardware is the magnet system. Since for whole-body proton studies one cannot exceed a frequency of about 5 MHz, the field required is a fairly modest 0.1 T, well within the design capabilities of conventional electromagnetic systems. However, if phosphorus imaging is contemplated, the alternative of a superconductive system is perhaps worthy of consideration. If the receiver bandwidth (and hence, noise) is to be kept to a minimum, the magnet homogeneity should be better than the natural linewidth of the nucleus under study. For the case of protons in biological tissues this is typically ~ 20 Hz, so that whole-body systems ideally require a homogeneity of a few ppm. The four-coil body magnets, as supplied by Oxford Instruments for example, can usually be aligned to give a homogeneity of about 10 ppm. Additional shims are necessary to improve this to the optimum value. Some techniques can operate at much lower homogeneities by the use of correspondingly greater field gradients, but some sacrifice of S/N may be incurred. In addition, the magnitude of gradient required by some methods such as echo-planar imaging may then become inordinately high. We therefore feel that the time-consuming business of magnet shimming is a worthwhile pursuit.

We have already commented on the low sensitivity of NMR. It is not surprising therefore that there is still scope for improvement in the receiver system, from the construction of more efficient probes to the design of pre-amplifiers with lower noise figures.

10.4. Applications of NMR Imaging

In Chapter 6 we considered a number of different NMR imaging regimes, ranging from human whole-body studies down to NMR microscopy. While the smaller size systems may well be of great interest to biochemists and food scientists, the major application of NMR imaging will undoubtedly be in the field of diagnostic medicine. Indeed, the subject has already advanced to the point where a number of research instruments are currently undergoing limited clinical evaluation. However, these studies are only in their preliminary stages, and we still await the proven medical application which will guarantee the acceptance of NMR as an important new imaging modality.

NMR imaging has the great advantage that it does not require the use of ionizing radiation. Biomagnetic effects have been dealt with in Chapter 9, and although it is always prudent to err on the side of caution, the indications are that the technique will be an extremely safe one.

We have discussed a number of potential applications to medicine in Chapter 7. There is every reason to expect that tissue contrast, both in terms of spin density T_1 and other parameters, will enable one to discriminate between normal and pathological states. In particular, large increases in T_1 are observed in the case of tumors. A smaller increase also accompanies the onset of ischemia, suggesting that myocardial infarctions may be observed through this effect. Contrast agents which alter relaxation times may have an important part to play here. Paramagnetic relaxation centers such as Mn^{2+} or even O_2 are an obvious possibility, but the field still remains to be fully explored.

In many NMR imaging systems fluids which are in motion are visible through their lack of response—a sort of inverse angiography—without the need for a contrast medium. However, if such discrimination proves insufficient, it should be possible in future to perform flow studies in which only the moving fluids are imaged.

Those conditions, which involve changes in water content such as hydrocephalus, pulmonary edema, etc., are obviously also good candidates for imaging studies. We may also envisage real-time and stroboscopic cardiac studies using the echo-planar technique.

We have stressed the exciting potential of ^{31}P imaging for the study of biochemistry *in vivo*. Whereas early NMR measurements of ATP concentrations in whole organs agreed reasonably well with conventional chemical assays, it has recently been demonstrated that this agreement was largely fortuitous. As perfusion techniques have improved to allow better oxygenation of tissues, so the ratio of ATP to inorganic phosphate measured by NMR has increased. In the best perfused system of all, a living animal, this ratio is so much greater that it may well require a reevaluation of our current theories of metabolic processes.

Thus, NMR imaging offers an exciting range of possibilities, and we feel confident that one or more of these will be shown to have an important clinical application within the next few years.

Notes Added in Proof

Note 4.1. Since this book was completed, a number of major developments in NMR imaging have been made. In most cases they represent substantial improvements in techniques already discussed at length in this chapter. In this category are the remarkable three-dimensional head images recently shown by Hinshaw,[1] the very high quality two-dimensional projection reconstruction images of Young *et al.*[2] and the two-dimensional Fourier transform images of Edelstein *et al.*[3]

Note 4.2. Improvements in the echo-planar technique have now allowed an experiment, only hinted at in Section 4.5.5.5, to be realized in practice. Ordidge *et al.*[4] have now produced the first ever real-time NMR moving images showing a beating heart in a live rabbit. Further theoretical discussion of echo-planar methods by other groups has also recently appeared.[5,6]

Note 6.1. Table 6.1 suggests that it should be possible to image physiological concentrations of ^{23}Na. DeLayre *et al.*[7] have recently published ^{23}Na images of a test tube phantom and an isolated perfused rat heart obtained by the projection reconstruction technique. The heart was visible as a

---

[1] W. S. Hinshaw, *Proc. Int. Symp. NMR Imaging* (N. Karstaedt, ed.). Bowman-Gray School of Medicine, Winston-Salem, 1982.

[2] I. R. Young, H. Clow, D. R. Bailes, M. Burl, D. J. Gilderdale, P. E. Walters, F. H. Doyle, J. M. Pennock, J. S. Orr, J. C. Gore, G. M. Bydder, and R. E. Steiner, *Lancet*, 11 July, p. 53 (1981).

[3] W. A. Edelstein, J. M. S. Hutchison, F. W. Smith, J. Mallard, G. Johnson, and T. W. Redpath, *Br. J. Radiol.* **54**, 149 (1981).

[4] R. J. Ordidge, P. Mansfield, M. Doyle, and R. E. Coupland, *Proc. Int. Symp. NMR Imaging* (N. Karstaedt, ed.). Bowman-Gray School of Medicine, Winston-Salem, 1982.

[5] M. M. Tropper, *J. Magn. Reson.* **42**, 193 (1981).

[6] L. F. Feiner and P. R. Locher, *Appl. Phys.* **22**, 257 (1980).

[7] J. L. DeLayre, J. S. Ingwall, C. Mallory, and E. T. Fossel, *Science* **212**, 235 (1981).

negative image (low signal intensity) since the ^{23}Na concentration in the cardiac tissue was much lower than that of the perfusate. Motional artifacts were avoided by gating the NMR acquisition from the aortic pressure wave. With the heart *in situ*, the higher concentration of ^{23}Na in blood should allow similar visualization of the chambers.

Note 7.1. The recent improvements in NMR image quality have surpassed all earlier expectations and could soon lead to a reassessment of the role of NMR as a diagnostic imaging modality.

Note 7.2. Since writing this chapter, a number of publications showing much improved image quality have appeared.[2,3,8-13] All images have been produced by methods described in Chapter 4.

Note 7.3. Some interesting studies have been made by the Massachusetts General Hospital group using liquid paraffin (or mineral oil) as an abdominal contrast agent.[14] This material has a short T_1 and is only digested rather slowly, thus showing up particularly well in the stomach and intestines.

Note 7.4. Preliminary experiments have now successfully demonstrated real-time moving images of the heart in a live rabbit using echo-planar imaging.[4] NMR cardiography in larger animals and humans is now likely to follow quickly.

Note 7.5. The recent advances in echo-planar imaging make generalized motion, including flow effects, directly observable in real-time moving images. Pure flow images, showing only the moving fluid, have also been demonstrated.[15]

Note 7.6. With the latest high-field NMR spectrometers it is becoming possible to unravel the proton spectra of proteins and other macromolecules in solution. In tissues, however, the wide range of structural components,

[8] R. C. Hawkes, G. N. Holland, W. S. Moore, and B. S. Worthington, *J. Comput. Assist. Tomogr.* **4,** 577 (1980).

[9] R. C. Hawkes, G. N. Holland, W. S. Moore, E. J. Roebuck, and B. S. Worthington, *J. Comput. Assist. Tomogr.* **5,** 605 (1981).

[10] R. C. Hawkes, G. N. Holland, W. S. Moore, E. J. Roebuck, and B. S. Worthington, *J. Comput. Assist. Tomogr.* **5,** 613 (1981).

[11] F. Smith, J. M. S. Hutchison, G. Johnson, J. R. Mallard, A. Reid, T. W. Redpath, and R. D. Selbie, *Lancet,* 10 Jan., p. 78 (1981).

[12] F. Smith, J. M. S. Hutchison, J. R. Mallard, G. Johnson, T. W. Redpath, R. D. Selbie, A. Reid, and C. C. Smith, *Br. Med. J.* **282,** 510 (1981).

[13] R. J. Ordidge, P. Mansfield, and R. E. Coupland, *Br. J. Radiol.* **54,** 850 (1981).

[14] See paper by J. Newhouse in *Proc. Int. Symp. NMR Imaging* (N. Karstaedt, ed.). Bowman-Gray School of Medicine, Winston-Salem, 1982.

[15] R. Rzedzian and P. Mansfield (to be published).

coupled with spectral lines broadened by magnetic susceptibility differences and other factors, makes the situation rather less hopeful. The high-resolution proton spectrum of muscle, for example, shows a large and very broad peak due to water with a much smaller resonance ~ 3.5 ppm upfield arising primarily from the $-CH_2-$ groups of fats. Although limited, it is nevertheless possible to extract some information from such results, the ratio of fatty to fibrous tissue for instance. Rather more promising are the recent results obtained with the ^{13}C nucleus.[16] Although it has a natural abundance of only 1.1%, the high concentrations in tissue allow excellent spectra to be obtained with a sensitivity similar to that of ^{31}P. Since the chemical shift range for ^{13}C is much greater than for ^1H (~ 300 ppm as compared with ~ 10 ppm) the resonances are clearly resolved and it is possible in muscle, for example, to measure the ratio of unsaturated to saturated fats. This might have important implications in the study of diseases such as muscular dystrophy. We should mention that if proton decoupling is used to simplify the spectra and enhance sensitivity the rf levels should be kept as low as possible in order to avoid tissue heating. Labeled ^{13}C compounds have also been extensively used for the study of metabolic pathways in culture cells.[17] Unfortunately the relatively high cost has tended to restrict the number of applications of this exciting technique. Compounds labeled with other NMR isotopes, for example ^{15}N, are also of interest.

Note 8.1. A disadvantage of the SSM of Fig. 8.21 is its low frequency cutoff, inherent in the use of a transformer-coupled quadrature phase splitter. Our original SSM, described in Mansfield *et al.*,[18] avoids this problem by using the computed cosine and sine Fourier transforms of the desired rf spectral distribution. This method works at zero as well as at higher frequencies.

Note 8.2. This calculation assumes an intermediate step in the three-dimensional reconstruction algorithm in which a series of two-dimensional projections is first produced. Since the final stage in reconstruction also involves some S/N enhancement it may well be possible to further reduce the ADC bit size.

Note 8.3. The advent of real-time moving images in NMR (see Note 4.2) now makes high-speed data handling essential. Most CPUs in the mini-computer range are really not fast enough for this type of imaging where a fast Fourier transform (FFT) is required in times less than 50 msec, depending on the number of data points transformed. Hard-wired and pro-

[16] R. E. Gordon, *Phys. Bull.* **32**, 178 (1981).
[17] See, for example, S. M. Cohen, P. Glynn, and R. G. Shulman, *Proc. Nat. Acad. Sci. USA* **78**, 60 (1981).
[18] P. Mansfield, A. A. Maudsley, and T. Baines, *J. Phys. E Sci. Instrum.* **9**, 27 (1976).

grammable systems capable of such high FFT speeds are now commercially available. The array processor (AP) is a flexible solution to the problem. Most APs require a host CPU to supervise its operation.

Note 8.4. The general form of F_0/Q_{min}, as plotted in Fig. 8.25, has been verified experimentally for saline solutions in both solenoidal and saddle coils.[19]

Note 9.1. An interesting report has recently appeared claiming to have observed effects of low-frequency magnetic fields on bacterial growth rate.[20,21] Cultures of *Escherichia coli* were grown in weak (1–20 G) alternating magnetic fields of square waveform at frequencies of 16.66 and 50.0 Hz. Periodic reductions in the mean bacterial generation time as a function of field strength were reported, and it is speculated that integral changes in the magnetic flux quanta coupling the individual cells may be responsible.

Note 9.2. Enzymes have been widely studied by NMR techniques.[22] Changes in enzyme conformation on addition of coenzyme or substrate are often visible through their effect on the observed chemical shifts. However, in cases where the enzyme has been studied at a number of frequencies there is no magnetic field dependence of the enzyme conformation, as witnessed by the linear dependence of the chemical shift dispersion on field strength. Other NMR parameters, such as the spin–lattice relaxation time, also show the expected frequency dependence. Thus there would appear to be no effect of the static magnetic field on enzyme conformation, and indeed kinetic parameters, as determined by NMR, agree well with those measured by other techniques.

Note 9.3. When calculating $j = \sigma E$ it is assumed that the current flow is continuous. In finite systems, as we have seen, surface charge accumulation may prevent continuous flow. When no other current pathways are present, current will flow to charge or discharge the capacitance of the cylinder surface in a characteristic time of $\epsilon\epsilon_0/\sigma$ which can be less than a microsecond. When charged, no further current flows. With a conducting cylinder surrounded by a nonconducting dielectric, however, we rather assume, as sketched in Fig. 9.6, that internal charges will dissipate, giving

[19] P. Mansfield, R. J. Ordidge, and D. Guilfoyle (to be published).

[20] E. Aarholt, E. A. Flinn, and C. W. Smith, *Phys. Med. Biol.* **26**, 613 (1981).

[21] F. Ivanho, *J. Human Evolution* **8**, 433 (1979).

[22] See for examples: R. A. Dwek, "NMR in Biochemistry: Application to Enzyme Systems." Clarendon Press, Oxford, 1973; K. Wüthrich, "NMR in Biological Research; Peptides and Proteins." North-Holland, Amsterdam, 1976; "NMR in Biology" (R. A. Dwek, I. D. Campbell, R. E. Richards, and R. J. P. Williams, eds.). Academic Press, New York, 1977.

rise to a sustained current density close to that calculated. However, the return paths in such current loops may carry quite different current densities.

Note 9.4. Therapeutic use of rf irradiation in physiotherapy is well established. However, there is a growing number of reports concerning other therapeutic uses of electric and magnetic fields of various frequencies and waveforms.[23-27] The power levels used in these cases are generally well below those required in diathermy. Dramatic effects are claimed using very low tissue current densities. Currents are produced by direct attachment of electrodes if dc, but similar biological effects have been reported using magnetically induced currents. Biomagnetic and bioelectric effects are claimed to accelerate tissue growth, especially useful in cases of nonunion bone fractures. surface wound healing and possibly organ and tissue regeneration. Some published guidelines for NMR clinical exposure are now available.[28]

[23] C. A. L. Bassett, A. A. Pilla, and R. J. Pawluk, *Clin. Orthop.* **124,** 128 (1977).

[24] C. A. L. Bassett, S. N. Mitchell, L. Norton, and A. A. Pilla, *Acta Orthop. Belgica* **44,** 706 (1978).

[25] R. B. Borgens, J. W. Vanable, and L. F. Jaffe, *Bioscience* **29,** 468 (1979).

[26] L. F. Jaffe, *in* "Membrane Transconduction Mechanisms" (R. A. Cone and J. E. Dowling, eds.). Raven, New York, 1979.

[27] H. Frölich, *Adv. Electron. Electron Phys.* **53,** 93 (1980).

[28] See: Exposure to NMR clinical imaging; a report by NRPB. *Radiography* **47,** 258 (1981).

Index

A

Abdomen, NMR imaging of, 225, 228
ADC, *see* Analog-to-digital converters
Adenosine diphosphate, 238
Adenosine triphosphate, 238
Age dependence, tissue water content and, 12–13
Air-cored electromagnets, 250–263
Amperian loop, 99
Analog-to-digital converters, 282–283
Anatomical detail, NMR imaging of, 219–233
Animals, imaging of, 209–211
ATP, *see* Adenosine triphosphate
Avantek amplifier, 278

B

Back projection
  FID and, 139
  in NMR imaging, 135–139
Barnacle muscle cells, transverse relaxation data from, 26
Biological systems
  self-diffusion coefficients for, 31
  water in, 10–31
Biological tissues, relaxation times in, 15–30, *see also* Relaxation times
Biomagnetic effects, 297–332, *see also* Magnetic field
  static mechanisms for, 300
    chemically induced magnetic polarization effects in, 308
    CNS effects in, 305
    fluid and gaseous diffusion effects in, 301
    genetic and enzymic effects in, 305
    kinetic effects in, 307
    magnetohemodynamic effects in, 304
    metabolic effects in, 309
    static hemomagnetic effects in, 303
    tunneling effects in, 307
Biomagnetics, safety recommendations in, 330–332, *see also* Static magnetic field
Biot–Savart expression, for static magnetic field, 264
Bloch equations, 87
  FID and, 45–46
  phenomenological, 41
  rf pulses and, 44
  solution to, 43–46
Blood flow, 236
  separators, 303
Breast cancer, NMR studies in, 243, 244

C

Camphor, transient nuclear signal from protons in, 178
Cancer diagnosis, NMR studies in, 243–245
Carbon, nuclear sensitivity and properties of, 176
Cardiac ejection fraction, 235
  output, 235
  studies, NMR imaging in, 234
Cardiac studies, high-speed echo-planar imaging in, 234–235
Carr–Purcell experiment, 30, 80–81, 83
Cathode-ray tube, in picture displays, 276, 286–289
Central nervous system effects, 305
Central processor unit, 282–285
Character recognition, 38
Chemical shift information, 127, 239, imaging of, 109–110
Coils, *see also* Solenoids
  crossed-ellipse, 271, 272
  design of, 252–257
  ellipsoidal, 252–254
  gradient, 266–275
  rectangular, 269
  saddle, 180, 268, 270
  transmitter, 266–271
  wire-wound, 262–263
Color display system, *see* Picture display
Computerized tomography, 217
Continuous offset distribution, in NMR imaging, 72–73
Contrast agents, *see* Paramagnetic impurity effects
Convolution filtering, in NMR imaging, 139
Cooley–Tukey algorithm, 57
CP, *see* Carr–Purcell experiment
CPU, *see* Central processor unit
Creatine, 238–240
Crossed-ellipse coil, 272